Encyclopaedia of Mathematical Sciences

Volume 5

Editor-in-Chief: R.V. Gamkrelidze

V.I. Arnol'd (Ed.)

Dynamical Systems V

Bifurcation Theory and Catastrophe Theory

With 130 Figures

Springer-Verlag
Berlin Heidelberg New York
London Paris Tokyo
Hong Kong Barcelona
Budapest

Title of the Russian edition:
Itogi nauki i tekhniki, Sovremennye problemy matematiki,
Fundamental'nye napravleniya, Vol. 5, Dinamicheskie Sistemy 5
Publisher VINITI, Moscow 1986

Mathematics Subject Classification (1991):
34A34, 34C35, 35B32, 35L67, 57R45, 58Fxx

ISBN 3-540-18173-3 Springer-Verlag Berlin Heidelberg New York
ISBN 0-387-18173-3 Springer-Verlag New York Berlin Heidelberg

Library of Congress Cataloging-in-Publication Data
Dynamical systems. (Encyclopaedia of mathematical sciences; v. 1–)
Translation of: Dinamicheskie sistemy, issued as part of the series:
Itogi nauki i tekhniki. Serii͡a Sovremennye problemy matematiki.
Vol. 2 edited by Ya. G. Sinai; v. 3, edited by V.I. Arnol'd;
v. 4, edited by V.I. Arnold, S.P. Novikov.
Vol. 4 has title: Symplectic geometry and its applications.
Includes bibliographies and index.
1. Mechanics, Analytic. 2. Mechanics, Celestial.
I. Anosov, D.V. II. Arnol'd, V.I. (Vladimir Igorevich), 1937–
III. Series: Encyclopaedia of mathematical sciences; v. 1, etc.
QA805.D5613 1994 531 87-20655
ISBN 0-387-17002-2 (U.S.: v. 3)
ISBN 0-387-17001-4 (v. 2: U.S.)

Typesetting: Asco Trade Typesetting Ltd., Hong Kong
SPIN 10005298 41/3140-543210—Printed on acid-free paper

List of Editors, Authors and Translators

Editor-in-Chief

R.V. Gamkrelidze, Russian Academy of Sciences, Steklov Mathematical Institute, ul. Vavilova 42, 117966 Moscow, Institute for Scientific Information (VINITI), ul. Usievicha 20a, 125219 Moscow, Russia

Consulting Editor

V.I. Arnol'd, Steklov Mathematical Institute, ul. Vavilova 42, 117966 Moscow, Russia

Authors

V.S. Afrajmovich, N. Novgorod State University, pl. Gagarina 23, 603078 N. Novgorod, Russia

V.I. Arnol'd, Steklov Mathematical Institute, ul. Vavilova 42, 117966 Moscow, Russia

Yu.S. Il'yashenko, Department of Mathematics and Mechanics, Moscow State University, 119899 Moscow, Russia

L.P. Shil'nikov, Institute for Applied Mathematics and Cybernetics, ul. Ul'janova 10, 603005 Nizhnij Novgorod, Russia

Translator

N.D. Kazarinoff†

Contents

Translator's Preface

In translating this volume, I am happy to thank Y.-H. Wan and James Boa for much help on technical points and P. Ashwin for the final check of Part I. I am particularly thankful to G. Wassermann for his careful reading of and many excellent suggestions for the translation of Part II.

N.D. Kazarinoff

Acknowledgement

Springer-Verlag would like to thank J. Joel, B. Khesin, V. Arnol'd and A. Paice for their mathematical and linguistic editing which was necessary after the untimely death of N.D. Kazarinoff. Without their efforts this book would have been delayed even longer.

Springer-Verlag, September 1993

I. Bifurcation Theory

V.I. Arnol'd, V.S. Afrajmovich,
Yu. S. Il'yashenko, L.P. Shil'nikov

Translated from the Russian
by N.D. Kazarinoff

Contents

Preface

The word "bifurcation" means "splitting into two". "Bifurcation" is used to describe any sudden change that occurs while parameters are being smoothly varied in any system: dynamical, ecological, etc. Our survey is devoted to the bifurcations of phase portraits of differential equations – not only to bifurcations of equilibria and limit cycles, but also to perestroikas of the phase portraits of systems in the large and, above all, of their invariant sets and attractors. The statement of the problem in this form goes back to A.A. Andronov.

Connections with the theory of bifurcations penetrate all natural phenomena. The differential equations describing real physical systems always contain parameters whose exact values are, as a rule, unknown. If an equation modeling a physical system is structurally unstable, that is, if the behavior of its solutions may change qualitatively through arbitrarily small changes in its right-hand side, then it is necessary to understand which bifurcations of its phase portrait may occur through changes of the parameters.

Often model systems seem to be so complex that they do not admit meaningful investigation, above all because of the abundance of the variables which occur. In the study of such systems, some of the variables that change slowly in the course of the process described are, as a rule, assumed to be constant. The resulting system with a smaller number of variables can then be investigated. However, it is frequently impossible to consider the individual influences of the discarded terms in the original model. In this case, the discarded terms may be looked upon as typical perturbations, and, accordingly, the original model can be described by means of bifurcation theory applied to the reduced system.

Reformulating the well-known words of Poincaré on periodic solutions, one may say that bifurcations, like torches, light the way from well-understood dynamical systems to unstudied ones. L.D. Landau, and later E. Hopf, using this idea of bifurcation theory, offered a heuristic description of the transition from laminar to turbulent flow as the Reynolds number increases. In Landau's scenario this transition was accomplished through bifurcations of tori of steadily growing dimensions. Later on when the zoo of dynamical systems and their bifurcations had significantly grown, many papers appeared, describing – mainly at a physical level – the transition from regular (laminar) flow to chaotic (turbulent) flow. The chaotic behavior of the 3-dimensional model of Lorenz for convective motions has been explained with the aid of a chain of bifurcations. This explanation is not included in the present survey since, to save space, bifurcations of systems with symmetry have not been included. Lorenz's system is centrally symmetric.

The theory of relaxation oscillations, which deals with systems in which the parameters slowly change with time (these parameters are called slow variables), closely adjoins the theory of bifurcations in which parameters do not change with time. In "fast-slow" systems of relaxation oscillations, a slowness parameter enters that characterizes the speed of change of the slow variables. When this parameter is zero, a fast-slow system transforms into a family studied in the theory of bifurcations, but at a nonzero value of the parameter specific phenomena arise which are sometimes called dynamical bifurcations.

In this survey, systematic use is made of the theory of singularities. The solutions to many problems of bifurcation theory (mostly of local ones) consist of presenting and investigating a so-called principal family – a kind of topological normal form for families of the class studied. The theory of singularities helps to guess at, and partially to investigate, principal families. This theory also describes the theory of bifurcations of equilibrium states, singularities of slow surfaces, slow motions in the theory of relaxation oscillations, etc.

We also note that finitely smooth normal forms of local families of differential equations are especially useful in the theory of nonlocal bifurcations. On one hand, these normal forms substantially simplify the presentation and investigation of bifurcations, and also simplify and clarify the proof and analysis of the results obtained. On the other hand, the nonlocal theory of bifurcations helps to select problems from the theory of normal forms that are important for applications. In our opinion, at the present time, the connection between the theory of normal forms and the nonlocal theory of bifurcations is not used often enough.

This survey includes, along with what is known, a series of new results, some of these are known to the authors through private communications. [Added in translation: The results mentioned below were new when the Russian text was written (1985). Now most of them have been published. The additional list of references is given after the main one and numbered.] Among these are eight new topics. The first is a complete investigation of bifurcations from equilibria in generic two-parameter families of vector fields on the plane with two intersecting invariant curves (the so-called reduced problem for two purely imaginary pairs, Sect. 4.5 and Sect. 4.6 of Chap. 1 (see Żołądek (1987)). The second is the construction of finitely smooth normal forms and functional moduli of the C^1-classification of local families of vector fields and diffeomorphisms (Yu.S. Il'yashenko and S.Yu. Yakovenko, Sect. 5.7–5.10 of Chap. 2 (see Il'yashenko and Yakovenko [3*, 4*])). The third is the construction of a topological invariant of vector fields with a trajectory homoclinic to a saddle with complex eigenvalues (Sect. 5.6 of Chap. 3). The fourth is the description of a generic two-parameter deformation of a vector field with two homoclinic curves at a saddle, in which the bifurcation diagram of the deformation contains a continuum of components. (D.V. Turaev and L.P. Shil'nikov [9*], Sect. 7.2 of Chap. 3). The fifth result is the definition of a statistical limit set as a possible candidate for the concept of a physical attractor (Sect. 8.2 of Chap. 3 (Il'yashenko [2*])). The sixth one is the description of connections between the theory of implicit equations and relaxation oscillations, and the normalization of slow motions for fast-slow systems with one or two slow variables (see Arnol'd's theorem in Sect. 2.2–2.7 of Chap. 4 and the related paper by Davidov [1*]). The seventh result is normalization of fast-slow equations, and the explicit form and investigation of systems of first approximation (Sect. 3.2–3.5 of Chap. 4; see the related paper by Teperin [8*]). The eighth and last one is the investigation of the delayed loss of stability in generic fast-slow systems as a pair of eigenvalues of a stable singular point of a fast equation crosses the imaginary axis (the birth of a cycle as a dynamical bifurcation (A.I. Nejshtadt, §4 of Chap. 4); see [6*, 7*]). We also point here to a conjecture on the bifurcations in generic multiple parameter families of vector fields on the plane that is closely related to Hilbert's 16th problem (Sect. 2.8 of Chap. 3).

Our survey, inevitably, is incomplete. We did not include in it the comparatively few works on local bifurcations in three-parameter families and on nonlocal bifurcations in two-parameter families; some relevant citations are, however, given in the References. In describing nonlocal bifurcations we limited ourselves to only those things which happen on the boundary of the set of Morse-Smale systems. The theory of such bifurcations is substantially complete, although it is not very well known; it is mostly due to works of the Gor'kij school, which often have been published in sources that are hard to obtain. That part of the boundary of the set of Morse-Smale systems on which a countable set of nonwandering trajectories arise is not yet fully explored; but Sect. 7 of Chap. 3 is devoted to this problem. For reasons of consistency of style we often formulate known results in a form different from that in which they first appeared.

Chap. 1 and 2 were written by V.I. Arnol'd and Yu.S. Il'yashenko. Chap. 3, in its final version, was written by V.S. Afrajmovich and Yu.S. Il'yashenko with the participation of V.I. Arnol'd and L.P. Shil'nikov. Sect. 1.6 of Chap. 2 was written by V.S. Afrajmovich. Sects. 1 and 2 of Chap. 4 were written by V.I. Arnol'd, Sect. 3, except for Sect. 3.7, by Yu.S. Il'yashenko. Sect. 3.7 was written by N.Kh. Rozov, Sect. 4 by A.I. Nejshtadt, Sect. 5 by A.K. Zvonkin; the authors sincerely thank them. The authors do not claim that the list of References is complete. In its organization we followed the same principles as in the survey by Arnol'd and Il'yashenko (1985). The symbol ▲ denotes the end of some formulations.

Chapter 1
Bifurcations of Equilibria

The theory of bifurcations of dynamical systems describes sudden qualitative changes in the phase portraits of differential equations that occur when parameters are changed continuously and smoothly. Thus, upon loss of stability, a limit cycle may arise from a singular point, and the loss of stability by a limit cycle may give rise to chaos. Such changes are termed bifurcations.

In Chap. 1 and 2 only local bifurcations are investigated, that is, bifurcations of phase portraits near singular points and limit cycles are considered.

In differential equations describing real physical phenomena, singular points and limit cycles are most often found in general position, that is, they are hyperbolic. However, there are special classes of differential equations where matters stand differently. Such classes are, for example, systems having symmetries related to the very nature of the phenomena investigated, and also Hamiltonian systems, reversible systems, and equations that preserve phase volume. Consider, for example, the one-parameter family of dynamical systems on the line with second-order symmetry:

$$\dot{x} = v(x, \varepsilon), \qquad v(-x, \varepsilon) = -v(x, \varepsilon).$$

A typical bifurcation of a symmetric equilibrium in such a system is the *pitchfork bifurcation* shown in Fig. 1 ($v = x(\varepsilon - x^2)$). In this bifurcation, from the loss of stability by a symmetric equilibrium, two new, less symmetric, equilibria branch out. In this process the symmetric equilibrium position continues to exist, but it loses its stability.

In typical one-parameter families of general (nonsymmetric) systems, pitchfork bifurcations do not occur. Under a small perturbation of the vector field $v(x, \varepsilon)$ above (although the breaking of symmetry may be ever so slight) the pitchfork in Fig. 1 changes into one of the four pairs of curves in Fig. 2. From these pictures it is evident that the phenomena occurring in response to a smooth, slow change of a parameter in an idealized, strictly symmetric system are qualitatively different from those in a perturbation of it. Therefore, it is necessary to take account of the influence of a slight breaking of symmetry when analysing bifurcations in symmetric systems, if such a break is generally possible. On the other hand, strictly symmetric models occur in some instances. Such is the case, for example, for normal forms (see § 3 below). In these cases it is necessary to investigate bifurcations of symmetric systems within the class of perturbations that do not break symmetry.

The degenerate cases which are avoidable by small generic perturbations of an individual system may become unavoidable when families of systems are studied. Therefore, in the investigation of degenerate cases, instead of studying an individual degenerate equation one should always consider the bifurcations that occur in generic families of systems that display a similar degeneracy in an

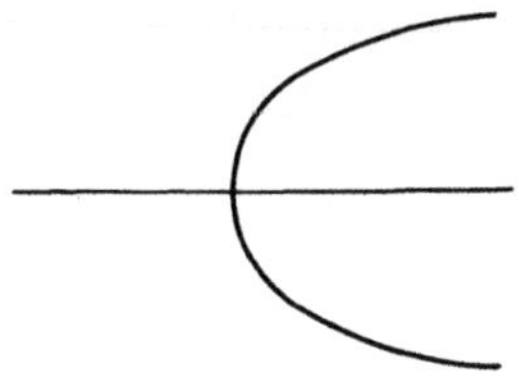

Fig. 1. Bifurcation of equilibria in a symmetric system

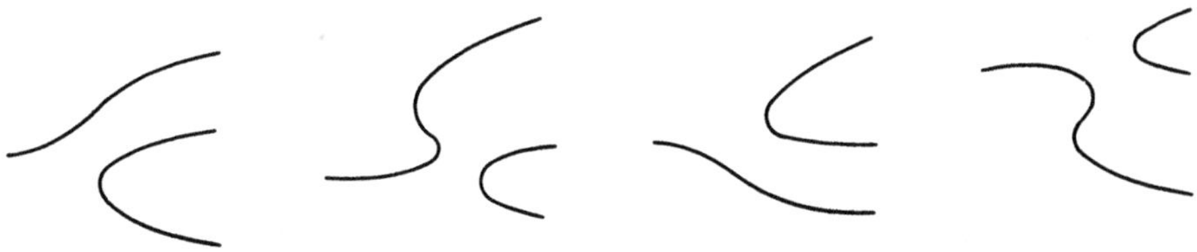

Fig. 2. Bifurcation of equilibria in a nearly symmetric system

unavoidable form. Technically, this investigation is carried out with the help of the construction of special, so-called versal, deformations; in some sense these contain all possible deformations.

§ 1. Families and Deformations

In this section the transversality theorem and the "reduction principle", which allows one to lower the dimension of phase space by "neglecting" inessential (hyperbolic) variables, are formulated.

1.1. Families of Vector Fields. We consider a family of differential equations, say,

$$\dot{x} = v(x, \varepsilon), \qquad x \in U \subset \mathbb{R}^n, \qquad \varepsilon \in B \subset \mathbb{R}^k.$$

The domain U is called *phase space*, B is called the *space of parameters* (or the *base* of the family), and v is called a family of vector fields on U with base B. Henceforth, unless stated otherwise, only smooth families will be considered (v is of class C^∞).

1.2. The Space of Jets. Let U and W be domains of the real, linear spaces $\mathbb{R}^n$ and $\mathbb{R}^m$, respectively. If we choose coordinate systems in $\mathbb{R}^n$ and $\mathbb{R}^m$, then the k-jet of a mapping $U \to W$ at a point x is the vector-valued Taylor polynomial at x with degree $\leqslant k$. Similarly, the set of all k-jets of mappings $U \to W$ is defined by $U \times$ {the space of m-component vector polynomials, of degrees no greater than k, in n variables, with constant terms in W}, and therefore it is a smooth manifold. The *manifold of k-jets of mappings* $U \to W$ is denoted by $J^k(U, W)$.

Analogously, $J^k(M, N)$ is the manifold of k-jets of mappings of a smooth manifold M into a smooth manifold N.

1.3. Sard's Lemma and Transversality Theorems. Consider a smooth mapping $f: U \to W$. A point x of U is *regular* if the image, under the derivative of f at x, of the tangent space at x is the whole tangent space to W:

$$f_*(x) T_x U = T_{f(x)} W.$$

The value of f at a *critical* (i.e., nonregular) point is called a *critical value*.

Sard's Lemma. *The set of critical values of a smooth mapping has Lebesgue measure zero.*

Definition. Two linear subspaces X and Y of a linear space L are *transversal* if their sum is the whole space: $X + Y = L$. [For example, two perpendicular planes in $\mathbb{R}^3$ are transversal, two perpendicular straight lines are not. Translator]

Everywhere in this subsection A and B denote smooth manifolds, and C is a smooth submanifold of B.

Definition. The mapping $f: A \to B$ is called *transversal to C at a point a* in A if either $f(a)$ does not belong to C or the tangent plane to C at $f(a)$ and the image, under the derivative of f at a, of the tangent plane to A at a are transversal:

$$f_*(a) T_a A + T_{f(a)} C = T_{f(a)} B.$$

Definition. The mapping $f: A \to B$ is *transversal* to C if it is transversal to C at each point of A.

Remark. If $\dim A + \dim C < \dim B$ and a mapping $f: A \to B$ is transversal to C, then the intersection $f(A) \cap C$ is empty.

We denote by $C^r(U, W)$ the space of r-smooth mappings of U into W.

The Weak Transversality Theorem for Domains in $\mathbb{R}^n$. *Let C be a smooth submanifold in W. The mappings $f: U \to W$ that are transversal to C form an everywhere dense countable intersection of open sets*[1] *in $C^r(U, W)$ (where $r >$* max $(\dim W - \dim U - \dim C, 0)$).

The Weak Transversality Theorem for Manifolds. *Let A be a compact manifold, and let C be a compact submanifold of a manifold B. Then the mappings $f: A \to B$ transverse to C form an open everywhere dense set in the space of all r-smooth mappings of A into B (where $r >$* max $(\dim B - \dim A - \dim C, 0)$).

Remarks. The closeness of two mappings is defined in terms of the C^r-norms of the functions determining them. If one of the manifolds A or C is not compact, then "open everywhere dense set" must be replaced by "residual set".

---

[1] Such intersections are sometimes called *thick sets* or *residual sets*.

Let M and N be smooth manifolds (or domains in vector spaces). Associated to each smooth mapping is its '*k-jet extension*' $j^k f: M \to J^k(M, N)$; the *k-jet* of the mapping f at x corresponds to a point x of M.

Thom's Transversality Theorem. *Let C be a proper submanifold of the space of k-jets $J^k(M, N)$. Then the set of mappings $f: M \to N$, whose k-jet extensions are transversal to C, forms a residual set in the space of mappings from M into N in the* C^r-topology (*where* $r \geqslant r_0(k, \dim M, \dim N)$, *for some function* r_0).

1.4. Simplest Applications: Singular Points of Generic Vector Fields. Everywhere in this subsection a "generic" field or family is a field or family from some residual subset of the corresponding function space. Vector fields are defined on domains of the space $\mathbb{R}^n$.

Theorem. *For a generic family of vector fields the set of singular points of the fields of the family forms a smooth submanifold in the direct product of phase space with the space of parameters.*

◀ The set of singular points of the fields of family has the form $\{(x, \varepsilon) | v(x, \varepsilon) = 0\}$. By Sard's lemma the set of critical values of the mapping v has measure zero. Consequently, there exists an arbitrarily small vector δ, for which $-\delta$ is a regular value of the mapping v. The set $\{v(x, \varepsilon) = -\delta\}$ is a smooth submanifold by the implicit function theorem. But this submanifold is the set of singular points of vector fields of the family $v(x, \varepsilon) + \delta$. ▶

The projection of the manifold of equilibria onto the space of parameters is a smooth mapping. The theory of singularities of smooth mappings (in particular, of projections) allows one to classify the critical points of generic mappings (and, consequently, also the bifurcations of equilibrium positions in generic families).

For example, if there is just one parameter, then a typical bifurcation is, modulo diffeomorphisms fibred over the axis of parameters, the same as in the family with equilibrium curve $\varepsilon = \pm x^2$ (birth or death of a pair of equilibria). If there are two parameters, then projection leads to one of the normal forms:

$$\varepsilon_1 = \pm x^2 \qquad \text{(a fold)},$$

$$\varepsilon_1 = x^3 \pm \varepsilon_2 x \qquad \text{(a Whitney pleat or cusp)}.$$

Theorem. *All the singular points of a generic vector field are nondegenerate* (*do not have zero eigenvalues*).

◀ Suppose v is a vector field with phase space U. Consider the mapping $v: U \to \mathbb{R}^n$, and suppose that a point 0 takes the role of the submanifold C. By the weak transversality theorem, a generic mapping v is transversal to C. This implies the nondegeneracy of the singular points of v. ▶

Theorem. *All the singular points of a generic vector field are hyperbolic.*

◀ Consider the one-jet extension of the mapping v from the phase space U to $\mathbb{R}^n$. The space $J^1(U, \mathbb{R}^n)$ consists of points of the form (x, y, A), where $x \in U$, $y \in \mathbb{R}^n$,

$A \in \mathrm{Hom}(\mathbb{R}^n, \mathbb{R}^n)$. The image of U under the action of the one-jet extension of v consists of the points $(x, v(x), \partial v/\partial x|_x)$. Denote by C the algebraic submanifold in $J^1(U, \mathbb{R}^n)$ consisting of points of the form $(x, 0, A)$, where the operator A has at least one eigenvalue on the imaginary axis. This algebraic manifold has codimension $n + 1$; it is not a smooth manifold, but it is the union of smooth, in general noncompact, manifolds of codimensions at least $n + 1$. The dimension of U is n. By the transversality theorem the image $j^1 v(U)$, for a vector field v in general position, does not intersect C. ►

1.5. Topologically Versal Deformations. We consider a family of differential equations, say,

$$\dot{x} = v(x, \varepsilon).$$

The germ of the field v at the point (x_0, ε_0) of the direct product of phase space with the space of parameters is called a *local family of vector fields* $(v; x_0, \varepsilon_0)$; the representatives of such germs are families of vector fields.[2]

A *topological equivalence* of local families $(v; x_0, \varepsilon_0)$ and $(w; y_0, \eta_0)$ is given by the germ of a homeomorphic mapping H between the direct product of phase space with the parameters spaces of the first family and the analogous product for second family; the germ is considered at the point (x_0, ε_0); $H(x_0, \varepsilon_0) = (y_0, \eta_0)$. The representative of the germ H is assumed to be fibered over the base of the family, that is, $H: (x, \varepsilon) \mapsto (y, \eta) = (H_1(x, \varepsilon), H_2(\varepsilon))$. The mapping $H(\cdot, \varepsilon)$ is, for each ε, a homeomorphism transforming phase curves of the vector field $v(\cdot, \varepsilon)$ in the domain of definition of H into phase curves of the family $w(\cdot, \varepsilon)$, and preserving the direction of motion. We note that for $\varepsilon \neq \varepsilon_0$ the point x_0 is not necessarily mapped onto y_0 by the mapping $H(\cdot, \varepsilon)$.

Weak equivalence of local families of vector fields is defined similarly; here the germ H need not be continuous: a representative of the germ H is a family of diffeomorphisms $H(\cdot, \varepsilon)$, defined in a common neighborhood of the point x_0, but not necessarily continuous in ε.

Two local families are *strongly equivalent* if they are equivalent, have a common base, and the conjugating homeomorphism H preserves values of the parameter: $H(x, \varepsilon) = (H_1(x, \varepsilon), \varepsilon)$. A local family $(u; x_0, \mu_0)$ is said to be *induced* from the local family $(v; x_0, \varepsilon_0)$ if there exists a germ at the point μ_0 of a continuous mapping φ of the parameter space (of u) into the parameter space (of v): $\mu \mapsto \varepsilon = \varphi(\mu)$ such that $u(x, \mu) = v(x, \varphi(\mu))$, $\varphi(\mu_0) = \varepsilon_0$. A local family $(v; x_0, \varepsilon_0)$ is called a *topologically orbitally versal* (or simply *versal*) *deformation* of the germ of the field $v_0 = v(\cdot, \varepsilon_0)$ at the point x_0 if every other local family containing the same germ is strongly equivalent to one induced from the given family.

A *weakly versal deformation* is defined in the same way, only the word "equivalence" is replaced by "weak equivalence".

---

[2] We emphasize the difference between a local family and a family of germs of vector fields: the field of a local family is defined in a neighborhood of x_0 which is independent of ε, ε sufficiently near to ε_0. The fields belonging to a family of germs do not have this property.

We now consider a group G of diffeomorphisms of a manifold M, for example the group of linear transformations of $\mathbb{R}^n$, the symmetry group S_2 of the plane generated by the reflection in a fixed straight line which passes through the origin, or the group $\mathbb{Z}_q$ of rotations of the plane by angles $2\pi p/q$. If, in the previous definitions, the germs of vector fields and the homeomorphisms are required to be G-equivariant, then we obtain the definition of a *G-equivariant versal deformation* of a G-equivariant germ of a vector field. (Recall that a vector field on M or its germ at a point 0 is G-equivariant if the field (germ) is mapped into itself by every mapping $g \in G$. The germ of a homeomorphism at $0 \in \mathbb{R}^n$ is G-equivariant if it commutes with all elements of the group G.)

The determination and investigation of versal deformations of a germ of a vector field are a way to represent in a condensed form the results of a very complete investigation of bifurcations of a phase portrait.

1.6. The Reduction Theorem. We consider a family of vector fields, depending on a finite-dimensional parameter ε. We assume that the field $v(\cdot, 0)$ has a singular point $x = 0$, and that the corresponding characteristic equation has s roots in the left half-plane, u in the right half-plane and c on the imaginary axis, counting algebraic multiplicity.

Definition. 1. The *saddle suspension* over a family (with s-dimensional stable[2a] manifold and u-dimensional unstable manifold; $s, u \geqslant 0$)

$$\dot{x} = w(x, \varepsilon)$$

is the family

$$\dot{x} = w(x, \varepsilon), \quad \dot{y} = -y, \quad \dot{z} = z, \quad x \in \mathbb{R}^c, y \in \mathbb{R}^s, z \in \mathbb{R}^u.$$

2. The *center manifold* of a local family $(v; 0, 0)$, $v(0, 0) = 0$, is the center manifold at $(0, 0)$ of the corresponding system:

$$\dot{x} = v(x, \varepsilon), \qquad \dot{\varepsilon} = 0.$$

The Reduction Theorem (Shoshitajshvili, 1975, Arnol'd, 1978). *A local family of vector fields* $(v; 0, 0)$, $v(0, 0) = 0$, *is topologically equivalent to the saddle suspension over the restriction of the family to its center manifold. (This restriction, denoted by* $(w; 0, 0)$, *is a local family with c-dimensional phase space, where c is the dimension of the center manifold of the germ* $v(\cdot, 0)$.) *If the local family* $(w; 0, 0)$ *is a versal deformation of the germ* $w(\cdot, 0)$, *then the original family* $(v; 0, 0)$ *is a versal deformation of the germ* $v(\cdot, 0)$. ▲

[See also this EMS, *Dynamical Systems* I]

Remark. The germ at zero of the system

$$\dot{\xi} = w(\xi, \varepsilon), \qquad \dot{\varepsilon} = 0$$

[2a] We use this bad notation in spite of the fact that the motion along the "stable" manifold is very unstable and hence this manifold should be called "unstable".

is topologically equivalent to the restriction of the system

$$\dot{x} = v(x,\varepsilon), \qquad \dot{\varepsilon} = 0$$

to its center manifold at 0; the conjugating homeomorphism preserves ε.

1.7. Generic and Principal Families. We begin with a definition. Consider a family of vector fields $v(\cdot, \varepsilon)$. Topological orbital equivalence (or weak equivalence) defines a partition of the parameter space into classes. This partition is called the *bifurcation diagram* of the family. If the kind of equivalence relation used in the construction of the bifurcation diagram is not indicated, then usual topological equivalence is understood.

A full topological study of the deformations of germs of vector fields at a singular point (in a case where it is possible to carry this out) is carried out according to the following plan.

1. Divide the class of undeformed germs, having some kind of degeneracy (for example, with a zero eigenvalue at the singular point) into two subclasses: typical and degenerate. In this class, the typical germs form an open, everywhere dense set and the degenerate germs form a subset of codimension 1 or higher. For example, in the class of germs $(ax^2 + \cdots)\partial/\partial x$ of vector fields on the line with a zero eigenvalue at the singular point, the typical germs are distinguished by the condition $a \neq 0$ and the degenerate germs by the condition $a = 0$.

2. Describe the *principal families* corresponding to the given class. These are standard families, playing the role of "topological normal forms" for the deformations of typical germs of the class studied. The germ, topologically equivalent to the undeformed germ from which we start, corresponds in the principal family to the zero value of the parameters.

3. Study the bifurcation diagrams of principal families and the phase portraits of the equations of these families. For the principal families described below, some neighborhood of zero in the base of the family is partitioned into a finite number of subsets (strata). The union of the open strata forms the complement to the bifurcation diagram. Any two fields that correspond to parameter values from the same stratum are topologically equivalent in some neighborhood of the origin in phase space, independent of the parameters.

4. For each stratum describe (up to homeomorphism) the phase portrait of the corresponding vector field.

The results of this study are summarized below in tables and illustrations. The dimension of the phase space of the equation, referred to in the tables, is equal to the dimension of the center manifold of the undeformed germ. The class of undeformed germs is listed in the first column of Table 1, its codimension ν is listed in the second column, typical germs are listed in the third column, the topological normal form of the undeformed germ is shown in the fourth column, the principal deformations are shown in the fifth column. The bifurcation diagrams and the corresponding phase portraits are shown in the figures and their corresponding numbers are given in the sixth column of Table 1. The typical and principal deformations for the classes considered below are related by the following.

1. In generic local ν-parameter families of vector fields only typical germs of the class are considered.

2. Any ν-parameter deformation of a typical germ of this class which is transversal to the class is equivalent to the saddle suspension[3] over one of the principal deformations and is versal.

3. A generic ν-parameter deformation of a typical germ of such a class is transversal to this class.

The principal families, their bifurcation diagrams, and phase portraits, which correspond to the simplest classes of typical germs, are described in the table below. Sometimes one considers a class of germs equivariant under the action of some group of symmetries. Then all deformations and conjugating mappings are considered to be equivariant under this action.

§ 2. Bifurcations of Singular Points in Generic One-Parameter Families

In generic one-parameter families of vector fields there are two types of nonhyperbolic singular points: one eigenvalue is equal to zero or two eigenvalues are purely imaginary and not zero, and the remaining eigenvalues of the singular point do not lie on the imaginary axis. In this section the versal deformations of such germs are described, and the phenomena of soft and hard loss of stability of an equilibrium are examined.

From here to the end of Chapter 2, unless stated otherwise, a 'generic family' is a family from some open, everywhere dense set in the space of families with the C^r-topology (r is any number, greater than or equal to the degrees of the vector fields giving the principal deformations).

2.1. Typical Germs and Principal Families

Theorem. *The class of all germs of vector fields at a nonhyperbolic singular point (having eigenvalues on the imaginary axis) has the form of the union of two sets of codimension* 1 *and a set of codimension greater than* 1 *in the space of all germs at the singular point. The first set of codimension* 1 *corresponds to a zero eigenvalue at the singular point, the second one to a pair of purely imaginary eigenvalues. The typical germs in each case reduce, on the center manifold, to the forms shown in* Table 1, *Rows* 1 *and* 2. *The deformations of such germs in generic one-parameter families are versal and are equivalent (up to a saddle suspension) to those principal deformations written down in* Table 1. *They are versal, and the equivalence is stable.*

[3] If there is no saddle suspension, then for uniformity we shall say that there is a suspension of the "trivial saddle" $\{0\}$.

Table 1

Class		Typical Germ				
(Description of the degeneracy)	ν	Normalized jet	Condition for typicality	Normalized germ	Principal families	Bifurcation diagrams & phase portraits
One zero eigenvalue	1	$\dot{x} = ax^2$	$a \neq 0$	$\dot{x} = x^2$	$\dot{x} = x^2 + \varepsilon$ (1^+) $\dot{x} = x^2 - \varepsilon$ (1^-)	Fig. 3
Pair of purely imaginary eigenvalues	1	$\dot{z} = z(i\omega + A\rho)$ $z \in \mathbb{C}, \rho = z\bar{z}$	Re $A \neq 0$	$\dot{z} = z(i \pm \rho)$	$\dot{z} = z(i + \varepsilon + \rho)$ (2^+) $\dot{z} = z(i + \varepsilon - \rho)$ (2^-)	Fig. 4
One zero and a μ-fold-degeneracy in the nonlinear terms: $A_{\mu+1}$	$\mu + 1$	$\dot{x} = ax^{\nu+1}$	$a \neq 0$	$\dot{x} = x^{\nu+1}$	$\dot{x} = \pm x^{\nu+1} + P_\nu(x, \varepsilon)$ $(3^\pm)$	Figs. 5, 6
Pair of purely imaginary eigenvalues and μ-fold degeneracy in nonlinear terms: $B_{\mu+1}$	$\mu + 1$	$\dot{z} = z\left(i\omega + \sum_1^\nu A_j\rho^j\right)$ Re $A_j = 0$, $j = 1, \ldots, \nu - 1$	Re $A_\nu \neq 0$	$\dot{z} = z(i \pm \rho^\nu)$	$\dot{z} = z(i \pm \rho^\nu) + P_\nu(\rho, \varepsilon))$ $(4^\pm)$	Figs. 7, 8
Two zero eigenvalues	2	$\dot{x}_1 = x_2$ $\dot{x}_2 = ax^2 + bx_1x_2 + cx_2^2$	$ab \neq 0$	$\dot{x}_1 = x_2$ $\dot{x}_2 = x_1^2$	$\dot{x}_1 = x_2$ $\dot{x}_2 = x_1^2 \pm x_1x_2 + \varepsilon_1 + \varepsilon_2x_2^2$ $(5^\pm)$	Figs. 9, 10

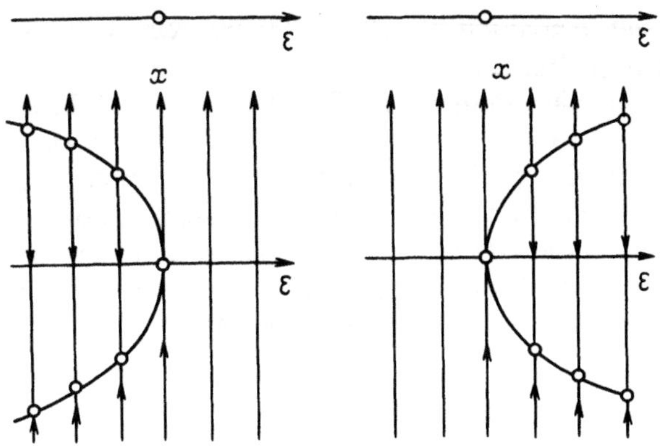

Fig. 3. Bifurcation diagrams and phase portraits for the principal families (1^+) and (1^-)

In Table 1

$$P_\nu(x, \varepsilon) = \varepsilon_1 + \varepsilon_2 x + \cdots + \varepsilon_\nu x^{\nu-1}. \tag{6}$$

Remarks. 1. The family (1^+) in Table 1 is obtained from the family (1^-) by changing the sign of the parameter. However, in general, after equivalent suspension, these families generate nonequivalent suspended families.

2. The families (2^+) and (2^-) in Table 1 are obtained from each other by reversing time: $t \mapsto -t$, by symmetry: $z \mapsto \bar{z}$, and by reversing the parameter: $\varepsilon \mapsto -\varepsilon$, but we study them individually, since in these families the loss of stability is accompanied by principally different phenomena.

2.2. Soft and Hard Loss of Stability. Consider the family (2^-) in Table 1. For $\varepsilon < 0$ the singular point 0 is asymptotically stable, and also for $\varepsilon = 0$. For $\varepsilon > 0$ it becomes unstable. However, for small $\varepsilon > 0$ a neighborhood of the critical point remains attracting: the phase curves originating on its boundary enter this neighborhood and remain forever inside it, only now they wind onto a limit cycle, not the critical point. This limit cycle is a circle of radius $\sqrt{\varepsilon}$. Physicists say that

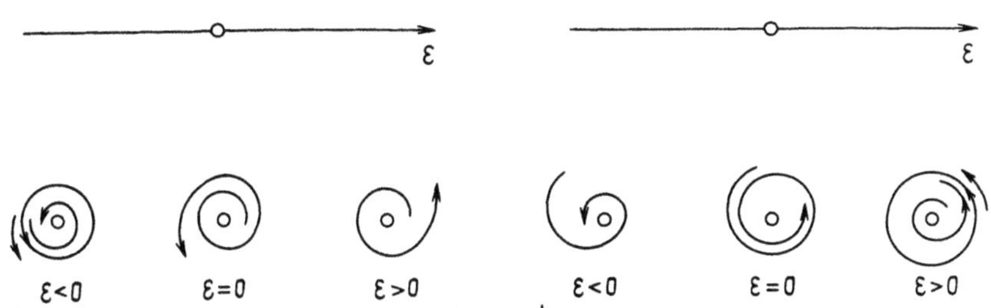

Fig. 4. Bifurcation diagrams and phase portraits for the principal families (2^+) and (2^-)

in this case a *soft generation of self-oscillations* occurs or that there is a *soft loss of stability* (Fig. 4b).

We now consider the family (2^+) in Table 1. For $\varepsilon < 0$ the singular point 0 is stable. However, as $\varepsilon \to 0$ its basin of attraction becomes small (has radius $\sqrt{-\varepsilon}$). For $\varepsilon \geq 0$ the singular point 0 is unstable; all the phase curves, except the equilibrium point, leave a neighborhood of the singular point for all sufficiently small $\varepsilon \geq 0$. This situation is known as a *hard loss of stability*: as ε increases past 0, the system jumps to another regime (a steady-state, a periodic, or a more complex regime, far from the studied equilibrium (Fig. 4a).

§ 3. Bifurcations of Singular Points in Generic Multi-Parameter Families with Simply Degenerate Linear Parts

In this section we consider the class of nonhyperbolic germs of vector fields having the same degeneracies in their linear parts as in the last section, and having additional degeneracies in their nonlinear parts.

3.1. Principal Families

Theorem. *Germs with one zero eigenvalue (respectively, with a pair of purely imaginary eigenvalues) are divided into an infinite number of adjacent*[4] *classes:*

$$A_1 \leftarrow A_2 \leftarrow \cdots$$

$$B_1 \leftarrow B_2 \leftarrow \cdots.$$

The class A_μ (respectively, B_μ) has codimension μ in the space of germs with singular point 0. *It is defined by the Taylor polynomial of a field of degree $\mu + 1$ ($2\mu + 1$) at the singular point: in suitable coordinates, the equation on the center manifold must take the form indicated in column* 3, *row* 3 (*row* 4), *of the table. The classes A_μ and B_μ are typically met in families depending on not less than μ parameters, and they are unremovable under small perturbations of such families. A generic family containing a germ of class A_μ is (up to a saddle suspension) stably locally topologically equivalent to the principal family shown in Table* 1 *and is, as is the principal family, a versal deformation of its most degenerate vector field. An analogous assertion holds for families, containing a germ of class B_μ, only the term "equivalence" must be replaced by "weak equivalence".* ▲

---

[4] Let A and B be two disjoint classes of germs of vector fields at a singular point 0. We say that the class B is adjacent to the class A (written $B \to A$) if for each germ v in the class B, there exists a continuous deformation taking that germ into the class A. More exactly, there exists a continuous family of germs $\{v_t \mid t \in [0, 1]\}$ such that $v_0 = v$ and v_t is a germ in the clas A for all $t \in (0, 1]$.

The classification of local μ-parameter families containing the germs of class B_μ up to usual, not weak, equivalence has functional moduli for $\mu \geqslant 4$. This phenomenon is discussed below in Sect. 5.11 of Chap. 2. For $\mu \leqslant 3$ "weak equivalence" for germs of class B_μ may, possibly, be replaced by the usual equivalence; for $\mu = 1$ this has been proved, see Sect. 2.1 above.

3.2. Bifurcation Diagrams of the Principal Families ($3^\pm$) in Table 1. The set of all singular points of any field in the families ($3^\pm$) forms a smooth submanifold in the product of phase space and the space of parameters. The bifurcation diagram of a principal family ($3^\pm$) (the set of values of the parameter at which some singular points of the family merge) is the set of coefficients of polynomials of degrees $\mu + 1$ having multiple roots. For $\mu = 1$ this set is a single point; for $\mu = 2$ it is a semicubical parabola (cusp) and for $\mu = 3$ it is a swallowtail (Fig. 5). Deformations of vector fields on the line with a degenerate singular point arise in the theory of relaxation oscillations as the equation of slow motions in a neigborhood of a point on the fold of the slow manifold (Sect. 2 of Chap. 4 below). Only the topological normal forms of such deformations are shown in Sect. 3.1. For applications, smooth normal forms are important as well; they are studied in § 5 of Chap. 2 and turn out to be very like the principal families ($3^\pm$).

For $\mu = 2$ the bifurcation diagram and the perestroika of the phase portraits in the family (3^-) are shown in Fig. 6.

3.3. Bifurcation Diagrams with Respect to Weak Equivalence and Phase Portraits of the Principal Families ($4^\pm$) in Table 1. The study of the bifurcation diagrams and the perestroikas of the phase portraits in the principal families ($4^\pm$) leads to the analogous problem for "factored" families, whose phase space is the one-dimensional ray $\rho \geqslant 0$ with coordinate $\rho = z\bar{z}$ as a factor. These factored families have the form

$$\dot{\rho} = 2\rho(\pm\rho^\mu + \varepsilon_1 + \cdots + \varepsilon_\mu \rho^{\mu-1}), \qquad \rho \geqslant 0. \tag{$7^\pm$}$$

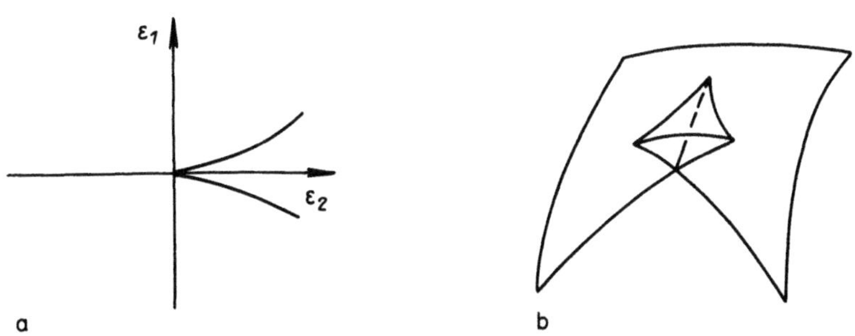

Fig. 5. Bifurcation diagrams for the principal families (3^-) for $\nu = 2$ and $\nu = 3$
a. A semicubical parabola. b. A swallowtail.

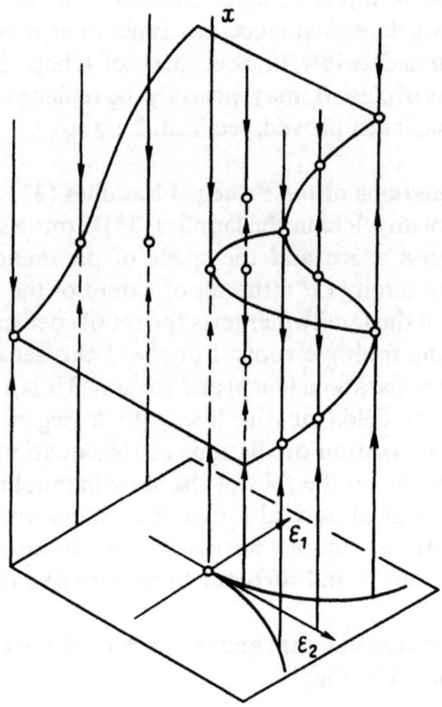

Fig. 6. Phase portraits for equations from the family (3^-) for $\nu = 2$

A limit cycle of the equations $(4^\pm)$, a circle $|z|^2 = \rho_0$, corresponds to a singular point $\rho_0 > 0$ of equation $(7^\pm)$. The stability characteristics of a corresponding point and cycle are the same; for the limit cycle this stability is, of course, orbital stability. The points of the bifurcation diagram of this family correspond to multiple cycles, or equivalently, to multiple singular points of the factored system $(7^\pm)$ (whose phase space is the positive semiaxis). In other words, the

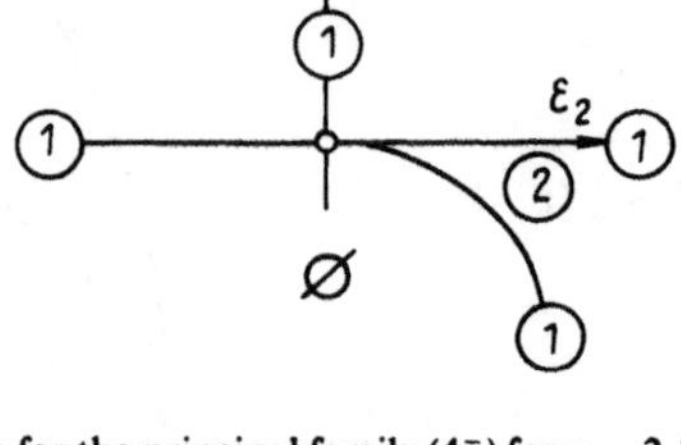

Fig. 7. The bifurcation diagram for the principal family (4^-) for $\nu = 2$. The number on the components of the bifurcation diagram indicates the number of cycles in an equation of the principal family corresponding to the parameter values on this component.

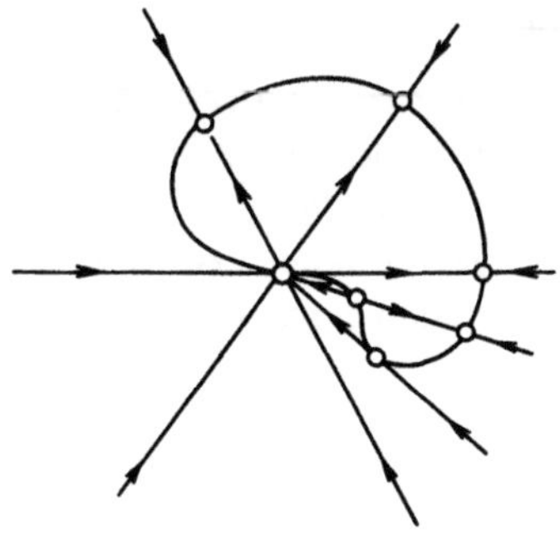

Fig. 8. Phase portraits of factored systems for the family (4^-), corresponding to a circle with center 0 on the plane of parameters

bifurcation diagram of the family ($4^{\pm}$) is the set of polynomials having non-negative multiple roots or a zero root. For example, for $\mu = 2$ this diagram consists of half of a semicubical parabola together with the straight line $\varepsilon_1 = 0$ (Fig. 7).

§ 4. Bifurcations of Singular Points of Vector Fields with a Doubly-Degenerate Linear Part

Not all bifurcations described in this section have been completely investigated.

4.1. A List of Degeneracies. In generic two-parameter families of vector fields the germs at a singular point having doubly degenerate linear parts are of exactly one of the following three types:

1. Two zero eigenvalues; the center manifold is two-dimensional; the corresponding block of the linear part is a nilpotent Jordan block.
2. One zero and a pair of purely imaginary eigenvalues; the center manifold is three-dimensional.
3. Two pairs of purely imaginary eigenvalues; the center manifold is four-dimensional.

A complete description of bifurcations has been found only for the first of these classes. For germs in the other two classes an analogous description seems to be impossible. The theory of normal forms leads to some auxiliary local families of plane equivariant vector fields that play the role of simplified models for the investigation of deformations of germs of these classes. The transfer of results from the auxiliary families to the original ones meets with some problems which are still unsolved. The study of the auxiliary systems involves very difficult problems concerning the bifurcation of limit cycles.

4.2. Two Zero Eigenvalues

Theorem (Bogdanov (1976)). *Generic two-parameter families of vector fields contain only germs with two zero eigenvalues at a singular point whose restrictions to the center manifold in suitable coordinates have the form shown in* Table 1 (line 5). *The deformations of such germs in generic two-parameter families are versal, and are stably equivalent to the principal deformations shown in* Table 1 (*up to a saddle suspension*).

We describe the bifurcations in the principal family (5^+). The bifurcation diagram divides the plane of $\varepsilon = (\varepsilon_1, \varepsilon_2)$ into four parts, denoted A, $B = B_1 \cup B_2 \cup B_3$, C, and $D = D_1 \cup D_2 \cup D_3$ in Fig. 10. The phase portraits corresponding to each of these parts of the ε-plane are shown in Fig. 10 . The branches of the bifurcation diagram correspond to systems with degeneracies of co-dimension 1. These are represented also in Fig. 10 (P, Q, R, and S).

Bifurcations in the principal family (5^-) are obtained from those shown for (5^+) by changing the signs of t and x_2.

4.3. Reductions to Two-Dimensional Systems. By the Reduction Theorem it is sufficient to study bifurcations of singular points with one zero eigenvalue and a pair of purely imaginary eigenvalues, or two pairs of purely imaginary eigenvalues, in three or four-dimensional spaces, respectively. Poincaré's method in this case leads to the following auxiliary problem. The family of equations $\dot{x} = v(x, \varepsilon)$ is transformed to the system

$$\dot{x} = v(x, \varepsilon), \qquad \dot{\varepsilon} = 0.$$

This system is reduced to the Poincaré-Dulac normal form by a transformation that preserves ε, and then terms of sufficiently high order in x (higher than 3 in the case of a zero together with a purely imaginary pair and higher than 5 in the case of two purely imaginary pairs) are neglected. The resulting polynomial vector field is invariant under the group of rotations, which is isomorphic to the torus of dimension equal to the number of purely imaginary pairs. The corresponding factored system is a family of equations on the plane, invariant with

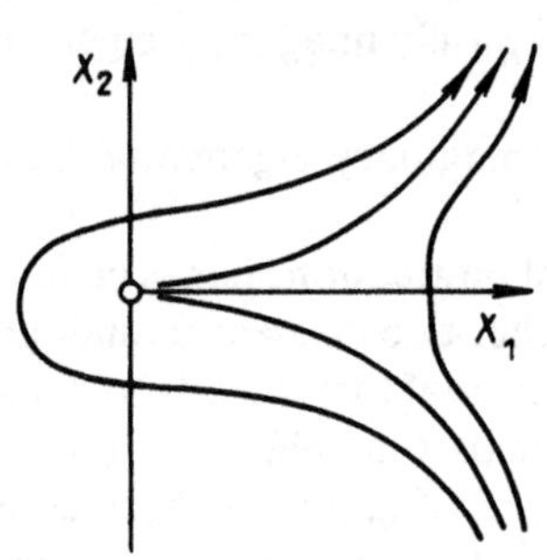

Fig. 9. The phase portrait of a vector field on the plane with nilpotent linear part and a generic nonlinearity

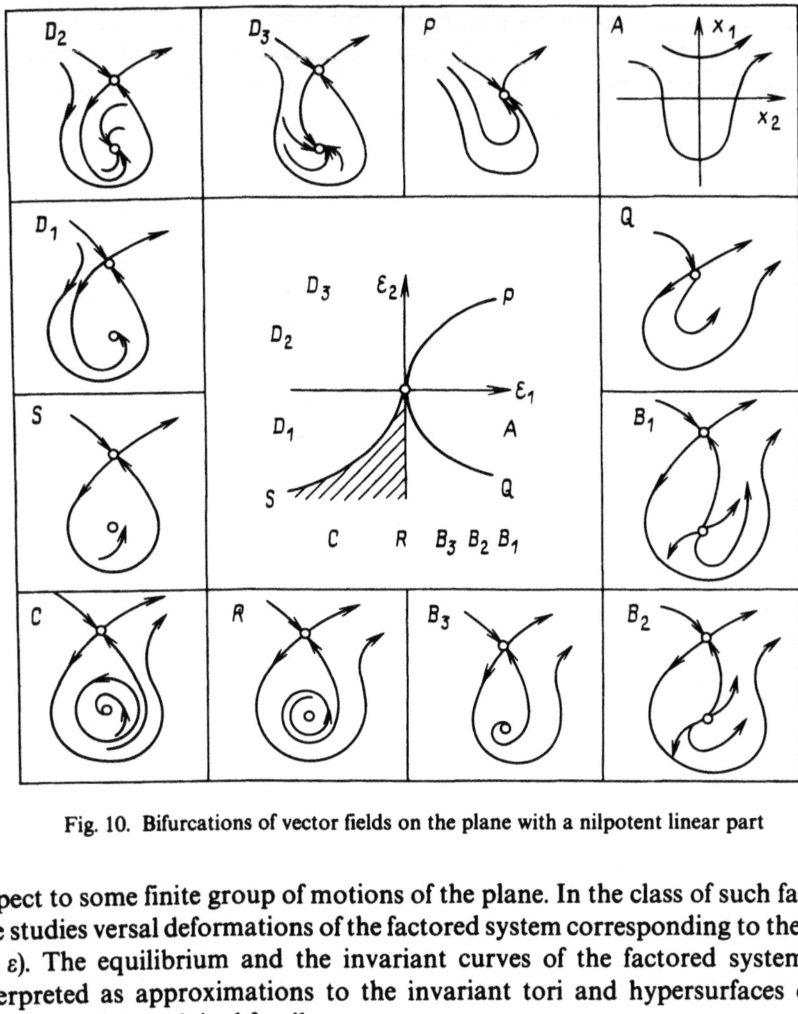

Fig. 10. Bifurcations of vector fields on the plane with a nilpotent linear part

respect to some finite group of motions of the plane. In the class of such families one studies versal deformations of the factored system corresponding to the germ $v(\cdot, \varepsilon)$. The equilibrium and the invariant curves of the factored systems are interpreted as approximations to the invariant tori and hypersurfaces of the equations of the original family.

As indicated above, neglecting higher-order terms in the above procedure is dangerous. For systems of the original family the existence of invariant tori corresponding to equilibria of the auxiliary factored systems is derived from the theorems of Krylov-Bogolyubov (N.N. Bogolyubov and Yu.A. Mitropol'skij, see ref. 17 in Arnol'd and Il'yashenko (1985)). Smooth tori corresponding to cycles of the original system seem to exist only for values of the parameters close to the curve of birth of cycles, but they may be destroyed earlier than the corresponding cycles disappear. For further discussion on this topic see Guckenheimer (1984) and the literature cited therein.

4.4. One Zero and a Pair of Purely Imaginary Eigenvalues. (Following H. Żołądek (written as Kh. Zholondek in *Math. in the USSR–Sbornik*) (1983)). The procedure described above transforms a deformation of a germ of a vector field with one zero and two purely imaginary eigenvalues at the singular point, into a family of equations invariant under the group $\mathbb{Z}_2$ of motions of the plane (x, r), generated by the symmetry $(x, r) \mapsto (x, -r)$. The germs with a zero and a pair of purely imaginary eigenvalues correspond to $\mathbb{Z}_2$-equivariant germs with zero linear parts on the plane.

Theorem. *In generic two-parameter families of $\mathbb{Z}_2$-equivariant vector fields on the plane, one finds only those germs with a zero linear part at the singular point and whose three-jets have the form given in Table 2 (the dots stand for neglected terms). Deformations of such germs in generic two-parameter, $\mathbb{Z}_2$-equivariant families are equivalent to principal deformations and are versal.*

Table 2. $\mathbb{Z}_2$-equivariant vector fields on the plane ($\mathbb{Z}_2$-symmetry $(x, r) \mapsto (x, -r)$)

Class	ν	Typical germ		Principal $\mathbb{Z}_2$-equivariant families	Bifurcation diagrams and phase portraits
		Normalized jet	Conditions for typicalness		
$\mathbb{Z}_2$-equivariant vector fields on the plane ($\mathbb{Z}_2$-symmetry $(x, r) \mapsto (x, -r)$)	2	$\dot{x} = ax^2 + br^2 + cx^3 + \dots$ $\dot{r} = 2dxr + \dots$	$abd \neq 0$ $c \neq 0$ for $b > 0$	$\dot{x} = \varepsilon_1 + \varepsilon_2 x + ax^2 + r^2 + x^3$ $\dot{r} = -2xr$ (8) $a = \pm 1$	Figs. 11, 12
				$\dot{x} = \varepsilon_1 + \varepsilon_2 x + ax^2 - r^2$ $\dot{r} = -2xr$ (9) $a \in \{-3; -1; 1\}$	Figs. 13, 14

Remarks. 1. A topological difference between the principal families (9) for $a = -1$ and $a = -3$ is observed only for the parameter equal to zero (see Fig. 13 and compare Fig. 14b with 14c, in which the structure of the set of 0-curves differs). A 0-curve is defined to be a phase curve with the origin as α or ω-limit set.

2. Equations in the family (9) from Table 2 do not have limit cycles in a sufficiently small neighborhood of the origin (in x, z, and ε). Equations from the family (8) in Table 2 have at most one cycle.

3. In investigating the family (8) in Table 2 for $a = -1$ it is important to pay attention to the form of the neighborhood:

$$(x, r) \in U_\delta = \{\tfrac{1}{2}x^2 + r^2 < \delta^2\}, \qquad \varepsilon_1^2 + \varepsilon_2^2 < \tfrac{1}{2}\delta^2.$$

The previous theorem is correct in this case for any sufficiently small δ.

The form of the neighborhood is important because a specific bifurcation takes place in the family under consideration: the exit of a limit cycle through the boundary of the domain U_δ for arbitrarily small values of the parameter. This bifurcation takes place on the curve N (Fig. 11 for $\alpha = -1$).

4. We emphasize yet again that in generic two-parameter families of $\mathbb{Z}_2$-equivariant vector fields, there are only those germs whose representatives in some neighborhood of the origin, common to all germs of the family, have no

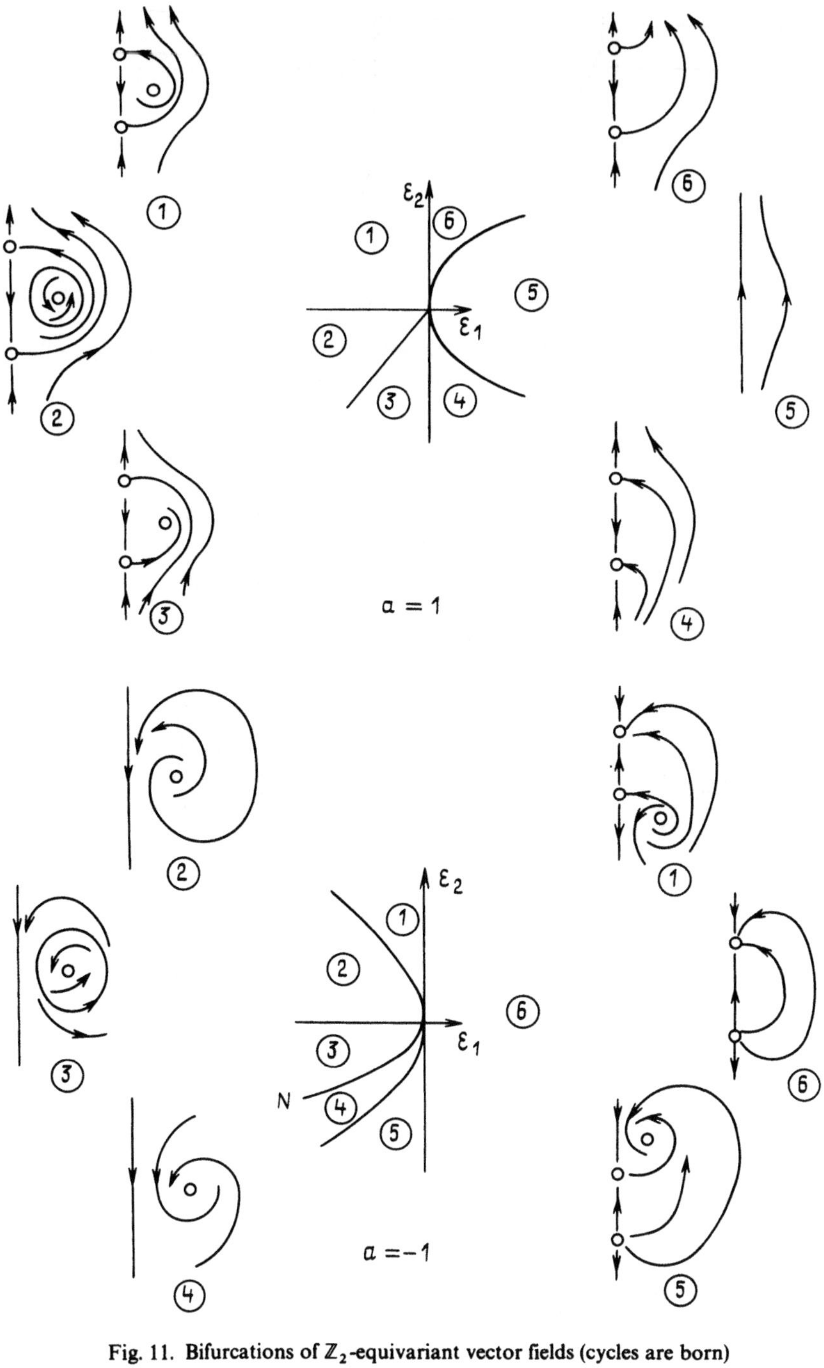

Fig. 11. Bifurcations of $\mathbb{Z}_2$-equivariant vector fields (cycles are born)

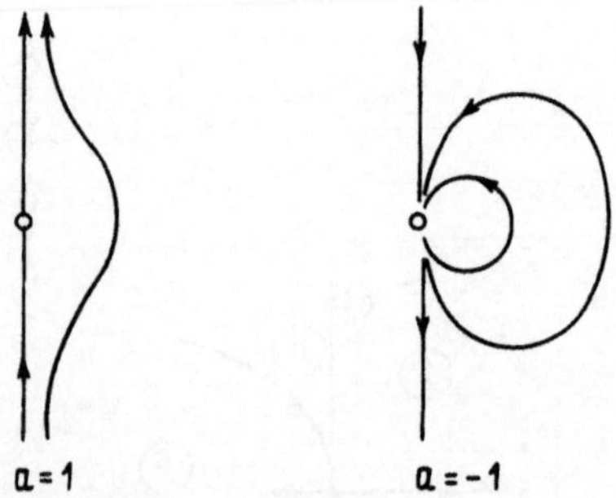

Fig. 12. Local phase portraits of principal $\mathbb{Z}_2$-equivariant vector fields, corresponding to the zero value of the parameter, whose bifurcation generates limit cycles

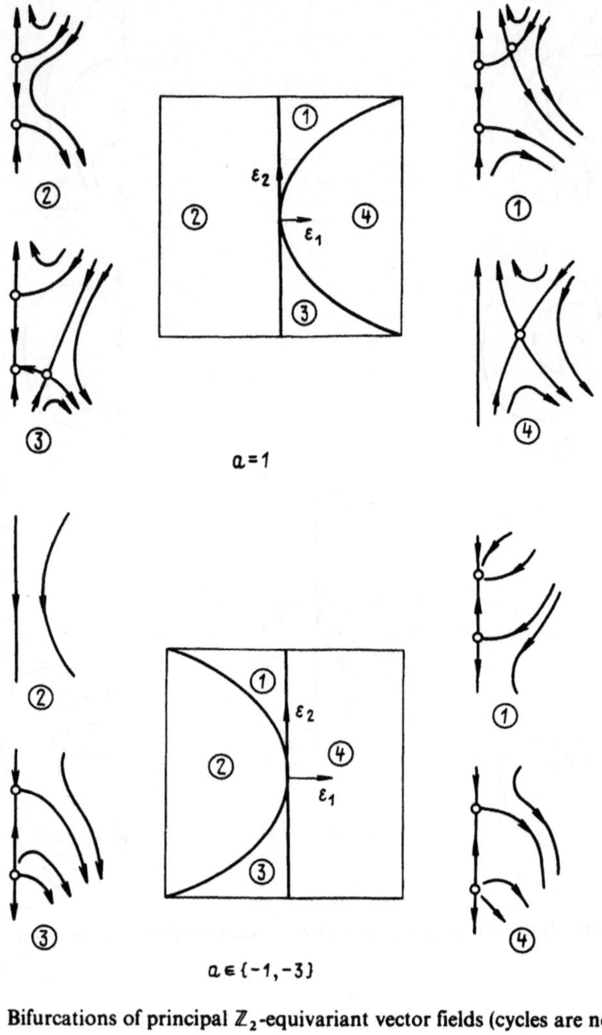

Fig. 13. Bifurcations of principal $\mathbb{Z}_2$-equivariant vector fields (cycles are not born)

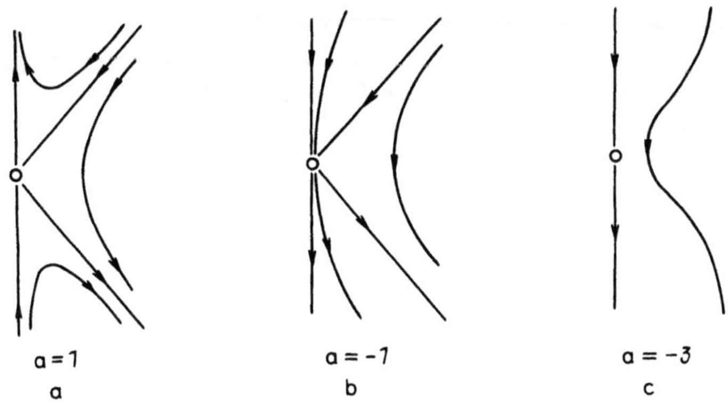

Fig. 14. Local phase portraits of principal $\mathbb{Z}_2$-equivariant vector fields, corresponding to the zero value of the parameter, whose bifurcations do not generate limit cycles

more than one limit cycle. This part of Żołądek's theorem is the richest in content and the most difficult to prove.

Analogous results (but without a proof of the theorem on the number of cycles) were obtained by N.K. Gavrilov (1978).

4.5. Two Purely Imaginary Pairs. We consider a vector field with two pairs of purely imaginary eigenvalues at a singular point 0 in the space $\mathbb{R}^4$. The reductions of Sect. 4.3 lead to the problem of studying the bifurcations of the phase portraits in generic two-parameter families in the quadrant $x \geqslant 0$, $y \geqslant 0$ (the vector field is tangent to the coordinate axes):

$$\dot{x} = xA(x, y),$$
$$\dot{y} = yB(x, y). \tag{10}$$

Systems of the form (10) also occur in ecology (models of Lotka-Volterra type), where the restrictions $x \geqslant 0$ and $y \geqslant 0$ are due to the actual meaning of the phase variables (the populations of predator and prey).

Comments. The two-dimensional system (10) is obtained from the four-dimensional system with two pairs of purely imaginary eigenvalues in the following way: x and y denote the squares of the moduli of the first and second complex coordinates (respectively) in the four-dimensional system after it is transformed into Poincaré-Dulac normal form . In the case of incommensurate frequencies (the ratio of the moduli of the purely imaginary eigenvalues being irrational), resonant terms are expressed through x and y; therefore the normal form admits a factorization up to the two-dimensional system (10).

The problem on vector fields in the first quadrant that arises from this is formally equivalent to a problem on vector fields in the plane that are even in both x and y. Indeed, denoting by x and y the squares of the moduli of the two complex coordinates, we transform the equation corresponding to the vector field to the equations (10).

Bifurcation in the generic two-dimensional families (10) was studied only recently (Żołądek (1985))[5]. The following results were obtained. In generic two-parameter families of systems of the form (10), the functions A and B are simultaneously zero at the origin only for discrete values of the parameters. We consider such a value of the parameters, say, the 0 value, and we write the system in the form:

$$\dot{x} = x(ax + by + \dots),$$
$$\dot{y} = y(cx + dy + \dots).$$

For generic systems of this form $ad \neq 0$. Rescaling x and y and, if necessary, changing the direction of time t, one sets $a = 1$ and $|d| = 1$. The sign of d plays an essential role. Consider the systems

$$\begin{aligned} \dot{x} &= x(x \pm by + \dots), \\ \dot{y} &= y(cx \pm dy + \dots), \end{aligned} \tag{11$^\pm$}$$

By an exchange of the variables $(x, y) \mapsto (y, x)$ one arrives at the condition that $b \geqslant c$ in the system (11^+); the same inequality can be obtained in the system (11^-) by reversal of time and by the same exchange of x and y.

Let $\Delta = bc - 1$. Systems (11^+) for which $bc\Delta = 0$, and systems (11^-) for which $b(b - 1)c(c - 1) = 0$ are called exceptional; they are not encountered in generic two-parameter families of equations of the form (10). The nonexceptional systems (11^+) and (11^-) for which $\Delta < 0$ are called systems of *easy type*; the rest of the nonexceptional systems (11^-) are said to be of *difficult type*.

Theorem. *In generic two-parameter families of systems of Lotka-Volterra type* (10) *there are only those deformations of systems of easy type which are topologically equivalent to one of the principal local families*:

$$\begin{aligned} \dot{x} &= x(\varepsilon_1 + x \pm by), \\ \dot{y} &= y(\varepsilon_2 + cx \pm y) \end{aligned} \tag{12$^\pm$}$$

with two parameters ε_2 and ε_2 (the topological equivalence preserves the first quadrant; time reversal is allowed). These deformations and their normal forms $(12^\pm)$ are topologically versal. The two families of systems $(12^\pm)$, that correspond to values of (b, c) in one "easy" connected component of the set of nonexceptional values are topologically equivalent. The principal families of easy type have no cycles in some neighborhood of the origin, independent of the parameters. Bifurcation diagrams and perestroikas of the phase portraits for such families are shown in Figs. 15a *and* 16a. ▲

In each of Figs. 15 and 16, bifurcation diagrams in the $(\varepsilon_1, \varepsilon_2)$-plane are pictured, under them are the phase portraits, below these is the partition of the

[5] Partial results were obtained in the references (Arnol'd (1972); Gavrilov (1980); Khorozov (1979); Guckenheimer (1984); Guckenheimer and Holmes (1983)), and by V.I. Shvetsov in his diploma thesis, Moscow State University, 1983, 15 pp.

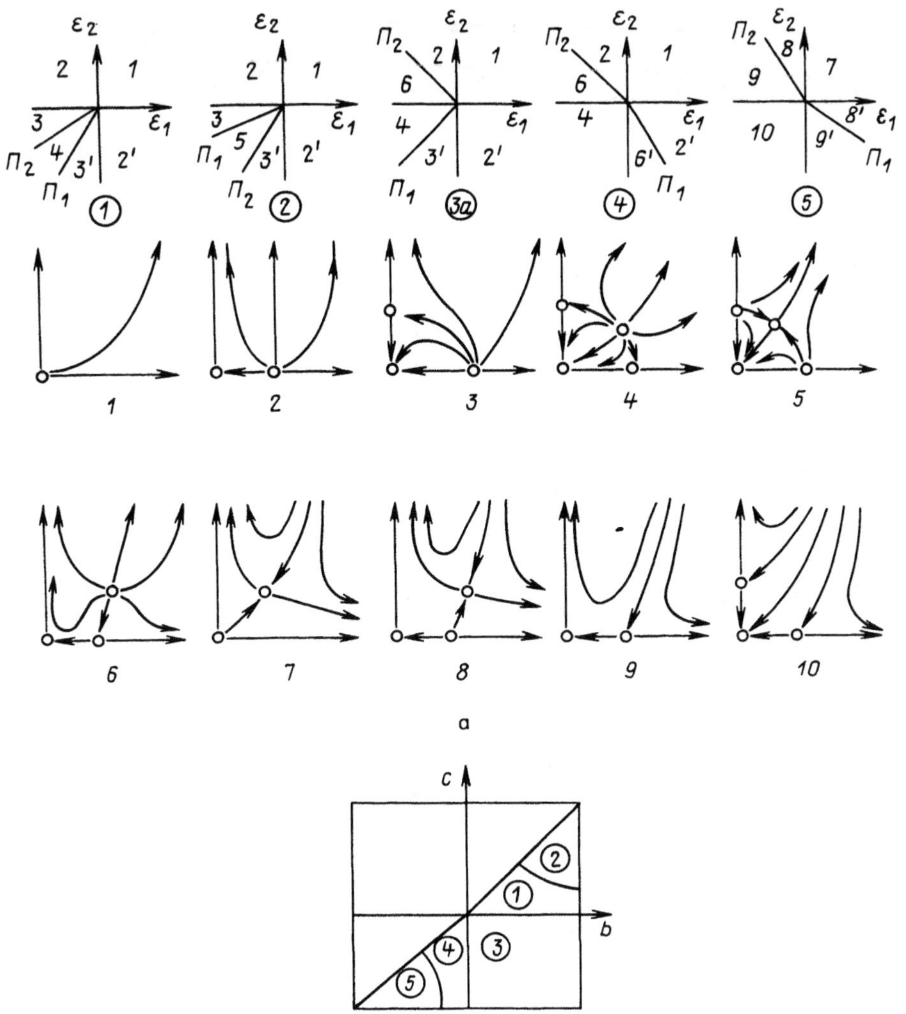

Fig. 15. a. Bifurcation diagrams and phase portraits for easy principal families (12^+) with $d > 0$, b. Partitioning of the half-plane of the parameters (b, c) for $b \geqslant c$

half-plane of the parameters (b, c) $(b \geqslant c)$ corresponding to classes of topologically equivalent "easy" families ($12^{\pm}$). The domains corresponding to difficult families are shaded. The numbers in the open sectors of the bifurcation diagrams correspond to the number of the phase portrait in the lower part the primes on $2'$, $3'$, etc. indicate that the corresponding phase portraits are obtained from 2, 3, ... by the symmetry $(x, y) \mapsto (y, x)$. If the axes of ε_1 and ε_2 (with the origin deleted) are crossed, then either singular points are born from the origin on the positive semi-axes x and y or the inverse process occurs. On passing through the ray Π_1 (resp., Π_2), from a singular point on the y-axis (resp., x-axis), a new one appears

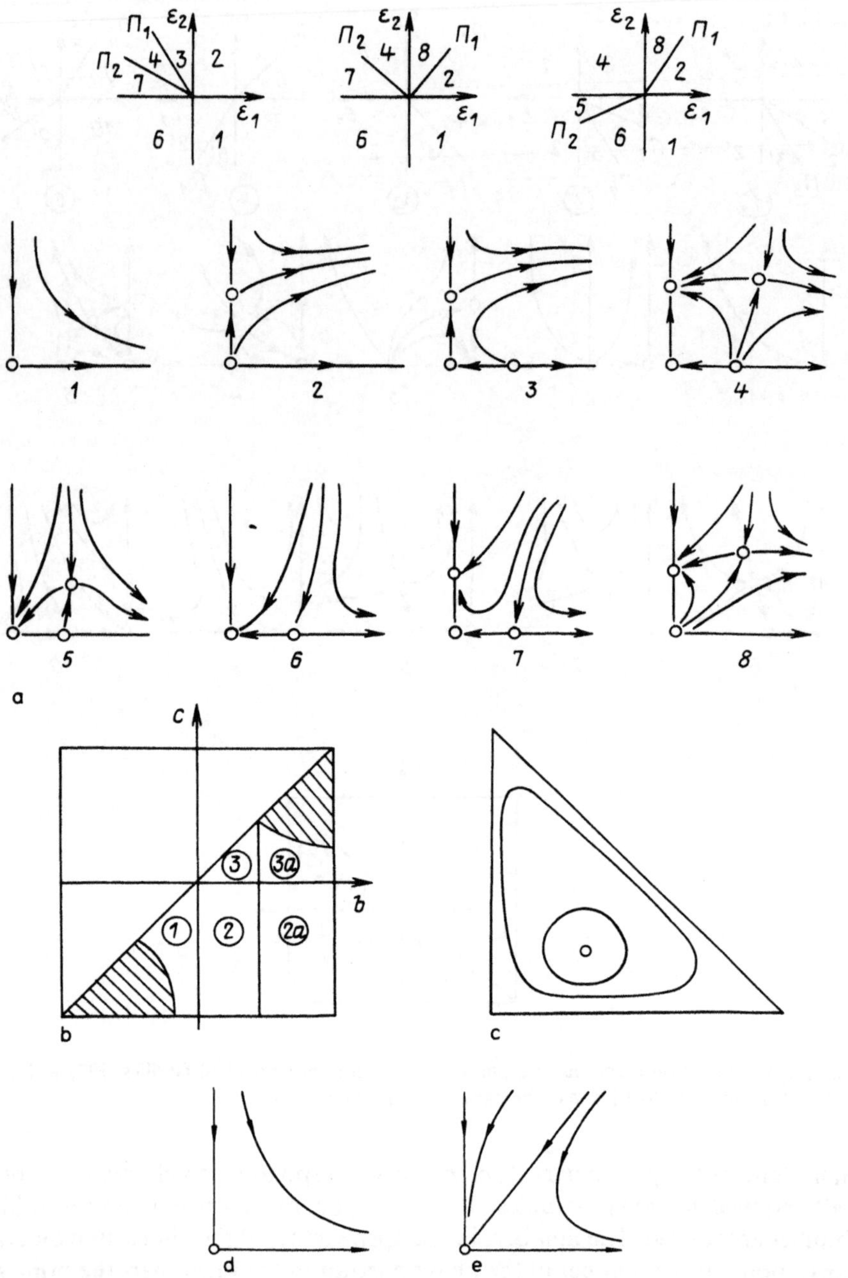

Fig. 16. a. Bifurcation diagrams and phase portraits for easy principal families (12^-) with $d < 0$, b. Partitioning of the half-plane of the parameters (b, c) for $b \geqslant c$. c. Level lines of the Hamiltonian H, corresponding to one of equations in the families (12^-) for $b < 0$, $c < 0$, $\Delta > 0$, d, e. Phase portraits of equations from easy principal systems, corresponding to a zero value of the parameter: d. for the regions 2, 3; e. for the regions 2a, 3a

strictly within the interior of the first quadrant or an existing one disappears from it. The easy families (12^-) of types 2 and 2a, and also of types 3 and 3a differ from each other only for the zero value of the parameter; the sets of 0-curves corresponding to degenerate systems are not equivalent (Figs. 16 d,e).

For each nonexceptional pair (b, c) belonging to one of the difficult components there exist arbitrarily small values of the parameters for which the equations (12^-) have a first integral and a continuous family of cycles. Such equations cannot be found in generic families with a finite number of parameters.

4.6. Principal Deformations of Equations of Difficult Type in Problems with Two Pairs of Purely Imaginary Eigenvalues (Following Żołądek)

Theorem. *A germ at the origin of a generic two-parameter family of difficult type may be transformed to the following "difficult principal family"*

$$\begin{aligned} \dot{x} &= x(\varepsilon_1 + x - by), \\ \dot{y} &= y(\varepsilon_2 + cx - y \pm f_2(x, y, \varepsilon_1)). \end{aligned} \tag{13}$$

The equations of this family have no more than one cycle in some neighborhood of the origin common to all members of the family. Here f_2 is a homogeneous polynomial of degree two in its three variables and with coefficients depending on b and c; its exact form is shown below. ▲

We stop here to give more details of the construction and investigation of difficult principal families. Changes of variables and multiplication by a positive function do not change the topology of the phase portrait. Therefore in the families (13) the only cubic terms remaining are "complementary" to those that can be annihilated by changes of the variables and time in the system ($11^\pm$). (The principal $\mathbb{Z}_2$-equivariant family in Sect. 4.4 was constructed in the same way.) In each of the "difficult" nonexceptional families (13), each time the parameters cross some curve with one endpoint at the origin of the ε-plane, a change of stability of the critical point takes place as a pair of eigenvalues crosses the imaginary axis and a periodic solution is born. The two families (13), which differ by only a change in sign of f_2, are topologically inequivalent: in one the loss of stability is soft and in the other it is hard (see § 2).

The domain of (x, y, ε)-space in which limit cycles of the "principal local family" (13) exist has the form of a narrow tongue extending to the origin. A change of time and of (x, y, ε) transforms the "difficult principal family" considered in this domain into an integrable equation with a small perturbation. We give this change of variables and perturbation in the case $b < 0$, $c < 0$, $\Delta > 0$. In this case the tongue in which we are interested is situated in the half-plane $\varepsilon_1 < 0$ of the $\varepsilon_1, \varepsilon_2$-plane. If $(\varepsilon_2/\varepsilon_1) + (c - 1)/(b - 1) = 0$, then the system ($12^-$) has a first integral (it is written down below after a change of scale). We make a change in the parameters so that one new parameter δ measures the perturbation from zero, and the product of δ by the second new parameter μ measures how far the system is away from being integrable:

$$\varepsilon_1 = -\delta, \qquad \varepsilon_2 = \delta(c-1)/(b-1) + \delta\mu \qquad (\delta > 0).$$

Then the change in scale of the variables and time

$$x = \delta u, \qquad y = \delta v, \qquad \tau = \delta t$$

takes the family (13) into the family

$$\begin{aligned} u' &= u(-1 + u - bv), \\ v' &= v[(c-1)/(b-1) + cu - v + \mu + \delta f_2(u, v, -1)]. \end{aligned} \tag{14}$$

For $(\delta, \mu) = (0, 0)$ the system is integrable; its first integral H and integrating factor m are:

$$H = (1/\beta)u^\alpha v^\beta\{-1 + u + [(1-b)/(1-c)]v\}, \qquad m = u^{\alpha-1}v^{\beta-1},$$

where

$$\alpha = (1-c)/\Delta, \qquad \beta = (1-b)/\Delta, \qquad \Delta = bc - 1.$$

In the case $\Delta = bc - 1 > 0$, we have $\alpha < 0$, $\beta > 0$, and the closed level curves of the Hamiltonian H fill the triangle in Fig. 16c. We denote the corresponding domain in the target space of H by σ. The phase curves of the system (14) are the integral curves of the equation

$$dH - \mu\omega_1 - \delta\omega_2 = 0,$$

where

$$\omega_1 = u^{\alpha-1}v^\beta\, du, \qquad \omega_2 = u^{\alpha-1}v^\beta f_2(u, v, -1)\, du.$$

The limit cycles of the perturbed system are both from the closed phase curve γ_h: $H = h$ of the unperturbed equation if the integral

$$I(c) = \int_{H \leqslant c} d\omega, \qquad \omega = \mu\omega_1 + \delta\omega_2,$$

has a simple zero for $h = c$. Suppose

$$f_2(x, y, \varepsilon_1) = \frac{(x + \varepsilon_1)^2}{\beta} - \frac{2(x + \varepsilon_1)by}{\beta + 1} + \frac{(by)^2}{\beta + 2};$$

$$I_1(c) = \int_{H \leqslant c} d\omega_1, \qquad I_2(c) = \int_{H \leqslant c} d\omega_2.$$

We obtain

$$I_1(c) = -\beta \int_{H \leqslant c} u^{\alpha-1}v^{\beta-1}\, du \wedge dv.$$

$$I_2(c) = -\int_{H \leqslant c} u^{\alpha-1}v^{\beta-1}z^2\, du \wedge dv, \qquad z = -1 + u - bv.$$

The functions I_1 and I_2 are linearly independent on the interval σ; therefore for suitable (δ, μ) the integral I has at least one simple zero on this interval. Generally, if k functions on an interval are linearly independent, then there exists a linear combination of them having $k - 1$ simple zeroes on this interval.

Theorem. *The integral I has no more than one zero on the interval σ.*

The difficult families (13) for $b > 0$ and $c > 0$ are studied analogously.

Remarks. If instead of the cubic terms $\pm yf_2$ in the system (13) we write arbitrary cubic terms that do not remove the system from the class (10): $xF_2(x, y)$, $yG_2(x, y)$, where F_2 and G_2 are homogeneous polynomials of degree 2, then the previous construction leads to a system of the form

$$dH - \omega = 0,$$

where

$$\omega = \mu\omega_1 + \delta\omega_2, \qquad \omega_1 = u^{\alpha-1}v^{\beta}\,du, \qquad \omega_2 = m(vG_2\,du - uF_2\,dv),$$
$$m = u^{\alpha-1}v^{\beta-1}. \tag{15}$$

The differential form ω is a linear combination of seven one-forms:

$$mv\,du,\ mv^3du,\ mv^2u\,du,\ mvu^2dv,\ mv^2u\,dv,\ mvu^2dv,\ mu^3dv.$$

The differentials of these forms span a four-dimensional linear space. Therefore, the space of integrals of linear combinations of these forms over any system of closed curves is at most four-dimensional. However, for the considered class of curves $\{H = h\}$, this space is two-dimensional, not four-dimensional; (and hence it is impossible to conclude that there exists a form ω of the class (15), whose integral has three zeros on σ as is asserted in (Guckenheimer and Holmes, (1983, p. 409)) [corrected in their 2nd edition. Translator].)

§ 5. The Exponents of Soft and Hard Loss of Stability

The exponents defined in this section give the speed with which the loss of stability occurs in generic ν-parameter systems of vector fields for $\nu \leqslant 3$.

5.1. Definitions. The space of germs of real vector fields at a singular point is divided into three parts: the domain of stability, the domain of instability, and the boundary of the domain of stability. This boundary consists of germs whose linearizations do not have eigenvalues lying strictly in the right half-plane, but have at least one eigenvalue on the imaginary axis.

Definition 1. A germ v of a vector field at a singular point 0, on the boundary of the domain of stability undergoes a *soft loss of stability* under a deformation

$$V = \{v_\varepsilon \mid \varepsilon \in B \subset \mathbb{R}^k, 0 \in B, v_0 = v\}$$

if there exists a neighborhood U of 0 and a family of neighborhoods, $\{U_\varepsilon | \varepsilon \in B \setminus \{0\}\}$, contracting to 0 as $\varepsilon \to 0$, such that:

1. The neighborhood U is absorbing for all fields v_ε: each positive semi-trajectory of a field v_ε originating in U remains there forever.

2. For $\varepsilon \neq 0$ all positive semi-trajectories originating in U eventually enter U_ε and remain therein.

Definition 2. A germ v of a vector field at a singular point 0 belonging to the boundary of the domain of stability undergoes *a hard loss of stability* under a deformation

$$V = \{v_\varepsilon | \varepsilon \in B \subset \mathbb{R}^k, 0 \in B, v_0 = v\}$$

if there exists a neighborhood U of the singular point 0 and a family of initial conditions x_ε, defined for all sufficiently small $\varepsilon \neq 0$, with $|x_\varepsilon| \to 0$ as $\varepsilon \to 0$, such that each positive semi-trajectory with initial condition x_ε leaves the neighborhood U forever.

Definition 3. A number κ is called the *exponent of soft* (*resp., hard*) *loss of stability of a germ* v if for any deformation of the germ v there exists a C (depending on the deformation and the metrics in the phase and the parameter spaces) such that each neighborhood U_ε from Definition 1 (resp., each initial condition x_ε of Definition 2) is contained in the ball $|x| \leqslant C|\varepsilon|^\kappa$. The least upper bound of such numbers κ is called the *maximal exponent of soft* (*resp., hard*) *loss of stability* of the germ.

Remarks. 1. The maximal exponent depends upon neither the deformations nor the metric.

2. An asymptotically stable germ always undergoes a soft loss of stability (Malkin (1952)).

3. The larger the exponent of soft loss of stability, the more slowly does the size of the attractor "replacing" the singular point grow as ε grows, and the more softly stability is lost.

The larger the exponent of hard loss of stability, the faster does the "dangerous zone" of initial conditions, from which solutions leave the fixed neighborhood, approach the equilibrium point as ε decreases, and the harder is the loss of stability.

4. The "dangerous zone" of initial conditions for a hard loss of stability can be very narrow; it is not easy to find this zone by calculations. In this there is a substantive difference between equations lying near the boundary of stability, on one hand, and equations having unstable linear parts, with large growth (large maximal real part of an eigenvalue), on the other hand.

Example. Bifurcations giving birth to cycles in a generic one-parameter family are accompanied by a soft or hard loss of stability with exponent 1/2 (see Sect. 2.2 and 2.3).

Formally, the maximal exponents κ_+ and κ_- are defined as follows. Let $v(x, \varepsilon)$ be a deformation of the germ $v(x, 0)$ with singular point 0, and let $\varphi_{x,\varepsilon}$ be a trajectory of the field $v(\cdot, \varepsilon)$ with initial condition x: $\varphi_{x,\varepsilon}(0) = x$. If the germ $v(\cdot, \varepsilon)$ is stable, then there exists an absorbing neighborhood U of the equilibrium. If the germ $v(\cdot, \varepsilon)$ is unstable, then there exists a neighborhood U of the equilibrium, for which one may find a positive semi-trajectory, having arbitrarily small initial condition, that leaves U. With this notation

$$\kappa_+ = \varlimsup_{\substack{\varepsilon \to 0 \\ \varepsilon \neq 0}} \sup_{\substack{x \in \partial U \\ t \in [0,\infty)}} \{\ln|\varphi_{x,\varepsilon}(t)|/\ln \varepsilon\}.$$

Let $E(x, \varepsilon)$ be the interval of $\mathbb{R}^+$ that is the "maximal set of definition of the negative semi-trajectory $\varphi_{x,\varepsilon}$ with values in U":

$$E(x, \varepsilon) = \{t \leqslant 0 \mid \varphi_{x,\varepsilon}(\tau) \in U \text{ for } t \leqslant \tau \leqslant 0\}.$$

Then

$$\kappa_- = \varlimsup_{\substack{\varepsilon \to 0 \\ \varepsilon \neq 0}} \sup_{\substack{x \in \partial U \\ t \in E(x,\varepsilon)}} \{\ln|\varphi_{x,\varepsilon}(t)|/\ln \varepsilon\}.$$

5.2. Table of Exponents

Theorem. *In generic three-parameter families there are only those germs of vector fields at a singular point that lie on the boundary of the domain of stability that belong to one of the classes listed in* Table 3 *below. If a germ is stable (resp. unstable), it undergoes a soft (resp. hard) loss of stability; the corresponding maximal exponents are also given in the tables.*

In Table 3, ν is the codimension of the degeneracy, and κ_+ and κ_- are the maximal exponents of soft and hard loss of stability, respectively. A blank space in the table means that in the class considered there are no stable germs (in generic

Table 3

ν	Class	κ_+	κ_-	Class	κ_+	κ_-
1	$W_1^{0;*}$	—	$\frac{1}{2}$	$W_2^{I;*}$	$\frac{1}{2}$	$\frac{1}{2}$
2	$W_1^{0;0}$	$\frac{1}{3}$	$\frac{1}{2}$	$W_2^{I;0}$	$\frac{1}{4}$	$\frac{1}{4}$
	$W_2^{J;*}$	—	3/4	$W_3^{0,I;*}$	—	$\frac{1}{2}$
	$W_4^{I,I;*}$	$\frac{1}{2}$	$\frac{1}{2}$	$W_1^{0;0,0}$	—	$\frac{1}{4}$
3	$W_2^{J;0}$	$\frac{1}{4}$	$\frac{1}{3}; \frac{1}{4}$	W_3^J	—	5/6
	$W_4^{I,J}$	—	$\geqslant \frac{1}{2}$ (3/4)	$W_3^{0,I;0}$	$\frac{1}{3}$	$\frac{1}{2}; \geqslant \frac{1}{3}$ $(\frac{1}{2})$
	A_1	$\frac{1}{4}$	$\frac{1}{4}$	A_2	$\frac{1}{4}$	$\frac{1}{4}; \geqslant \frac{1}{4}$ $(\frac{1}{4})$
	A_3	—	1	A_4	—	1
	A_5	$\frac{1}{2}$	$\frac{1}{2}$	$W_5^{0,I,I}$	—	$\frac{1}{2}$
	$W_6^{I,I,I}$	$\frac{1}{2}$	$\frac{1}{2}$	$W_2^{I;0,0}$	1/6	1/6

The classes $A_1, \ldots, A_5$ were defined in Arnol'd and Il'yashenko (1985, Sect. Chap. 3).

three-parameter families). The classes listed in the table are defined in Arnol'd and Il'yashenko (1985, Sect. 5, Chap. 3). We recall some notation. The subscript on a particular class $W_\cdot^\cdot$ indicates the dimension of the center manifold; the superscript before the semicolon indicates the degeneracy of the linear part: a 0 means that there is a single 0 eigenvalue, an I means that there is a pair of purely imaginary eigenvalues, a J means that there is a nilpotent Jordan block, the order of which is given by the dimension of the center manifold. An asterisk after the semicolon in the superscript of a W symbolizes the absence of a degeneracy in the nonlinear terms, while the number of 0's there is equal to the number of degeneracies of the nonlinear terms.

Remarks. 1. The conjectured values of the maximal exponents are shown in parentheses.

2. The maximal exponents of soft and hard loss of stability need not coincide for all germs of a single class (in generic three-parameter families). Cases of such noncoincidence occur in the classes $W_2^{J;0}$ and, possibly, in $W_3^{0,I;0}$.

3. The long horizontal lines in the table separate classes of different codimensions.

Dangerous and safe zones on the boundary of the domain of stability were first studied by N.N. Bautin (1949).

This section consists of an exposition of results of L.G. Khazin and Eh. Eh. Shnol' (1981, 1991).

Chapter 2
Bifurcations of Limit Cycles

Bifurcations of phase portraits in the neighborhood of a cycle are completely described by bifurcations of the corresponding monodromy transformations. Therefore the basic objects of study in this chapter are the bifurcations of germs of diffeomorphisms at a fixed point. Local families of germs of diffeomorphisms, their equivalence and weak equivalence, and induced and versal deformations of such germs are defined just as for germs of vector fields (see Sect. 1.5 of Chap. 1). Analogs of the Reduction Theorem are also true for germs of diffeomorphisms at a fixed point; see Sect. 1.6 of Chapter 1 and Arnol'd and Il'yashenko (1985, Sect. 2.4 of Chap. 6). The restriction of a germ of a diffeomorphism to its center manifold is called the *reduced germ* of the diffeomorphism. We note that a reduced germ can change orientation, even if the original germ did not: for example, $x \mapsto \operatorname{diag}(1; -1; 2; -\frac{1}{2})x$, $x \in \mathbb{R}^4$. The bifurcations of germs of diffeomorphisms are described below, and then the results obtained are translated into the language of differential equations.

In § 5 below the "finitely smooth" theory is set forth in detail. The normal forms of local families of vector fields and diffeomorphisms studied there are those into which these families can be transformed by a finitely smooth change of coordinates in phase space. These normal forms are useful in the theory of nonlocal bifurcations and relaxation oscillations.

In § 6 the theory of Feigenbaum is discussed, mainly for multi-dimensional mappings.

§ 1. Bifurcations of Limit Cycles in Generic One-Parameter Families

Limit cycles, having one (Floquet) multiplier[6] equal to ± 1 or having a complex conjugate pair $\exp(\pm i\omega)$ on the unit circle, occur in generic one-parameter families. The remaining multipliers do not lie on the unit circle.

It also is useful to study two-parameter families of bifurcations with a complex conjugate pair of multipliers passing through the unit circle: perestroikas, which seem to be nonlocal from a one-parameter point of view, become tractable by local methods if one considers the problems to be two-parameter ones (see Sect. 1.5 below).

1.1. Multiplier 1

Definition. A principal one-parameter deformation of a germ of a diffeomorphism of the line at a fixed point with multiplier 1 is one of the two families

$$x \mapsto x + x^2 + \varepsilon \qquad (1^+)$$

and

$$x \mapsto x + x^2 - \varepsilon. \qquad (1^-)$$

The saddle (with s-dimensional stable and u-dimensional unstable manifold, $s \geqslant 0$, $u \geqslant 0$) *suspension over the family*

$$x \mapsto w(x, \varepsilon)$$

is the family

$$(x, y, z, u, v) \mapsto (w(x, \varepsilon), \tfrac{1}{2}y, -\tfrac{1}{2}z, 2u, -2v), \quad (y, z) \in \mathbb{R}^s, \quad (u, v) \in \mathbb{R}^u.$$

Theorem. *In generic one-parameter families of diffeomorphisms of the line with a fixed point having multiplier* 1, *only those germs occur which under a homeomorphism take the form* $x \mapsto x + x^2$. *The deformations of such germs in generic families are equivalent to the principal deformations and are versal.*

The assertion of the theorem also holds for families of diffeomorphisms of $\mathbb{R}^n$ if one replaces each principal deformation by its saddle suspension.

---

[6] We recall that the multipliers of a limit cycle are the eigenvalues of the Poincaré mapping on a disc into itself transversal to the cycle.

Conditions for Genericity. 1. The reduced germ of a diffeomorphism with multiplier 1 has the form

$$x \mapsto x + ax^2 + \cdots, \quad x \in (\mathbb{R}, 0), \quad a \neq 0.$$

2. The family $(1^{\pm})$ is transversal to the manifold of germs defined by condition 1 above.

Remarks. The proof of the theorem is not simple. Moreover, an unexpected "rigidity theorem" is stated below. Any germ of a smooth diffeomorphism of the line $x \mapsto x + ax^2 + \cdots$, $a \neq 0$, can be represented as a germ of a time-one shift along the phase curves of a smooth uniquely defined field v, the so-called generating field: $v(x) = ax^2 + \cdots$.

Theorem (Newhouse, Palis, Takens (1983)). *The homeomorphism conjugating two generic one-parameter deformations of germs*

$$f: x \to x + ax^2 + \cdots, \quad g: x \to x + bx^2 + \cdots, \quad x \in (\mathbb{R}, 0), \quad ab \neq 0,$$

that correspond to the zero value of the deformation parameter ε, *is smooth in* x *for* $\varepsilon \neq 0$ *and conjugates the generating fields of the germs* f *and* g.

The bifurcations of orbits of diffeomorphisms in the principal family (1^+) are shown in Fig. 17. As ε moves to the right from zero, the fixed point vanishes, and as ε moves to the left from zero, the fixed point splits into two hyperbolic fixed points: one attracting and the other repelling. This perestroika becomes, in

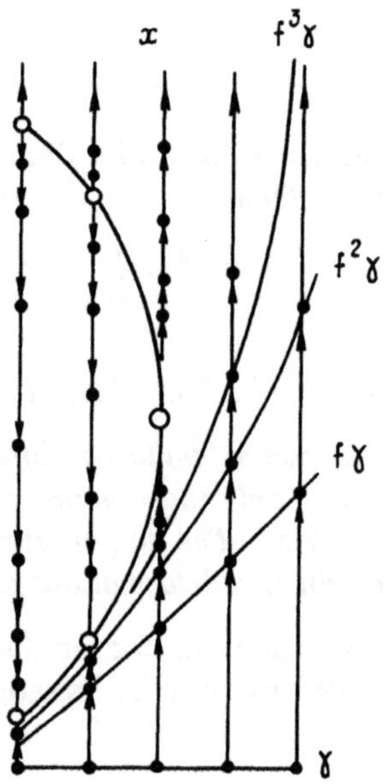

Fig. 17. Orbits of the groups of iterates of germs of diffeomorphisms of the family (1^+)

the corresponding family of differential equations on the plane, an approach of two periodic orbits to each other, one stable and the other unstable, which at the moment they join ($\varepsilon = 0$) form a semi-stable orbit that disappears as ε increasing passes through 0.

1.2. Multiplier -1 and Period-Doubling Bifurcations

Definition. *A principal one-parameter deformation of a germ of a diffeomorphism of the line at a fixed point with multiplier* -1 is one of the two families $\{f_\varepsilon\}$:

$$f_\varepsilon\colon x \mapsto (-1+\varepsilon)x \pm x^3. \qquad (2^\pm)$$

Theorem (Arnol'd (1978), Newhouse *et al.* (1983)). *In generic one-parameter families of diffeomorphisms of the line at a fixed point having multiplier* -1, *only those germs occur which under a homeomorphic change of coordinates take the form of one of the germs* $x \mapsto x + x^3$ *or* $x \mapsto x - x^3$. *The deformations of such germs in generic families are equivalent to the principal deformations, and are versal.*

The assertion of the theorem holds for families of diffeomorphisms of $\mathbb{R}^n$ as well if one replaces each principal deformation by its saddle suspension.

Conditions for Genericity. 1. The reduced germ of a diffeomorphism with multiplier -1 has the form $x \mapsto f(x)$, where

$$f^2\colon x \mapsto x + ax^3 + \cdots, \qquad a \neq 0.$$

2. The family $(2^\pm)$ is transversal to the manifold of germs defined by condition 1.

Remarks. A detailed proof of the above theorem does not exist in the literature, although this theorem is simpler than the proceeding one and is proved by the same methods; see Newhouse *et al.* (1983).

In the family (2^+) a soft loss of stability takes place as ε is increased through 0. Namely, for $\varepsilon \leqslant 0$ the fixed point 0 of the germ f_ε is stable; but for $\varepsilon > 0$ it loses stability, and a stable cycle of period 2 arises: a pair of points, close to $\pm\sqrt{\varepsilon}$, are permuted by the diffeomorphism f_ε. For the diffeomorphism $f_\varepsilon^2 \equiv f_\varepsilon \circ f_\varepsilon$ each of these points is fixed and stable. Under the assumption that for $\varepsilon \leqslant 0$, all the other multipliers lie inside the unit circle, a soft loss of stability by a limit cycle corresponds to this perestroika. For $\varepsilon > 0$, the original cycle becomes unstable and a stable limit cycle of approximately double the period appears at a distance of order $\sqrt{\varepsilon}$. (see Fig. 18).

Feigenbaum (1978) discovered that in generic one-parameter families of diffeomorphisms a change of the parameter within a finite interval may cause an infinite number of period-doublings. For concrete mappings infinite doubling sequences were numerically found several years before this by two ecologists: A.P. Shapiro (1974) and R.M. May (1975).

The phenomenon of Feigenbaum period-doubling cascades is described in detail in Sect. 6 below.

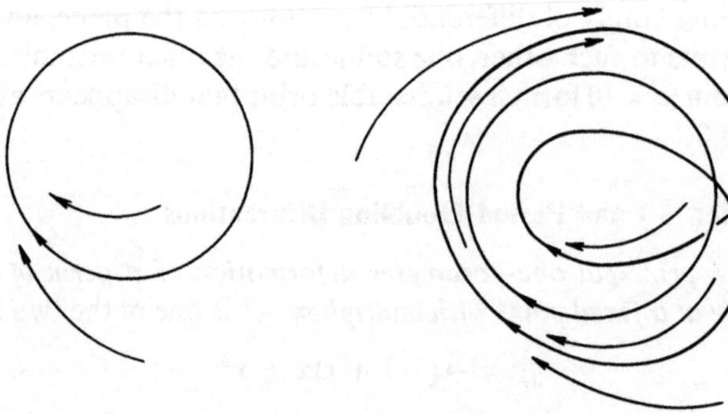

Fig. 18. A period doubling bifurcation

1.3. A Pair of Complex Conjugate Multipliers. Deformations of germs of diffeomorphisms with a pair of complex conjugate multipliers have a topological invariant along the unit circle (the argument of the multiplier of modulus 1). Even in the class of germs with a pair of multipliers $\exp(\pm i\omega)$ (ω fixed), versal deformations depending upon a finite number of parameters have not been constructed and, probably, do not exist.

In generic one-parameter families, there are germs with a pair of multipliers $\exp(\pm i\omega)$ that satisfy the following condition of genericity: by a change of coordinates the corresponding reduced germ takes the form

$$z \mapsto e^{i\omega}z + az|z|^2 + O(|z|^4). \tag{3}$$

Only deformations of such reduced germs are considered below.

A frequency ω "commensurable with 2π" ($\omega/2\pi = p/q$ with p and q positive integers) is called *a resonance of order q*. A resonance is called *strong* if its order is at most 4.

Conditions for Genericity

1. Absence of strong resonance: $\omega \neq 2\pi p/q$ for any $q \leq 4$, (3a)

2. $\operatorname{Re} a \neq 0$. (3b)

Everywhere in this subsection we assume that there is no strong resonance; a strong resonance appears unavoidably only for two (and more) parameters.

A deformation of the germ (3), with the aid of a change of coordinates depending upon the parameters, takes the form

$$z \mapsto g^1_{v(z,\varepsilon)}z + O(|z|^4), \tag{4}$$

where g^1_v is a shift by unit time along a phase curve of the flow v, where:

$$v(z, \varepsilon) = z[i\omega + \lambda(\varepsilon) + A(\varepsilon)\rho], \tag{5}$$

$$\rho = z\bar{z}, \qquad \lambda(0) = 0, \qquad A(0) = a, \qquad \text{and} \qquad \operatorname{Re} a \neq 0.$$

For generic families Re $\lambda'(0) \neq 0$. As ε passes through 0 a limit cycle is born in the family of equations

$$\dot{z} = v(z, \varepsilon). \tag{6}$$

It is a circle with center 0 and radius proportional to $\sqrt{\varepsilon}$ (see Sect. 2.2 of Chapter 1). Consequently, in the family (4), if the higher order terms $O(|z|^4)$ are discarded, as the parameter passes through 0, a smooth curve (a circle) is born, which the diffeomorphism rotates by an angle depending upon ε (since the field $v(\cdot, \varepsilon)$ is invariant under rotations).

The bifurcations in the original family are substantially more complicated. An invariant curve homeomorphic to a circle does indeed arise, but it is not smooth. The restriction of the diffeomorphism to the invariant curve is not necessarily equivalent to a rotation. The rotation number of the diffeomorphism on the invariant curve depends on the parameter and converges to $\omega/2\pi$ as the parameter converges to the critical value 0.

Theorem (Nejmark (1959), Marsden and McCracken (1976), Sacker (1964, 1965)). *Consider a local family of diffeomorphisms* $(f; 0, 0)$:

$$f(z, \varepsilon) = e^{i\omega + \lambda(\varepsilon)}z + a(\varepsilon)z\rho + O(\rho^2), \qquad \lambda(0) = 0.$$

Suppose the germ $f(\cdot, \varepsilon)$ *satisfies the genericity conditions* (3a,b), *and suppose the family satisfies the following transversality condition*:

$$\operatorname{Re} \lambda'(0) \neq 0.$$

Then in the local family $(f; 0, 0)$ *an invariant curve homeomorphic to a circle surrounding the origin is born as* ε *passes through* 0 *to the right if* Re $\lambda'(0)$ Re $a < 0$ *(to the left if the inequality is reversed). This curve, generally speaking, is finitely smooth, but its degree of smoothness tends to infinity as* $\varepsilon \to 0$.

Theorem (Newhouse *et al.* (1983)). *If two generic one-parameter deformations of germs of diffeomorphisms* $(\mathbb{R}^2, 0) \to (\mathbb{R}^2, 0)$ *with a pair of complex multipliers on the unit circle are topologically equivalent, then the multipliers of the germs being deformed coincide.*

This theorem follows from the topological invariance of the rotation number for a diffeomorphism of the circle.

1.4. Nonlocal Bifurcations in One-Parameter Families of Diffeomorphisms. On the invariant curve of a diffeomorphism in the family (4), its rotation number changes with changes of the parameter. If the rotation number is irrational, the orbits formed by iterates of a germ of the diffeomorphism are everywhere dense on the invariant curve; if it is rational, then in a generic family, as distinct from its reduced form (4), (5), (6), there arise a finite number of long-period cycles (the period is equal to the denominator of the irreducible fraction defining the rotation number).

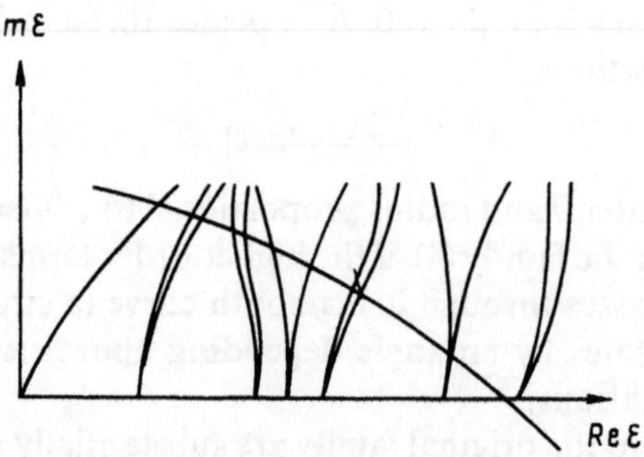

Fig. 19. The bifurcation diagram of the family (7) of diffeomorphisms and the corresponding family of differential equations. The base of the one-parameter family and the real axis are shown as thick curves.

It is convenient to study this phenomenon by considering a two-parameter family of diffeomorphisms in which the parameters are proportional to the logarithm of the complex multiplier: its real and imaginary parts are the two (real) parameters of the family. After a change of coordinates such a family has the form

$$z \mapsto e^{i\varepsilon}z + A(\varepsilon)z\rho + O(|z|^4), \tag{7}$$

if ε varies over a neighborhood of any value ω on the interval $[0, 2\pi]$ not containing "points of strong resonance" ($\varepsilon \neq 2\pi p/q$ for $1 \leqslant q \leqslant 4$). Suppose $\operatorname{Re} A(\omega) < 0$. Then an invariant curve is born as the parameter ε passes from the lower to the upper half-plane near to the point ω. It can be proved[7] that this curve depends on ε in a finitely-smooth way as ε passes through the intersection of some neighborhood of the point ω with the upper half-plane. The set of values of the parameter ε for which the rotation number of the diffeomorphism (7) on its invariant curve equals p/q is called the resonance domain of p/q.

The resonance domain of p/q lies in the upper half-plane and approaches the real axis in the upper half-plane at the point $2\pi p/q$ in a narrow tongue: its boundary curves intersect like two parabolas of degree $(q-2)/2$ (see Fig. 19). The position of these zones is reminiscent of the position of resonance zones of families of diffeomorphisms of the circle given by trigonometric polynomials; that is, it recalls a problem of Mathieu type in the sense of Arnol'd (1983); see Fig. 11 of Arnol'd and Il'yashenko (1985).

A generic one-parameter family induced from (7) intersects a countable number of resonance zones on any interval (however small) containing a real value of the parameter, different from a strong resonance. As the parameter passes

[7] This is easy to accomplish with the help of the considerations in Marsden and McCracken (1976, Chapter 6). However, it seems that an explicit formulation of the result and its proof is absent from the literature.

through this value a countable number of cycles are born and die, the periods of which become larger the closer the parameter approaches to the real axis (see Fig. 19); (V.S. Kozyakin; p. 283 in Arnol'd (1978)).

1.5. Nonlocal Bifurcations of Periodic Solutions. Suppose that in three-dimensional phase space for a generic one-parameter family of differential equations a stable limit cycle exists at the zero value of the parameter ε, with a pair of (Floquet) multipliers $e^{\pm i\omega}$ on the unit circle (stability may be attained by reversal of time if necessary). Since this is a generic one-parameter family, one can assume that $\omega \neq 2\pi p/q$ for any $q \leqslant 4$. Then, as ε passes through 0 in the direction corresponding to the passage of the multipliers from the interior of the unit circle to its exterior, an invariant torus of thickness $\sqrt{\varepsilon}$ arises close to the limit cycle. On this torus an infinite number of limit cycles with long periods are born and die as the parameter varies. As ε moves further from 0, the torus loses smoothness and may turn into a strange attractor, as is described below.

1.6. Bifurcations Resulting in Destructions of Invariant Tori. Suppose that at $\varepsilon = 0$, a limit cycle loses stability to a newly born invariant torus in a generic two-parameter family of C^k-vector fields ($k \geqslant 4$). Then, as was shown above, resonance tongues exist in the plane of parameters; these tongues correspond to the presence of nondegenerate limit cycles of the vector field on the torus. Moreover, the torus itself is a union of the unstable manifolds of saddle cycles with the stable cycles.

There is great interest in clarifying the scenarios leading away from a periodic regime, which corresponds to the presence of a stable cycle on the torus, to a regime of aperiodic oscillations, which may correspond to a strange attractor. In the first place, it is important to do this because numerical and laboratory investigations, and even investigations in nature, of a large number of physical and other problems (Couette flow, convection in a horizontal layer of a viscous fluid, oscillations in radio and VHF generators, etc.) show that the birth of stochastic oscillations as a two-dimensional torus (on which the rotation number is rational) is destroyed is a widely occurring phenomenon. Before the invariant torus disintegrates, it must lose smoothness, nevertheless remaining for some time a topological submanifold of phase space. It is convenient to demonstrate the ways stability is lost by investigating a mapping of an annulus into itself, which for the initial values of the parameters contains a smooth invariant curve. The concrete form of the mapping is immaterial; for example it may be as in Afrajmovich and Shil'nikov (1983) or as in Sect. 8.5 of Chap. 3. Hence, we show only the geometrical picture (Fig. 20). In this drawing the annulus is shown as a rectangle, the left and right sides of which are identified and which consists of points of the stable manifold of a fixed point on the boundary: a saddle-node in Fig. 20e, and a saddle for all the other drawings. In the beginning (Fig. 20a) the invariant curve is smooth. In the cases Fig. 20b, c, d, the invariant curve is still continuous but already has no tangent at the stable fixed point N. The stable

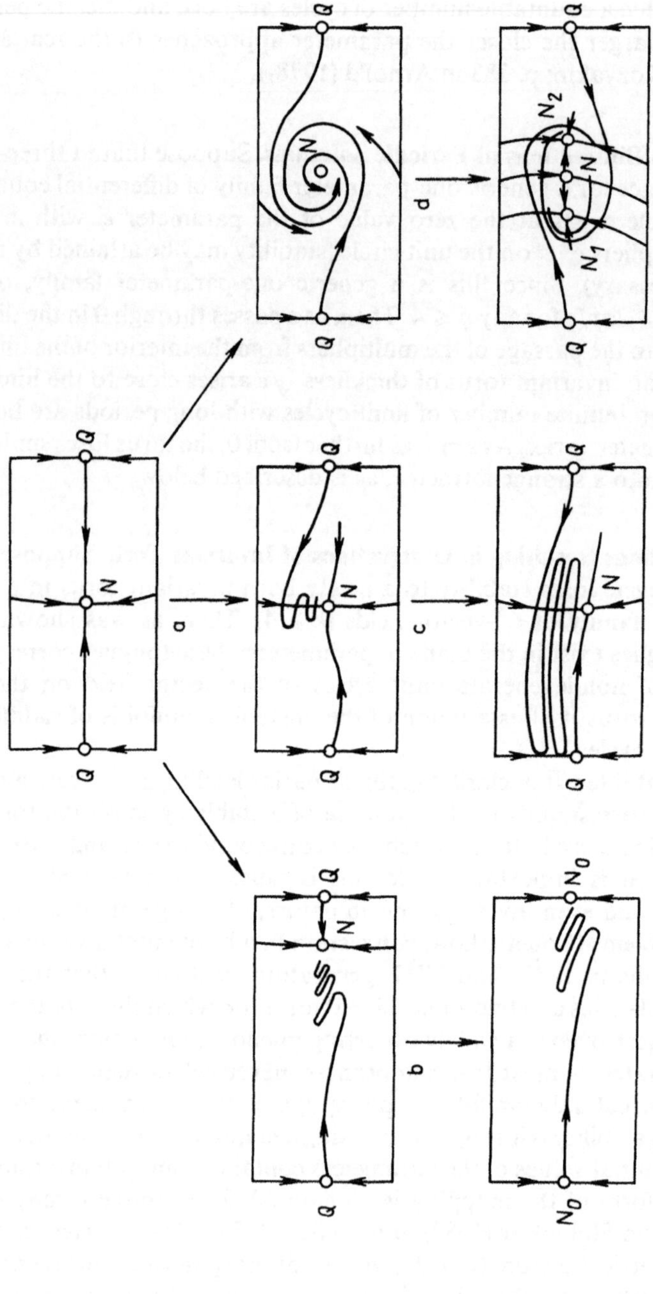

Fig. 20. Scenarios for break-up of a two-dimensional invariant torus

fixed point is a node in the cases Fig. 20b and Fig. 20c; moreover, in case Fig. 20b, the unstable manifold of the saddle Q does not, unlike case Fig. 20c, intersect the nonleading manifold of the node N (this is the invariant manifold of the node corresponding to the eigenvalue with largest modulus). In case Fig. 20d, N is a stable focus with complex multipliers.

The remaining drawings illustrate further possible changes in the phase portrait. In Fig. 20e, the moment of formation of an s-critical saddle node[8] is shown; its disappearance leads to the birth of a strange attractor. In Fig. 20f, the first simple tangency of the unstable and stable manifolds of the point Q is shown. At this moment, and for further changes of the parameters leading to the birth of homoclinic points of a transversal intersection, the attractor in the annulus becomes strange. In Fig. 20g, a period-doubling bifurcation of the fixed point N has already occured, and a stable trajectory with twice the period has arisen (the closed invariant curve has disappeared). Upon further changes in the parameters, a cascade of period-doubling bifurcations can be realized, and a Feigenbaum attractor may arise. In addition to these scenarios, the point N may lose its stability in still another way, for example, a closed invariant curve may arise for which the same scenario may occur as originally.

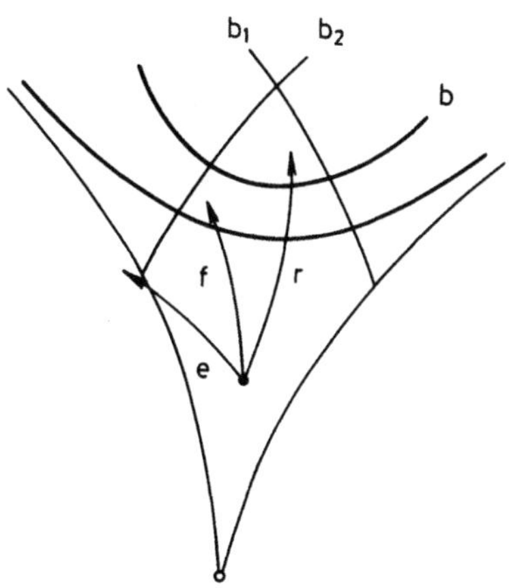

Fig. 21. Bifurcation curves corresponding to the perestroika of an invariant torus

In Fig. 21, a typical bifurcation diagram in a resonant tongue is given. At the point $\varepsilon = 0$ the diffeomorphism is as shown in Fig. 20a. The sequences of bifurcations corresponding to varying ε along the curves e, f, and r are shown, respectively, from left to right in the three columns of Fig. 20 . The bifurcation curves b_1 and b_2 correspond to the formation of points of simple tangency on

[8] An s-critical saddle-node is defined and its bifurcations are studied in Sect. 4 of Chap. 3.

each of the rays $W_Q^u \setminus Q$, and the bifurcation curve b corresponds to a change in stability of the point N.

For one and the same system, the loss of stability at a fixed point may take place differently in different resonance tongues.

§ 2. Bifurcations of Cycles in Generic Two-Parameter Families with an Additional Simple Degeneracy

This section begins with a list of degeneracies that occur in generic two-parameter families of germs of diffeomorphisms at a fixed point, corresponding to isolated values of the parameters. The bifurcations of fixed points with multipliers $+1$ or -1 and with an additional degeneracy in the nonlinear terms remind one of bifurcations of singular points with eigenvalue 0. In contrast, in the case of a pair of complex conjugate multipliers with an additional degeneracy in the nonlinear terms, along with the appearance of closed invariant curves, the bifurcations lead to completely new effects.

2.1. A List of Degeneracies

1°. One multiplier 1 with an additional degeneracy in the nonlinear terms.

2°. One multiplier -1 with an additional degeneracy in the nonlinear terms.

3°. A pair of complex (nonreal) multipliers on the unit circle with an additional degeneracy in the nonlinear terms.

4°. One multiplier $+1$ with multiplicity two; the linear part at 0 is equivalent to the Jordan block

$$\begin{pmatrix} 1 & 1 \\ 0 & 1 \end{pmatrix}.$$

5°. One multiplier 1 and one -1.

6°. One multiplier -1 of multiplicity 2.

7°. A pair of multipliers $e^{\pm i\omega}$, $\omega = 2\pi p/q$, $q = 3$ or 4.

8°. A trio of multipliers: $e^{\pm i\omega}$ and $+1$.

9°. A trio of multipliers: $e^{\pm i\omega}$ and -1.

10°. Two pairs of multipliers $e^{\pm i\omega_1}$ and $e^{\pm i\omega_2}$.

The cases 5° to 7° are called cases of strong resonance.

The cases 8° and 10° lead, in some sense, to the investigation of bifurcations from equilibrium with one zero and a pair of purely imaginary eigenvalues and with two purely imaginary pairs, respectively. As far as we know, specific investigations of bifurcations of fixed points of diffeomorphisms in the cases 8°–10° have not been carried out.

In this section deformations of germs of the first three types with degeneracies in the nonlinear terms are investigated.

2.2. A Multiplier +1 or −1 with Additional Degeneracy in the Nonlinear Terms

Definition 1. *A principal ν-parameter deformation of a germ of a diffeomorphism of the line at a fixed point with multiplier* +1 is one of the two families:

$$x \mapsto x \pm x^{\nu+1} + \varepsilon_1 + \varepsilon_2 x + \cdots + \varepsilon_\nu x^{\nu-1}. \qquad (8^\pm)$$

Definition 2. *A principal ν-parameter deformation of a germ of a diffeomorphism of the line at a fixed point with multiplier* −1 is one of the two families:

$$x \mapsto -x \pm x^{2\nu+1} + \varepsilon_1 x + \varepsilon_2 x^3 + \cdots + \varepsilon_\nu x^{2\nu-1}. \qquad (9^\pm)$$

"Theorem." *In generic μ-parameter families of germs of diffeomorphisms at a fixed point one finds, for* $\nu \leqslant \mu$, *only those germs with multiplier* 1 *(or* −1) *and one-dimensional center manifolds near to which the families are locally weakly equivalent to saddle suspensions over one of the principal families* $(8^\pm)$ *(respectively,* $(9^\pm)$). The case $\nu = \mu$ *corresponds to isolated points in the parameter space. These local families are weakly versal.*

Remark. The word "theorem" was put inside quotation marks above because, as far as we know, a proof has not been published. For $\mu \geqslant 3$ the classification of families of diffeomorphisms described in the "theorem" up to (usual) topological equivalence has functional moduli (see Sect. 5.11 in this chapter).

2.3. A Pair of Multipliers on the Unit Circle with Additional Degeneracy in the Nonlinear Terms (see Chenciner (1981) and Chenciner and Iooss (1979, Sect. 6–13)). Following Chenciner (1981), we consider a germ of a diffeomorphism $f_0: (\mathbb{R}^2, 0) \to (\mathbb{R}^2, 0)$ with a pair of nonreal multipliers on the unit circle and an additional degeneracy: condition (3b) of Sect. 1.3 is violated.

A change of coordinates and parameters reduces a generic two-parameter deformation of this germ to the form:

$$z \mapsto N_{\varepsilon,a}(z) + O(|z|^5); \qquad N_{\varepsilon,a} = g^1_{v_{\varepsilon,a}}$$

Here

$$v_{\varepsilon,a}(z) = \nu z + z(\tilde{A}|z|^2 + B|z|^4), \qquad \nu = i\omega + \varepsilon + i\alpha(\varepsilon),$$

$$\tilde{A} = ia_0 + a + ia_1(\varepsilon, a), \qquad B = B(\varepsilon, a), \qquad \operatorname{Re} B(0,0) < 0;$$

the last inequality can be satisfied by reversing time if necessary. The space of the parameters (ε, a) divides into three domains (Fig. 22), corresponding to one, two, or no closed invariant curves of the field $v_{\varepsilon,a}$ and the mapping $N_{\varepsilon,a}$ (these curves are circles). The curve Γ on which the two invariant circles come together and disappear is given by the equation

$$4\varepsilon b + a^2 = 0, \qquad b = \operatorname{Re} B, \qquad a \geqslant 0,$$

and resembles a branch of the parabola

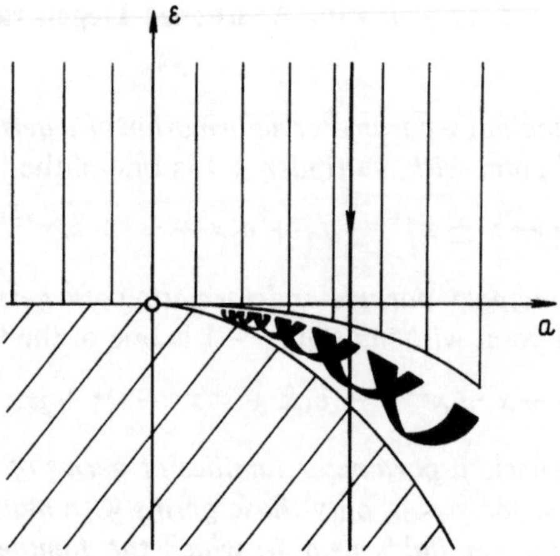

Fig. 22. Zones, colored black, of existence of closed invariant curves near the curve Γ. The zones where the perturbed map has as many closed invariant curves as the unperturbed map is shaded.

$$a^2 = -4\varepsilon b(0, 0), \qquad a \geqslant 0$$

(compare this with Fig. 7).

The following theorem compares the behavior of the normalized mappings $N_{\varepsilon,a}$ with the mappings arising under a generic deformation of the germ f_0. For some values of the parameters one observes similarity, but for others one observes sharp differences between the geometric properties of the perturbed and unperturbed mappings.

Theorem (Chenciner (1981)). *Consider the family f of germs of diffeomorphisms*

$$f_{\varepsilon,a}\colon z \mapsto N_{\varepsilon,a}(z) + O(|z|^5).$$

Let the number ω satisfy Siegel's condition: for some positive constants C and δ and for each rational p/q the inequality $|\omega - p/q| \geqslant Cq^{-(2+\delta)}$ holds. Then for any natural number k there exists a neighborhood U of zero and a neighborhood W of the "parabola" $\Gamma\backslash\{0\}$ bounded by the curve ∂U and two curves tangent at 0 such that:

1. *For each pair (ε, a) in the set $U\backslash W$ the mapping $f_{\varepsilon,a}$ has the same number of closed invariant curves as $N_{\varepsilon,a}$; these curves are C^k-smooth.*

2. *Inside of the neighborhood W and there exists a Cantor set "close to Γ" such that at each of its points (ε, a) the mapping $f_{\varepsilon,a}$ has a unique closed invariant curve. Moreover, such a point (ε, a) is a vertex of a double funnel (colored black in Fig. 22). For all values (ε', a') from the left (right) half of the funnel the mapping $f_{\varepsilon',a'}$ has an attracting (repelling) closed invariant curve.*

Chenciner (1985, Ref. 10) asserts also that in the domain W there exist values of (ε, a) arbitrarily close to zero for which the mapping $f_{\varepsilon,a}$ has arbitrarily many periodic points and homoclinic curves in any neighborhood of zero. Similar effects had been observed previously for germs of diffeomorphisms of the plane, but only in the presence of degeneracies of infinite codimension.

We note in conclusion that information about monodromy transformations translates in a standard way into the language of differential equations: fixed points and periodic points correspond to closed orbits, invariant topological circles correspond to invariant tori or Klein bottles, etc.

§ 3. Bifurcations of Cycles in Generic Two-Parameter Families with Strong Resonances of Orders $q \neq 4$

Generic diffeomorphisms with two multipliers that are roots of unity probably have no finite-parameter versal deformations. In this section instead of families of diffeomorphisms we consider families of vector fields, for which the time-one shifts along trajectories approximate the original families of diffeomorphisms.

Thus, we consider a family of time-one shifts along trajectories of $\mathbb{Z}_q$-equivariant vector fields as a simplified model of a two-parameter family of diffeomorphisms close to a resonance $e^{i\omega}$, $\omega = 2\pi p/q$. The normal forms of families of such fields are described in Sects. 3.3 and 3.4 . Although these simplified families of shifts are not equivalent to the original families of diffeomorphisms, they have more or less the same properties as the original families.

In other words, we limit ourselves to the investigation of bifurcations in the factor-system of the simplified normal form of a family of equations in the neighborhood of a cycle. The interpretation of the results in terms of the original system requires additional work, since even topologically, the bifurcations in the original system and the system in simplified normal form are not always the same (see, for example Sect. 3.5). We begin with the construction of auxiliary families of vector fields on the plane, approximating the monodromy transformations (linearizations of the Poincaré map) of cycles in the case of a strong resonance.

3.1. The Normal Form in the Case of Unipotent Jordan Blocks. A germ of a diffeomorphism at a fixed point on the plane with unipotent linear part can be realized as a monodromy transformation of a periodic differential equation with nilpotent linear part:

$$\dot{x} = Jx + f(x, t), \qquad x \in (\mathbb{R}^2, 0), \qquad t \in S^1 = \mathbb{R}/2\pi\mathbb{Z},$$

$$J = \begin{pmatrix} 0 & 1 \\ 0 & 0 \end{pmatrix}, \qquad f(0, t) \equiv 0, \qquad \partial f/\partial x|_{(0,t)} \equiv 0.$$

Such an equation can be made autonomous (independent of t) with a formal change of variables periodic in t:

$$\dot{x} = Jx + \tilde{f}(x), \qquad \tilde{f}(0) = 0, \qquad \partial\tilde{f}/\partial x|_0 = 0.$$

By a similar change of variable, deformations of the preceding nonautonomous equation are brought to the form

$$\dot{x} = A(\varepsilon)x + F(x, \varepsilon), \qquad A(0) = J, \qquad F(x, 0) = \tilde{f}(x).$$

Moreover, in a neighborhood of $(x, \varepsilon) = (0, 0)$ there exists a smooth change of variables taking the original deformation into the above autonomous family, except for the addition of a remainder-germ which is flat (all its derivatives are zero at $(x, \varepsilon) = (0, 0)$). This "almost autonomous" deformation is little studied; on the other hand, if one disregards the flat remainder-germs, the resulting deformations of germs of vector fields with nonzero nilpotent linear parts at a singular point on the plane can be investigated in detail. These deformations are described in Sect. 4.2 of Chap. 1.

Analogously, neglecting flat remainder-germs in the remaining cases of strong resonance, deformations of diffeomorphisms can be put into the form of deformations of shift-maps along the phase curves of a vector field so that a shift and a deformation are equivalent relative to a finite symmetry group acting on the phase space. For the pair of multipliers $+1$ and -1 this group is S_2 and is generated by the symmetry $(x, r) \mapsto (x, -r)$; for a pair $\exp(\pm 2\pi i p/q)$ this group is $\mathbb{Z}_q$ and is generated by a rotation of angle $2\pi/q$.

The reduction of the problem for deformations of germs of diffeomorphisms to a problem for equivariant deformations of germs of vector fields (in cases of strong resonance $\omega = 2\pi p/q$ or a pair of multipliers $+1$ and -1) can be carried out with the help of averaging in Seifert and Möbius foliations, as is described below.

3.2. Averaging in the Seifert and the Möbius Foliations. We consider the differential equation

$$\dot{z} = i\omega z, \qquad \omega = p/q, \qquad t \in \mathbb{R}/2\pi\mathbb{Z} = S^1, \qquad z \in \mathbb{C}$$

in the product space $S^1 \times \mathbb{C}$. The partition of the extended phase space $S^1 \times \mathbb{C}$ into the integral curves of this equation is called a Seifert foliation of type p/q. All solutions of this equation, except the trivial solution, are $2\pi q$-periodic, and a rotation of the z-plane by $2\pi p/q$ takes each integral curve into itself.

Suppose v is an arbitrary vertical (tangent to the fibers $\{t\} \times \mathbb{C}$) vector field in the product $S^1 \times \mathbb{C}$, fibered over S^1. We average it with respect to time along integral curves of the previous equation. By this we mean that the field v defines a field $\tilde{v}$ on the universal covering space $\mathbb{R} \times \mathbb{C}$ of the space $S^1 \times \mathbb{C}$, periodic under shifts of 2π along $\mathbb{R}$. Fix an initial section, say $\{t_0\} \times \mathbb{C}$. The total space of the bundle $\mathbb{R} \times \mathbb{C}$ is mapped onto this section so that each phase curve of the field

$$i\omega z\partial/\partial z + \partial/\partial t$$

goes into its point on the initial section. This mapping carries the vectors of the vector field $\tilde{v}$ into the initial section. At each point of the initial section arises a vector periodic in t. Averaging it with respect to t, we obtain a vector of the averaged field at the point of the plane $\mathbb{C}$ considered.

This operation is called *averaging the original field in the Seifert foliation.* An arbitrary vector field v is transformed by averaging in the Seifert foliation into a $\mathbb{Z}_q$-equivariant vector field on the plane.

We also consider the product of a Möbius strip with a line, obtained by identifying the points (t, x, r) and $(t + 2\pi, x, -r)$ in $\mathbb{R}^3$. The partition of this space into the integral curves of the equations

$$\dot{x} = 0, \qquad \dot{r} = 0$$

is called the Möbius foliation. This foliation is a "linear approximation" for studying a limit cycle with multipliers $+1$ and -1. Averaging along this foliation yields S_2-equivariant vector fields on the plane, the deformations of which are described in Sect. 4.4 of Chap. 1.

3.3. Principal Vector Fields and their Deformations

Definitions. 1. Equations of the form

$$\dot{z} = Az|z|^2 + B\bar{z}^{q-1}, \qquad z \in \mathbb{C},$$

and the corresponding vector fields on the plane are called *principal singular* $\mathbb{Z}_q$*-equivariant equations and fields* for $q > 2$.

2. The two-parameter family $v_\varepsilon = \varepsilon z + v_0$, where the parameters are the real and imaginary parts of ε, is called the *principal deformation of the principal singular* $\mathbb{Z}_q$*-equivariant vector field* v_0 $(q > 2)$.

3. The equations

$$\ddot{x} = ax^3 + bx^2y \qquad (q = 2), \qquad \ddot{x} = ax^2 + bxy \qquad (q = 1)$$

and the vector fields on the phase plane $(x, y = \dot{x})$ given by these equations are called *principal singular* $\mathbb{Z}_q$*-equivariant equations* and *fields* for $q = 2$ and $q = 1$, respectively.

4. A deformation, produced by the addition of the terms $\alpha x + \beta y$ $(q = 2)$ and $\alpha + \beta x$ $(q = 1)$ to the right-hand sides of these second order equations is called *a principal deformation of a principal singular field* for $q = 2$ and $q = 1$.

Thus, the list of principal deformations of principal singular $\mathbb{Z}_q$-equivariant vector fields is as follows

$$\dot{z} = \varepsilon z + Az|z|^2 + B\bar{z}^{q-1}, \qquad q \geqslant 3,$$

$$\ddot{x} = \alpha x + \beta y + ax^3 + bx^2y, \qquad q = 2,$$

$$\ddot{x} = \alpha + \beta x + ax^2 + bxy, \qquad q = 1.$$

Here z, ε, A and B are complex variables and x, y, α, β, a, and b are real, the parameters of the deformations are denoted by Greek letters, and $y = \dot{x}$.

3.4. Versality of Principal Deformations

"Theorem". *For each q all principal singular $\mathbb{Z}_q$-equivariant vector fields can be classified as degenerate or nondegenerate so that:*

1) *In generic two-parameter families of germs of singular $\mathbb{Z}_q$-equivariant vector fields at zero there are only those germs with nilpotent linear parts that are equivalent to one of the nondegenerate principal fields.*

2) *The corresponding local families are equivalent to principal deformations and are versal.*

3) *Degenerate fields form the union of a finite number of open submanifolds in the space of principal singular fields.*

4) *Nondegenerate fields form the union of a finite number of open connected domains.*

5) *The principal deformations of germs of nondegenerate fields within each connected component are topologically equivalent.*

The word "theorem" is in quotation marks because the theorem has been proved only for $q \neq 4$ (Arnol'd (1978, 1977); Khorozov (1979)). For $q \neq 4$ the conditions for nondegeneracy can be written explicitly:

$$a \neq 0, \quad b \neq 0 \quad \text{for } q = 1, 2; \qquad \operatorname{Re} A \neq 0, \quad B \neq 0 \quad \text{for } q = 3 \text{ and } q \geqslant 5.$$

The bifurcation diagrams and perestroikas of the phase portraits are illustrated above in Fig. 10 for $q = 1$, in Fig. 23 for $q = 2$ (we obtain $b < 0$ by reversing time if necessary), and in Figs. 24 and 25 for $q = 3$ and $q = 5$ (we obtain $\operatorname{Re} A < 0$ by reversing time if necessary).

3.5. Bifurcations of Stationary Solutions of Periodic Differential Equations with Strong Resonances of Order $q \neq 4$. In each principal singular $\mathbb{Z}_q$-equivariant family, for some values of the parameters forming curves in the ε-plane, separatrix polygons arise. A time-one shift along the phase curves of a field of a family approximates the qth-iterate of the monodromy transformation of a limit cycle losing its stability as a pair of multipliers passes through a strong resonance. Fixed points of the qth-iterate of the monodromy transformation and $2\pi q$-periodic cycles of a periodic equation correspond to singular points of the fields of a family; the incoming and outgoing separatrices of saddles are the stable and unstable manifolds of fixed points. Once two separatrices of singular points intersect, they must coincide in their entirety. This is false for invariant curves of fixed points of diffeomorphisms. These curves generally intersect transversally, but for diffeomorphisms of a generic one-parameter family these curves may be tangent for some values of the parameters. Such a tangency is called homoclinic or heteroclinic, depending upon whether the tangency is of invariant curves belonging to the same or different singular points. Consider the value of the parameter of a principal $\mathbb{Z}_q$-equivariant family corresponding to a vector field with a separatrix loop (homoclinic loop) (the cases $q = 1, 2$), or a separatrix polygon (the cases $q = 2, 3, 4$). One should expect that there exists a nearby value of the parameter of the family of periodic differential equations to which there corresponds either a homoclinic or a heteroclinic tangency of invariant manifolds of fixed points of the qth-iterate of the monodromy transformation. The bifurca-

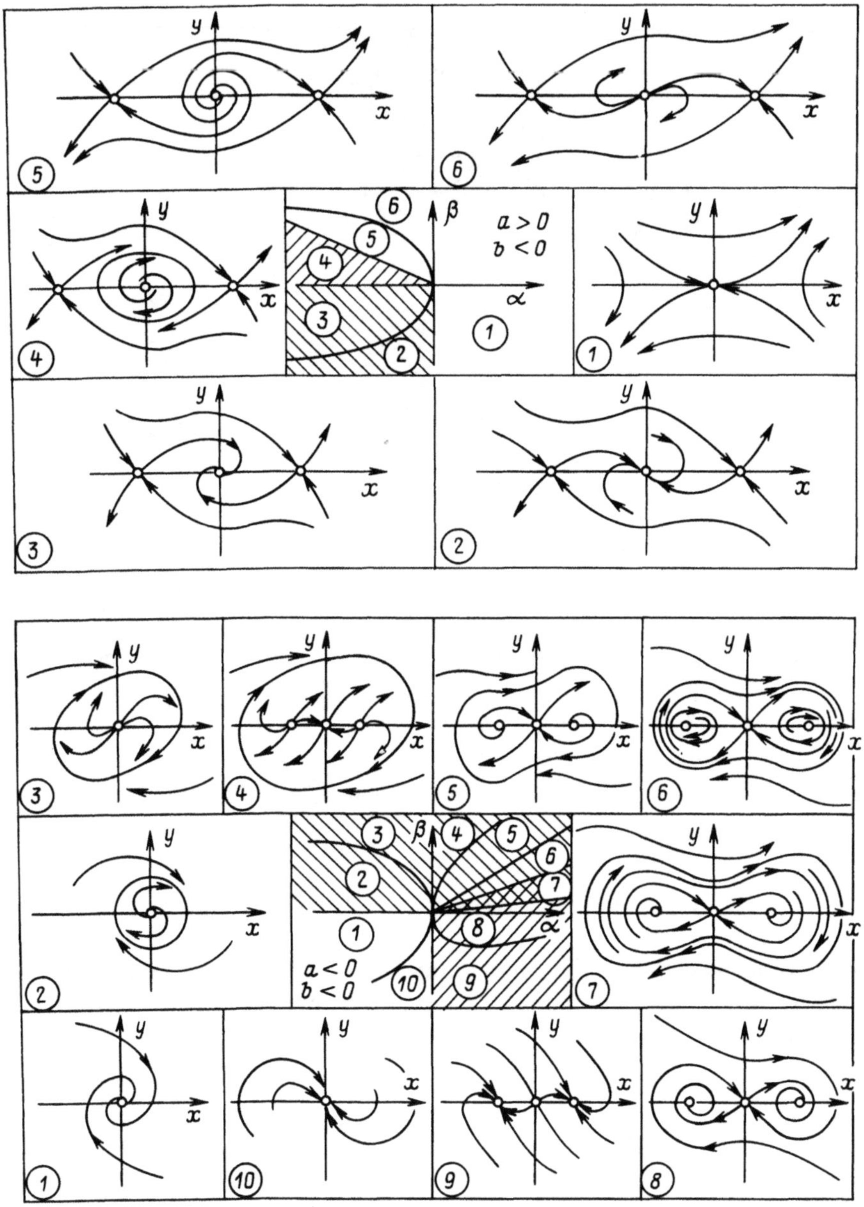

Fig. 23. Bifurcations in the principal $\mathbb{Z}_2$-equivariant family

tions of such diffeomorphisms are described in Sect. 6 of Chap. 3. Here we only remark that, as a rule, nontrivial hyperbolic sets arise at such bifurcations.

Conjecture. In generic two-parameter families of vector fields in which a loss of stability of a limit cycle occurs when passing through a strong resonance, there

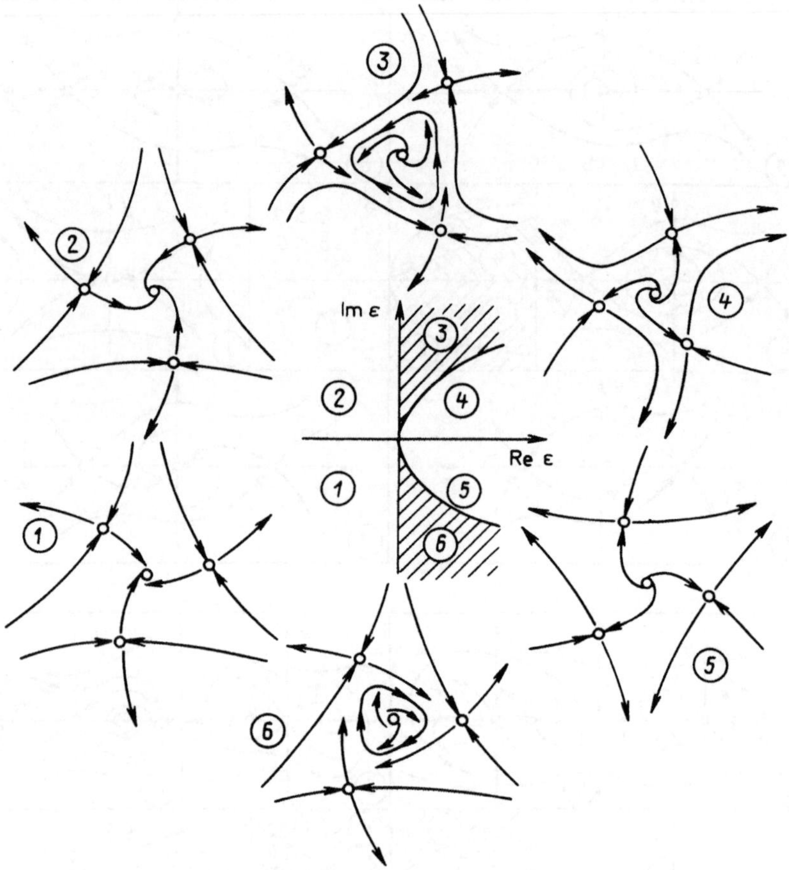

Fig. 24. Bifurcations in the principal $\mathbb{Z}_3$-equivariant family

are vector fields with nontrivial hyperbolic sets. The parameter values that correspond to such fields form very narrow tongues as they approach the critical value of the parameter.

Remarks. 1. As far we know, this conjecture is not proved, although statements near to it were given long ago (see Arnol'd (1978, § 21f)).

2. The union of a hyperbolic set arising at a homoclinic tangency, and all the trajectories which are attracted to it, in general has measure zero in phase space. However, there is a set of positive measure that is the union of trajectories which spend arbitrarily long times near the hyperbolic set (compared with the period of a cycle; from the point of view of a physical observer this time may be considered infinite). Therefore it follows that, upon loss of stability by a limit cycle near a strong resonance, one expects chaos to arise.

3. Consider a one-parameter family in which a limit cycle loses stability as a pair of multipliers pass through the unit circle close to the point -1. As the parameter changes the family may undergo the following sequence of changes:

Fig. 25. Bifurcations in the principal $\mathbb{Z}_5$-equivariant family

the stable cycle softly loses stability with the birth of a stable torus, which rapidly develops a pinch, so that the form of the meridian of the torus approaches a figure eight; in approaching the center of the figure eight (where an unstable cycle is found), the attracting set, almost contracting to the figure-eight meridian, disintegrates near a homoclinic separatrix (Yu.I. Nejmark).

In this case, a phase trajectory winds around one and then another half of the destroyed torus, jumping in an apparently random fashion from one side to the other.

§4. Bifurcations of Limit Cycles for a Pair of Multipliers Crossing the Unit Circle at $\pm i$

For the study of the loss of stability by a cycle with multipliers near to $\pm i$, it is necessary to study the $\mathbb{Z}_4$-equivariant family of equations

$$\dot{z} = \delta z + Pz|z|^2 + Q\bar{z}^3. \tag{10}$$

The bifurcations of the phase portraits in this family are described below.

4.1. Degenerate Families. Here we study those fixed values of P and Q for which nongeneric bifurcations take place in the family (10) with the parameter $\delta \in \mathbb{C} \setminus 0$.

Lemma. *For $PQ \neq 0$, the family* (10), *with parameter $\delta \in \mathbb{C}\backslash 0$, is equivalent (perhaps after reversal of time) to the family induced from the family*

$$\dot{z} = \varepsilon z + Az|z|^2 + \bar{z}^3 \tag{11_A}$$

with parameter ε: $|\varepsilon| = 1$, where $A = a + bi$, $a \leqslant 0$, $b \leqslant 0$.

◂ The equality $|\varepsilon| = 1$ can be obtained by multiplying t by a nonzero constant. Changing the sign of the time and Im z, we obtain the inequalties $a \leqslant 0$, $b \leqslant 0$. Sending $z \mapsto cz$, we get the equality $Q = 1$. Each transformation preserves the results of the previous ones, and together they transform the family (10) into (11_A). ▸

We say that those values of A for which there are degeneracies with codimension 2 or higher in the one-parameter family (11_A) are degenerate values. The degenerate values of A found up to now are shown in Fig. 26; those shown by solid curves were found analytically; those shown by dashed curves were found numerically.

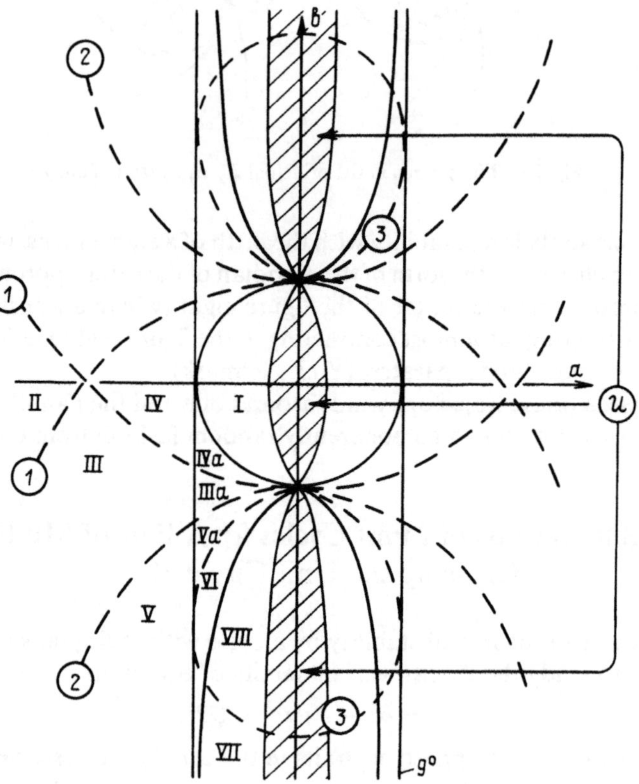

Fig. 26. The set of values of A, corresponding to degenerate principal $\mathbb{Z}_4$-equivariant families (shown with solid and dashed curves). The shaded regions illustrate values of A for which limit cycles in the families (11_A) have been investigated.

4.2. Degenerate Families Found Analytically. These families are classified in Table 1 below. The equations of the components of the set of degenerate values of A are given in the first column of Table 1; degeneracies with codimensions higher than 1 are described in the second column. The values of α ($\varepsilon = e^{i\alpha}$) at which degeneracies occur are given in the third column (sometimes α is given implicitly).

Table 1

Component	The degeneracies	Values of α
$a^2 + b^2 = 1$	Degenerate singular points are born at infinity	$\lvert a \sin \alpha - b \cos \alpha \rvert = 1$
$b = \pm(1 + a^2)/(1 - a^2)^{1/2}$	Nontrivial singular points are nonelementary (the linearization operator is nilpotent).	$\lvert a \sin \alpha - b \cos \alpha \rvert = 1$
$\lvert a \rvert = 1$	Two bifurcations of codimension 1 occur simultaneously: nonzero singular points become degenerate, and the singular point 0 changes stability.	$\alpha = \pm\pi/2$
$a = 0$	The equation is Hamiltonian	$\alpha = \pm\pi/2$

4.3. Degenerate Families Found Numerically. These correspond to the union of the three dashed curves in Fig. 26. If A belongs to curve 1 or 2, then one of the equations in the family (11_A) has a complex cycle (a separatrix polygon) having four singular points that are saddle-nodes at its vertices; moreover, the center manifold of one singular point is the stable (or unstable) manifold of another (Fig. 27 a,b). If A belongs to curve 3, then one of the equations of the family (11_A) has a complex cycle with four singular points that are saddles (Fig. 27c), and the principal term of the monodromy transformation of this cycle is linear (the monodromy transformation is defined on exterior semi-transversals).

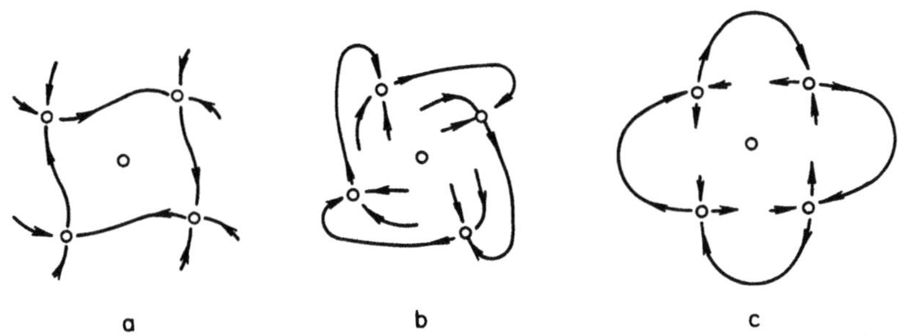

Fig. 27. Degeneracies corresponding to: a) the curve 1, b) the curve 2, c) the curve 3 of Fig. 26.

4.4. Bifurcations in Nondegenerate Families. The connected components into which the curves of degenerate values of A divide the third quadrant are numbered in Fig. 26. In Fig. 28 the sequence of bifurcations occurring in the family (11_A) is shown for A belonging to the domain VIII. The sequences of perestroikas known to occur in the remaining domains were given by Arnol'd (1977, 1978) and Berezovskaya and Khibnik (1979).

In the domains, whose numbers have the letter '*a*' attached, the sequences of perestroikas seem to be identical, with a single exception. In the families (11_A) corresponding to one of two such regions, first the nonzero singular points disappear, and then a limit cycle, surrounding 0, disappears at the origin; in the families corresponding to the other domain, the order of these events is reversed. The curves 1, 2, 3 and the degeneracies connected with them were predicted by Arnol'd (1977, 1978), and were investigated by Berezovskaya and Khibnik (1979, 1980).

4.5. Limit Cycles of Systems with a Fourth Order Symmetry. Limit cycles of systems (11_A) that are nearly Hamiltonian were investigated by Nejshtadt (1978). Namely, he showed that there exists a neighborhood U of the imaginary axis with the points $A = \pm i$ excluded (shaded in Fig. 26) and having the following property: For each point A in this neighborhood, the equations in the family (11_A) have no more than two limit cycles; each of the cases of 0, 1, and 2 cycles is realized.

Remarks. The following questions are open:

1. Do there exist degenerate values of A besides those indicated above?
2. In nondegenerate families (11_A) do perestroikas occur besides those specified in Arnol'd (1977, 1978)?
3. How many limit cycles can equation (10) have?

§ 5. Finitely-Smooth Normal Forms of Local Families

A family of differential equations may be reduced to a normal form by an analytic or C^∞ transformation only in exceptionally rare cases. Useful information often can be extracted, however, from a finitely-smooth reduction. For example, a C^1-smooth reduction permits one to follow directions of invariant manifolds, etc. Finitely-smooth normal forms of families are used to normalize the equations of fast motions in the theory of relaxation oscillations (see Sect. 2.1 of Chap. 4) and also in the investigation of nonlocal bifurcations in Chap. 3.

5.1. A Synopsis of Results. Integrable finitely-smooth normal forms have been obtained for deformations of germs of vector fields at a hyperbolic fixed point or on a hyperbolic cycle, under the assumption that the linearizations of corresponding germs are nonresonant or have at most a simple resonance. One can also obtain a finitely-smooth normal form for a versal deformation of a germ of a vector field with one zero eigenvalue at a singular point.

Fig. 28. The sequence of perestroikas that automatically occur in the principal $\mathbb{Z}_4$-equivariant family corresponding to values of the parameter A from a subdomain of the domain VIII

At this point, positive results, i.e., normal forms given by simple formulas, are exhausted. Already a deformation of the germ of the mapping

$$x \mapsto x + x^2 + \cdots$$

has a functional invariant, even in the C^1-smooth classification: two deformations with different functional invariants are not C^1-smoothly equivalent. Matters are analogous for deformations of other nonhyperbolic germs of diffeomorphisms at fixed points, or vector fields on cycles found in generic one-parameter families. The C^1-smooth classification of deformations of germs of vector fields at a singular point with a pair of purely imaginary eigenvalues also has functional invariants.

General theorems about finitely-smooth normal forms that are not necessarily integrable are given in Sect. 5.3.

5.2. Definitions and Examples

Definition 1. A deformation of a germ of a vector field at a singular point is called *C^k-smoothly (orbitally) versal* if any deformation of this germ is C^k-smoothly (orbitally) equivalent to one induced from the original germ.

Definition 2. A deformation of a germ of a vector field at a singular point is called *finitely-smoothly (orbitally) versal* if for any k there exists a representative germ that is a C^k-smooth (orbitally) versal deformation of this germ.

Finitely-smooth (orbitally) versal deformations of a vector field on a cycle or finitely-smooth (orbitally) versal deformations of a diffeomorphism at a fixed point are defined analogously.

Remark. A finitely-smooth versal deformation is *arbitrarily smooth*, but it is not *infinitely smooth.* The reason is that the higher the degree of smoothness of a diffeomorphism conjugating an arbitrary deformation with the deformation induced from the versal deformation, the smaller the domain of variation of parameters. This situation is analogous to that of the smoothness of a center manifold: for a smooth vector field the center manifold is arbitrarily smooth, but it is not infinitely smooth: the higher the requirement of smoothness, the smaller the neighborhood of the singular point on the center manifold where this smoothness holds.

Example 1. Consider the system

$$\begin{aligned} \dot{\varepsilon} &= 0 \\ \dot{x} &= x^2 - \varepsilon, \\ \dot{y} &= -y + f(x, y)(x^2 - \varepsilon), \end{aligned} \qquad (12)$$

The center manifold of this system is two-dimensional. We investigate its intersection with the planes $\varepsilon = \text{const}$. The system (12) is obtained by adding the equation $\dot{\varepsilon} = 0$ to the system of the last two equations in (12). For $\varepsilon > 0$, this two-dimensional system has two singular points: the saddle S_ε: $(\sqrt{\varepsilon}, 0)$ and the node N_ε: $(-\sqrt{\varepsilon}, 0)$; the ratio α of their eigenvalues is equal to $1/2\sqrt{\varepsilon}$.

The intersection of the center manifold of (12) with a plane $\varepsilon = \text{const.}$ contains a (smooth) separatrix of the saddle S_ε and a phase curve that enters the node N_ε. But, for nonintegral α, exactly two smooth invariant phase curves enter the node and the rest of the phase curves have only a finite number of one-sided derivatives at the point $(-\sqrt{\varepsilon}, 0)$ (this number is $[\alpha]$, the integer part of α). Therefore, if we choose the function $f(x, y)$ so that in the system (12) we separate the separatrix of the saddle from the smooth invariant manifolds of the node, then the center manifold of the system (12) is not smooth. The smoothness of that part, contained in the strip $|\varepsilon| < \varepsilon_0$, is no greater than $1/(2\sqrt{\varepsilon_0})$ and goes to infinity as $\varepsilon_0 \to 0$.

Example 2. Consider the deformation of the germ of the vector field at a saddle point on the plane given by the system:

$$\dot{x} = A(\varepsilon)x + \cdots,$$
$$\dot{\varepsilon} = 0, \qquad x \in \mathbb{R}^2, \qquad \varepsilon \in \mathbb{R}^m. \tag{13}$$

If the ratio α of the eigenvalues of the operator $A(0)$ is negative and irrational, then the formal normal form of this system has the form

$$\dot{x} = A(\varepsilon)x$$
$$\dot{\varepsilon} = 0.$$

However, since the eigenvalues of the operator $A(0)$ have different signs (the singular point 0 is assumed to be a saddle), the ratio of eigenvalues of $A(\varepsilon)$ admits rational values on any interval of variation of the parameters (if the deformation is generic). Therefore, there exist arbitrarily small values of ε for which the formal normal form of the equation

$$\dot{x} = A(\varepsilon)x + \cdots$$

contains nonlinear (resonant) terms. Consequently, there does not exist a C^∞-change of variables transforming the original system into a family of linear equations.

However, the smaller the base of the family, the higher the order of the resonances in the equations of the family. A resonance of high order does not prevent C^k-equivalence of the system with its linear part, unless the order of the resonance is sufficiently large compared to k (see references 3–5 to Sternberg in Hartman (1964)). For irrational α the local family (13) is finitely-smoothly equivalent to a linear one. Therefore the relationship of finitely-smooth equivalence is quite natural for the study of normal forms of local families.

5.3. General Theorems and Deformations of Nonresonant Germs

Theorem 1. (G.R. Belitskij (1978, 1979)). *A smooth germ of a diffeomorphism at a hyperbolic fixed point has a C^k-smoothly versal deformation having finitely many parameters for any k. This deformation is C^k-smoothly equivalent to a* *polynomial deformation. If the multipliers of the germ form a multiplicatively nonresonant n-tuple* $\lambda = (\lambda_1, \ldots, \lambda_n)$:

$$\lambda_j \neq \lambda^s, \quad \text{for each } j \in \{1, \ldots, n\}, \quad s \in \mathbb{Z}^n_+,$$

$$s = (s_1, \ldots, s_n), \quad |s| \overset{\text{def}}{=} s_1 + \cdots + s_n \geqslant 2,$$

then a versal deformation of the germ is equivalent to a linear one:

$$x \mapsto A(\varepsilon)x.$$

The analogous result holds for differential equations.

Remarks. 1. The family $\{A(\varepsilon)\}$ is a versal deformation of the operator $A(0)$; such deformations were found by Arnol'd (1972).

2. In applications, changes of variables of a moderate degree of smoothness are often used. Therefore requirements on germs are separately given below which allow one to estimate the degree of smoothness of changes of variables. The following theorem is valid for deformations of arbitrary germs, not only hyperbolic ones.

We consider germs of diffeomorphisms at a fixed point:

$$(x, y) \mapsto (x', y'), \qquad x' = A^c x + \cdots, \qquad y' = A^h y + \cdots$$

(the superscripts c and h stand for "center" and "hyperbolic", respectively); all eigenvalues of A^c lie on the unit circle, and they all lie outside of it for A^h. The variables $y = (y_1, \ldots, y_h)$ are called *hyperbolic* and the eigenvalues of the operator A^h are the *multipliers corresponding to the hyperbolic variables.*

Theorem 2a (Takens (1971)). *Consider a germ of a diffeomorphism at a fixed point for which the moduli of the multipliers corresponding to the hyperbolic variables form a nonresonant n-tuple. Then for any k there exists a representative of the germ that is* C^k*-equivalent to the following*:

$$f: (x, y) \mapsto (f_0(x), A(x)y); \tag{14}$$

here $y = (y_1, \ldots, y_h)$ *is the set of hyperbolic variables,* $y = 0$ *is the center manifold,* x *is a chart on the center manifold, and* f_0 *is the germ of a homeomorphism, whose multipliers are all of modulus one.*

This theorem can be strengthened. For each k we define a "forbidden order of resonance" $N(k)$ in the following way: let $|\lambda_1| \leqslant \cdots \leqslant |\lambda_s| < 1 < |\lambda_{s+1}| \leqslant \cdots \leqslant |\lambda_h|$. We set

$$\beta(k) = \left[\frac{\ln|\lambda_1| - k(\ln|\lambda_h|)}{\ln|\lambda_s|} + k\right] + 1,$$

$$N(k) = \left[\frac{\ln|\lambda_h| - \beta(k)(\ln|\lambda_1|)}{\ln|\lambda_{s+1}|} + \beta(k)\right] + 1.$$

Theorem 2b (Takens (1971)). *Consider a germ of a diffeomorphism at a fixed point for which the moduli of the multipliers corresponding to the hyperbolic* *variables form an n-tuple which does not satisfy resonance relations of order $N(k)$ or less, that is,*

$$|\lambda_j| \neq |\lambda|^s \qquad \text{for } |s| \leqslant N(k).$$

Then the diffeomorphism is C^k-smoothly equivalent to the diffeomorphism (14). *The analogs of Theorems 2a and 2b hold for differential equations.*

In particular, consider a germ of a vector field at a fixed point for which the real parts of the eigenvalues corresponding to the hyperbolic variables form a nonresonant set. For any k there exists a representative of the germ that is C^k-smoothly equivalent to the following:

$$\dot{x} = w(x),$$

$$\dot{y} = A(x)y,$$

where y is the set of hyperbolic variables and x is a chart on the center manifold. ▲

These results can be called "theorems on finitely-smooth suspensions of saddles" and are finitely-smooth analogs of the reduction principle; see Arnol'd (1978) and Arnol'd and Il'yashenko (1985). They are less general than the latter result, however: they place arithmetic requirements on the multipliers (or eigenvalues of the singular point) which are not in the Reduction Theorem. We now turn our attention to deformations of hyperbolic resonant germs.

Definition. The center manifold of the system $\dot{x} = v(x, \varepsilon)$, $\dot{\varepsilon} = 0$, corresponding to the family $\dot{x} = v(x, \varepsilon)$, is called *a center manifold of the local family of equations at the point* (0, 0).

Theorem 3. a) *Consider a family of vector fields at a singular point (germs of diffeomorphisms at a fixed point, or of periodic vector fields on a cycle). Then for each natural number k there exists an $N = N(k)$, such that all representatives of the N-jet of the family are C^k-equivalent on its center manifold.*

b) *Let* $\alpha = \max_{j=1,\dots,h}|\lambda_j|/\min_{j=1,\dots,h}|\lambda_j|$. *Then one can choose $N(k)$ to be the integer* $N(k) = 2[2(k+1)\alpha] + 2$, *where* $[a]$ *denotes the integer part of the number a.*

Remark. We emphasize that all the representatives referred to in this theorem are germs of families on a common center manifold, the N-jets of which coincide at all points of the center manifold.

Theorem 3a for germs of diffeomorphisms easily follows from the theorem of Belitskij (1978, Theorem 2.3.2) in which an estimate on $N(k)$ is given that is somewhat weaker than the one given above.

Theorem 3b was proved by V.S. Samovol (1982), and he also independently obtained the proof of Theorem 3a. These results are applied below to generic one-parameter deformations of hyperbolic germs; for these deformations one can successfully write down integrable normal forms.

5.4. Reduction to Linear Normal Form. From the preceding theorems we immediately obtain the following

Corollary. Let k be an arbitrary natural number. If the eigenvalues of a hyperbolic germ of a vector field of a singular point do not satisfy a resonance relation of order $N(k)$ or less, then a versal deformation of the germ is C^k-smoothly equivalent to a versal deformation of its linear part. In other words, a C^k-smooth change of variables reduces any deformation of the germ to a family of linear vector fields.

We note that the quantity $N(k)$ depends upon the number α that measures the dispersion of the real parts of the eigenvalues at the singular point. The following theorem requires only the absence of any resonance of order 2.

Theorem (E.P. Gomozov (1976)). *Suppose the multipliers of a hyperbolic germ of a diffeomorphism at a fixed point do not satisfy any relation of the form*

$$|\lambda_i| = |\lambda_j| \cdot |\lambda_k| \qquad \text{for } |\lambda_j| < 1 < |\lambda_k|.$$

Then any smooth deformation of this germ is C^1-equivalent to the linear part of the deformation.

5.5. Deformations of Germs of Diffeomorphisms of Poincaré Type.

We recall that a germ of a diffeomorphism at a fixed point is of *Poincaré type* if its multipliers lie on one side of the unit circle (either all inside or all outside the unit circle).

Theorem (N.N. Brushlinskaya (1971). *A versal deformation of a germ of a diffeomorphism of Poincaré type at a fixed point is equivalent to a polynomial family of diffeomorphisms depending upon $d + m$ parameters. Here d is the number of parameters of a versal deformation of the linear part of the original germ, and m is the number of resonance relations satisfied by the multipliers of this linear part. If the deformation is smooth (analytic), then the normalizing change of variables is also smooth (analytic).*

Analogous theorems hold for germs of vector fields at a singular point or on a cycle.

5.6. Deformations of Simply Resonant Hyperbolic Germs

Definition 1. An n-tuple Λ of complex numbers $\lambda \in \mathbb{C}^n$ is called *k-resonant* (*multiplicatively k-resonant, periodically k-resonant*) if the number of generators of the additive group generated by the set of vectors $\{r \in \mathbb{Z}^n_+ \mid (r, \lambda) = 0\}$ (resp., $\{r \in \mathbb{Z}^n_+ \mid \lambda^r = 1\}$, or $\{r \in \mathbb{Z}^n_+ \mid (r, \lambda) \in 2\pi i\mathbb{Z}\}$) is equal to k. If $k = 1$, a k-resonant set is said to be *simply resonant*. A linear vector field with spectrum λ, and also a linear diffeomorphism or a periodic differential equation

$$\dot{x} = Ax, \qquad (t, x) \in S^1 \times \mathbb{R}^n, \qquad S^1 = \mathbb{R}/\mathbb{Z},$$

with the operator A having spectrum λ, is called *k-resonant* if the set λ is k-resonant.

Definition 2. If all resonance relations on the spectrum of the linear part of a vector field at a singular point (the linear part of a diffeomorphism or of a periodic differential equation at a fixed point) are consequences of one relation

$$(r, \lambda) = 0, \tag{15}$$

respectively,

$$\lambda^r = 1, \tag{16}$$

or

$$(r, \lambda) + 2\pi i l = 0, \tag{17}$$

then the field (respectively, the diffeomorphism or the periodic differential equation) is called *strongly simply resonant.*

Definition 3. Let the operator A be the linear part of a strongly simply resonant vector field at a singular point, a diffeomorphism at a fixed point, or a periodic differential equation with autonomous linear part, and suppose λ is the spectrum of A. A real resonant monomial that corresponds to the operator A is defined in the following way.

Let $z_1, \ldots, z_n$ be coordinates in a Jordan basis for A; and in addition let the collection of conjugate coordinate functions on $\mathbb{R}^n$ correspond to the conjugate eigenvalues of A. We call Re z^r, resp. $\mathrm{Re}(z^r e^{2\pi i l t})$, the *resonance monomial* corresponding to the operator A in the first two cases, resp. in the third case. We shall say that the first of these monomials corresponds to the relation (15) or (16), and the second to the relation (17).

Definition 4. a. A family

$$w(x, \varepsilon) = Xg(u, \varepsilon), \qquad X = \mathrm{diag}\, x \tag{18}$$

is called *a principal family of germs of strongly simply resonant vector fields at a singular point,* if u is the resonance monomial corresponding to (15) and g is a vector polynomial in u whose coefficients are parameters of the family. We denote this set of parameters by ε.

b. A family

$$f(x, \varepsilon) = g_w^1(x, \varepsilon)$$

is called *a principal family of germs of strongly simply resonant diffeomorphisms at a fixed point,* if w is the family of vector fields from Definition 4a, g_w^1 is a time-one shift along the phase curves of the field w, and u is thc resonance monomial corresponding to (16).

c. A family (18) is called *a principal family of germs of strongly simply resonant periodic vector fields on a cycle,* if u is the resonance monomial corresponding to relation (17).

Theorem. *Let v be a strongly resonant hyperbolic germ of a vector field at a singular point. Then*

a) *For each natural number k, any smooth deformation of the germ v is C^k-smoothly equivalent to one induced from the principal family* (18), *in which g is a vector polynomial of degree $N(k)$, where $N(k)$ is the same as in Theorem 3.*

b) *If the germ v does not belong to an exceptional subset of infinite codimension, then any smooth deformation of this germ is finitely-smoothly equivalent to one induced from the principal family* (18); *the degree of the polynomial g in this family depends upon the undeformed germ and not upon the smoothness of the conjugating diffeomorphism.*

c) *The exceptional subset in this theorem is the same as in the the theorem of Ishikawa on formal finite determinacy for vector fields; see Arnol'd and Il'yashenko* (1985, Sect. 3.4 of Chap. 3).

The analog to part a) of this theorem is true for germs of diffeomorphisms at a fixed point and periodic vector fields on a cycle.

Theorem. *Only those germs of saddle resonant vector fields (a resonance* $p\lambda_1 + q\lambda_2 = 0$, *p and q relatively prime natural numbers) that are smoothly orbitally equivalent to the germ*

$$\dot{x} = x(1 \pm u^\mu + a_0 u^{2\mu}),$$
$$\dot{y} = -yp/q$$

occur in generic smooth finite-parameter families of vector fields on the plane.

A generic deformation of such a germ is finitely-smoothly orbitally equivalent to a deformation induced from a principal one:

$$\dot{x} = x(1 + P_{\mu-1}(u, \varepsilon) \pm u^\mu + au^{2\mu}),$$
$$\dot{y} = -yp/q,$$

and is finitely-smoothly orbitally versal. Here $u = x^p y^q$ *is a resonance monomial,* $(\varepsilon, a) = (\varepsilon_1, \ldots, \varepsilon_\mu, a) \in \mathbb{R}^{\mu+1}$ *is the multi-dimensional parameter of the family, and*

$$P_{\mu-1}(u, \varepsilon) = \varepsilon_1 + \varepsilon_2 u + \cdots + \varepsilon_\mu u^{\mu-1}. \tag{19}$$

Remark. A theorem on formal finite determinacy analogous to the theorem of Ichikawa was proved recently by M.B. Zhitomirskij (private oral communication). Results for periodic vector fields analogous to the results above follow easily from his theorem. The analog of part b) of the next to last theorem is probably also true for both cases.

We next turn to the investigation of deformations of nonhyperbolic germs.

5.7. Deformations of Germs of Vector Fields with One Zero Eigenvalue at a Singular Point

Definition. A family of germs given by the equation

$$\dot{x} = \pm x^{\nu+1} + P_{\nu-1}(x, \varepsilon) + ax^{2\nu+1} \tag{20}$$

is a principal $(\nu + 1)$-*parameter deformation of a germ of a vector field on the line.* Here $P_{\nu-1}(x, \varepsilon) = \varepsilon_1 + \varepsilon_2 x + \cdots + \varepsilon_\nu x^{\nu-1}$; the undeformed germ corresponds to the parameter value $\varepsilon = 0$, $a = a_0 \in \mathbb{R}$.

Remark. ν-parameter principal families are parametrized by one discrete (equal to ± 1) and one continuous parameter (equal to a_0). Different principal families are not finitely smoothly equivalent on the line if the conjugating diffeomorphism preserves orientation.

Theorem. *A generic μ-parameter family of vector fields on the line in a neighborhood of each degenerate singular point may be transformed by a change of variables and parameters into one of the principal families* (20) *for $\nu + 1 \leqslant \mu$ or into the family*

$$\dot{x} = \pm x^{\mu+1} + P_{\mu-1}(x, \varepsilon) + a(\varepsilon)x^{2\mu+1}. \tag{21}$$

The corresponding change of variables is analytic, smooth, or finitely smooth if the original family is analytic, smooth, or finitely smooth. More precisely, for any natural number k there exists an $N(k)$ such that if the original family is of class $C^{N(k)}$, then the normalizing change of variables may be chosen to be of class C^k.

This theorem in the analytic and (infinitely) smooth cases, was proved by V.P. Kostov (see Math. USSR Izvestija vol. 37 (1991) No. 3, pp. 525–537), and for the finitely smooth case, by S.Yu. Yakovenko (1985) (see Yu.S. Il'yashenko and S.Yu. Yakovenko [3*]). The proof in the analytic case has been published by Kostov (1984).

Corollary. Let v be a germ of a smooth vector field at a singular point with eigenvalue 0 and a one-dimensional center manifold. Let the multiplicity of this singular point be $\mu + 1$, and let the real parts of its nonzero eigenvalues form a nonresonant n-tuple. Germs with these properties are found in generic families that depend on at least μ parameters. A deformation of such a germ in a generic smooth $(\mu + 1)$-parameter family is finitely-smoothly equivalent to the principal deformation

$$\dot{x} = \pm x^{\mu+1} + P_{\mu-1}(x, \varepsilon) + a(\varepsilon)x^{2\mu+1}$$
$$\dot{y} = A(x, \varepsilon)y.$$

The undeformed germ corresponds to $\varepsilon = 0$, $a = a_0$, where a_0 is some real constant.

This corollary follows from the last assertion of the above theorem and from the general theorem on finitely smooth saddle suspensions for differential equations (Sect. 5.3).

Remarks. 1. A principal deformation depends upon a $(\mu + 1)$-dimensional parameter $(\varepsilon_1, \ldots, \varepsilon_\mu, a)$ and a functional parameter A.

2. The corollary becomes false if one replaces finitely smooth by analytic or infinitely smooth in its conclusion.

This corollary allows us to normalize an equation of fast motion near a generic fold point of a slow manifold (Sect. 2 of Chapter 4).

5.8. Functional Invariants of Diffeomorphisms of the Line. Functional invariants arise in the C^1-classification of mappings of the line that have more than one hyperbolic fixed point (G.R. Belitskij and others). Consider a diffeomorphism of an interval having two hyperbolic fixed points, one attracting and one repelling. In a neighborhood of each of these points, the diffeomorphism is included in a smooth flow in a unique way.

In other words, a germ of a diffeomorphism at a fixed point is a germ of the time-one map along phase curves of a unique C^1-smooth vector field. Both fields arising close to the fixed points are extended by the diffeomorphism to the whole interval between the singular points. The quotient-space of this interval under the action of the diffeomorphism is diffeomorphic to a circle. On this circle two vector fields without singular points arise for which the circle is a cycle with period 1. Therefore, on the circle there are two charts defined uniquely modulo shifts, the times of the motions corresponding to each of the fields. The function transforming one chart into the other generates a functional modulus of the original diffeomorphism; namely, this transition function is a diffeomorphism of the circle: $t \mapsto t + \varphi(t)$.

Translations in the image, and in the preimage, transform this diffeomorphism in the following way:

$$t \mapsto t + \psi(t), \qquad \psi(t) = \varphi(t + a) + b - a;$$

where a and b are constants. Choosing suitable a and b, one can achieve the equality

$$\bar{\varphi} \equiv \int_{S^1} \varphi = 0.$$

The functional invariant for the C^1-smooth classification of diffeomorphisms of an interval with two hyperbolic fixed points is an equivalence class of diffeomorphisms of the circle of the form

$$t \mapsto t + \varphi(t), \qquad \bar{\varphi} = 0.$$

The equivalence relation is: $\varphi \approx \psi \Leftrightarrow \varphi(t + a) = \psi(t)$, for some a.

5.9. Functional Invariants of Local Families of Diffeomorphisms. Consider the local family of diffeomorphisms of the line

$$(f; 0, 0); \qquad f(x, \varepsilon) = f_\varepsilon(x): x \mapsto x - \varepsilon + ax^2 + \cdots, \qquad a \neq 0. \tag{22}$$

For $\varepsilon > 0$, the mapping f_ε has two hyperbolic fixed points. As shown in Sect. 5.8 above, the finitely smooth classification of such mappings has a functional modulus which is a diffeomorphism of the circle into itself. An equivalence class of germs in ε at 0 of families of diffeomorphisms (23) of the circle corresponds to the local family (22):

$$\Phi(f) = \{\Phi_\varepsilon: S^1 \to S^1\}, \qquad \Phi_\varepsilon = \mathrm{id} + \varphi_\varepsilon, \qquad \bar{\varphi}_\varepsilon = 0; \qquad \varepsilon > 0; \tag{23}$$

for $\varepsilon \leqslant 0$, by definition, we assume $\Phi_\varepsilon = \mathrm{id}$ $(\varphi_\varepsilon \equiv 0)$.

From the C^k-smoothness of the local family f it follows that the corresponding family Φ is C^k-smooth. Two families Φ and Ψ of the form (23) are equivalent if there exists a function a such that

$$\varphi_\varepsilon(t + a(\varepsilon)) = \psi_\varepsilon(t), \tag{24}$$

where a is a C^k-smooth function and $a \equiv 0$ for $\varepsilon \leqslant 0$.

Theorem (S.Yu. Yakovenko (1985)). 1. *To each C^k-smooth local family* (22) *there corresponds an equivalence class of germs at 0 in ε of smooth, equivalent families of diffeomorphisms of the circle of the type* (23) *with the equivalence relation* (24). 2. *Each such class is realized as a functional invariant of some local family* (22). 3. *If the functional invariants and the multipliers of the fixed points, considered as functions of the parameter, coincide for two C^k-smooth families, then the families are C^1-smoothly equivalent for $k \geqslant 8$.*

The analogue of this theorem, with C^k and C^1 replaced by C^∞, is proved in full detail in a paper by Il'yashenko and Yakovenko [4*].

5.10. Functional Invariants of Families of Vector Fields. The C^1-smooth classification of deformations of germs of vector fields at a singular point with a pair of pure imaginary eigenvalues also has functional invariants. If we restrict a family to its center manifold, we obtain a (finitely smooth) deformation of the germ of a vector field on the plane with linear part corresponding to a center. The monodromy transformation corresponding to the deformed germ has two hyperbolic fixed points (at those values of the parameter that correspond to cycles of the deformed equation): one is singular, the other belongs to a cycle. The functional invariant of the C^1-classification of such transformations was constructed above.

The functional invariant of the classification of generic one-parameter deformations of germs of diffeomorphisms with multiplier -1 is constructed analogously.

The classification theorems for these local families are also proved in the paper by Il'yashenko and Yakovenko [4*].

5.11. Functional Invariants of Topological Classifications of Local Families of Diffeomorphisms of the Line (Following Roussarie (1986)). There exists a continuum of topologically inequivalent three-parameter deformations of the germ of the diffeomorphism

$$(\mathbb{R}, 0) \to (\mathbb{R}, 0): x \mapsto x + ax^4 + \dots.$$

Theorem (Roussarie (1986). *For a generic smooth three-parameter deformation of the germ of the diffeomorphism of the line*

$$f: (\mathbb{R}, 0) \to (\mathbb{R}, 0), \qquad x \mapsto x + ax^4 + \cdots, \qquad a \neq 0,$$

there exists a functional invariant: a one-parameter family of classes of equivalent diffeomorphisms of the circle; the equivalence relation is the same as in Sect. 5.8. For μ-parameter deformations of the germ

$$x \mapsto x + ax^{\mu+1} + \cdots, \qquad a \neq 0,$$

the functional invariant is a $(\mu - 2)$-parameter family of classes of equivalent diffeomorphisms of the circle ($\mu \geqslant 3$).

Roussarie (1986) does not describe the set of all families of diffeomorphisms of the circle arising as functional invariants of local families of diffeomorphisms of the line; however, he does show that this set has the cardinality of the continuum.

We give a sketch of the proof of the theorem. Suppose $\{f_\varepsilon\}$ is a generic three-parameter deformation of a germ f. The fixed points of the diffeomorphism f_ε merge if and only if ε belongs to a surface that is diffeomorphic to a swallowtail. We suppose that this diffeomorphism has been carried out; then the surface on which the fixed points merge is a swallowtail. To generic points on the swallowtail, there correspond diffeomorphisms with one fixed point of multiplicity two; the remaining fixed points, if any, are simple. To points on the curve Γ of self-intersection of the swallowtail, there correspond diffeomorphisms with two fixed points of multiplicity two. The curve Γ will be the base of a family of diffeomorphisms of the circle; this family will form the invariant of the deformation $\{f_\varepsilon\}$.

For $\varepsilon \in \Gamma$ the diffeomorphism f_ε corresponds to a functional invariant that is a class of equivalent diffeomorphisms of the circle onto itself. Namely, the germ of the diffeomorphism f_ε is generated at each one of its two semi-stable fixed points by the germ of a vector field: the germ of the diffeomorphism is the time-one shift along the phase curves of the vector field. The germ of each of the generating vector fields is uniquely defined by the diffeomorphism f_ε. Both fields are extended, with the help of f_ε, to the whole interval between the fixed points of the diffeomorphism, and on this whole interval they generate f_ε. Thus two vector fields on this interval are constructed that commute with a diffeomorphism of the interval onto itself that has no fixed points. Such a pair of vector fields generates a diffeomorphism of the circle onto itself, which is defined up to translations in the image and in the preimage, as was described in Sect. 5.8. Two diffeomorphisms of the circle onto itself are equivalent if they take the form

$$t \mapsto t + \varphi(t), \quad t \mapsto t + \psi(t), \quad \text{where } \varphi(t + a) = \psi(t) + b \text{ for some } a \text{ and } b.$$

The family of such classes of equivalent diffeomorphisms of the circle that we have constructed for the map f_ε ($\varepsilon \in \Gamma$) is the desired functional invariant of the deformations $\{f_\varepsilon \mid \varepsilon \in (\mathbb{R}^3, 0)\}$.

We now prove that the functional invariants of equivalent deformations coincide. If two families are equivalent, then the surfaces (swallowtails) in parameter space that correspond to the diffeomorphisms of both families having nonhyperbolic fixed points coincide. Suppose f_ε and g_ε are diffeomorphisms from the two families corresponding to a value of the parameter on the curve of self-intersection Γ of the swallowtail. There exists a rich set of homeomorphisms, conjugating f_ε and g_ε; most of them do not map the corresponding generating fields onto one another.

Let H be a homeomorphism conjugating the families $\{f_\varepsilon\}$ and $\{g_\varepsilon\}$. From the Rigidity Theorem (Sect. 1.1 of Chap. 2) it follows that for $\varepsilon \in \Gamma$ the homeomorphism $H(\cdot, \varepsilon)$ maps the generating fields of the diffeomorphism f_ε into the generating fields of the diffeomorphism g_ε. Consequently, the functional invariants of f_ε and g_ε coincide. ►

Remarks. 1. The Rigidity Theorem attaches some degree of smoothness to the conjugating mapping, which by definition was only a homeomorphism.

Therefore, one can carry out the same construction, as for smooth mappings in Sect. 5.9, for mappings that realize only topological, and not smooth, equivalence of families of diffeomorphisms.

2. The continuous dependence of the conjugating homeomorphism on the parameter is crucial for the Rigidity Theorem. Therefore, weakly equivalent deformations of germs of diffeomorphisms of the line do not generate functional invariants (see Sect. 2.2 of Chap. 2).

Corollary. 1. The topological classification of three-parameter deformations of vector fields with limit cycles of multiplicity four (degeneracy of codimension three) has functional invariants. To be convinced of this, it is necessary to carry Roussarie's Theorem over to the case of a family of monodromy transformations.

2. The topological classification of four-parameter deformations of the germ of a vector field on the plane with two pure imaginary nonzero eigenvalues and, additionally, a three-foldly degenerate nonlinear part (in brief, a germ of the class B_4; see Sect. 3.1 of Chap. 1) has functional invariants.

Actually, the corresponding family of monodromy transformations has a two-parameter subfamily, consisting of diffeomorphisms with two fixed points of multiplicity two. Therefore to study multiparameter deformations of vector fields on the plane it is helpful to weaken "equivalence" to "weak equivalence".

§ 6. Feigenbaum Universality for Diffeomorphisms and Flows

One of the possible sequences of bifurcations of an attractor, often occurring in systems depending upon one parameter, is a sequence of period-doublings of a stable cycle. This sequence of bifurcations happens on a finite interval of variation of the parameter and takes the system from a stable periodic regime to chaos.

6.1. Period-Doubling Cascades. A sequence of period-doubling bifurcations in one-parameter families happens in the following way. A cycle that is initially stable loses stability as a multiplier passes through -1. At that moment, in a generic family of systems, a stable cycle of twice the period (at the moment of bifurcation) branches off from it; the new-born cycle closes after two circuits around the cycle that has lost stability (Sect. 1.2). Upon further change of the parameter the new cycle also undergoes period-doubling, giving birth to an attracting cycle of twice the last period (approximately quadruple the original), and then this cycle doubles in turn, etc. It turns out that, in a generic family, this whole infinite cascade of doublings takes place on a finite interval of variation of the parameter. Moreover, the intervals between consecutive period-doublings decrease asymptotically in a geometric progression. The common ratio is universal: it does not depend upon the family considered, that is, it is the same for all generic families. It equals $1/4.6692\ldots$; $4.6692\ldots$ is called the Feigenbaum constant (Vul, Sinai and Khanin (1984)).

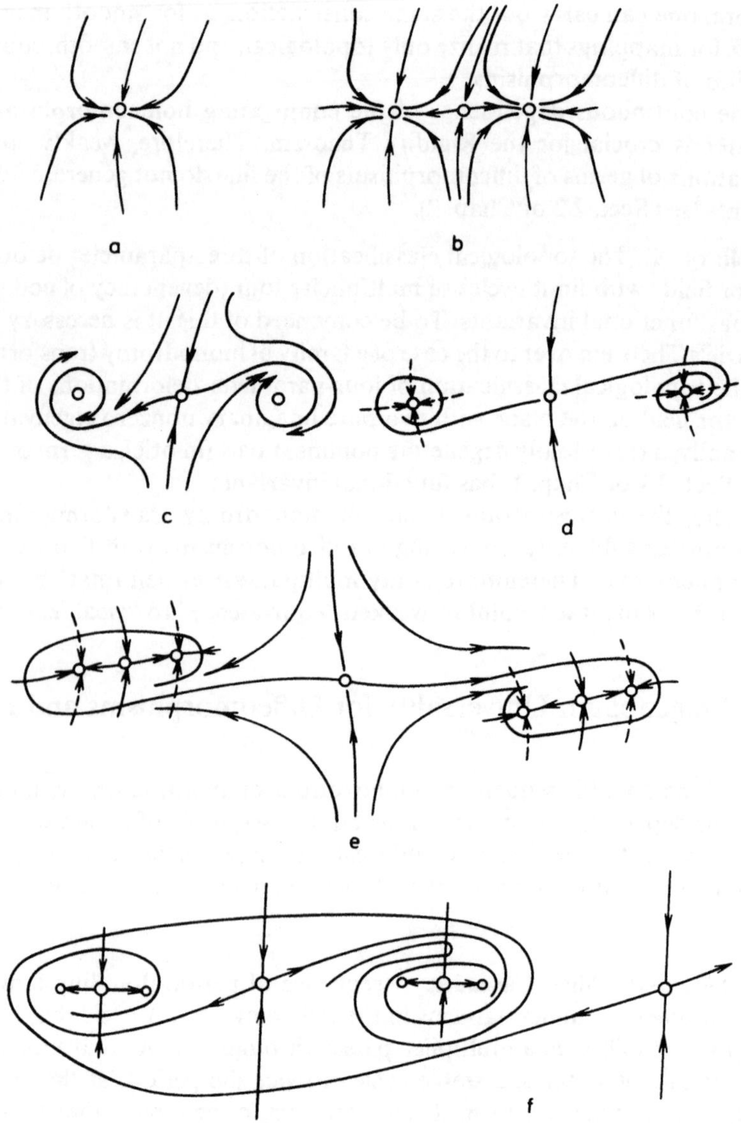

Fig. 29. Three successive period-doubling bifurcations for a diffeomorphism of the plane. The bifurcations occur going from (a) to (b), from (d) to (e), and from (e) to (f). Perestroikas of fixed points of the square of the diffeomorphism are shown in (c) and (d). In (d), the solid curves are invariant curves of the diffeomorphism and the dotted curves are invariant curves of its square; the diffeomorphism acts like an involution on these curves. In (e), invariant curves of the diffeomorphism are shown as solid curves and invariant curves of its fourth power are shown as dotted curves. The curves in (f) are curves related to the 16th power of the diffeomorphism. The unstable manifold of each saddle fixed point contains in its closure the unstable manifolds of all the saddle fixed points born in the next bifurcation. Only part, the "center" and "left", of the set of fixed points of the 16th power of the diffeomorphism is shown in (f).

6.2. Perestroikas of Fixed Points. Analogous cascades of period doublings are observed in generic families of diffeomorphisms: a fixed point, stable for values of the parameter less than the first critical value, loses stability as a multiplier passes through -1, giving birth to a stable cycle of period 2; then this cycle loses stability, giving birth to a stable cycle of period 4, etc. The intervals between successive bifurcations decrease, just as for continuous time systems.

The linearization of a diffeomorphism at a fixed point that loses stability as a multiplier passes through -1 cannot have real eigenvalues for all values of the parameter: otherwise, one of the multipliers of the fixed point would pass through the origin, and the mapping would cease to be a diffeomorphism. The perestroikas of the fourth power of a diffeomorphism of the plane for two successive period-doublings are shown in Fig. 29.

6.3. Cascades of *n*-fold Increases of Period. In two-parameter systems, cascades of period triplings, quadruplings, quintuplings, etc. occur in the same generic way. In these cases the common ratio of the geometric progression, defining the sequence of parameter values at bifurcations, is a complex number such that the bifurcation values of the parameters lie asymptotically on a logarithmic spiral (in a suitable Euclidean structure of the parameter plane). For period-tripling this number is equal to $(4.600\ldots + 8.981\ldots i)^{-1}$. Calculations show that for cascades of bifurcations caused by passage of a pair of multipliers through a resonance $\exp(\pm 2\pi i p/q)$, the universal common ratio is approximately equal to $C(p, q)/q^2$. Because of this, with the growth of multiplicity of period increases, bifurcation events take place ever more rapidly (see refs. 56, 57, 58 in Vul, Sinai and Khanin (1984)).

6.4. Doubling in Hamiltonian Systems. Period-doubling cascades also occur in Hamiltonian systems, but they look somewhat different. In this case a period-doubling bifurcation occurs if for a change of the parameter an elliptic periodic trajectory becomes hyperbolic[9], but along-side it appears a period-doubled, elliptic, periodic trajectory (Fig. 30). The universal common ratio for period doubling in Hamiltonian systems is equal to 1/8.72...(see refs. 54, 55 in Vul, Sinai and Khanin (1984)).

We next describe the mechanism by which a period-doubling cascade arises for diffeomorphisms. We first recall some results from the one-dimensional theory (Vul, Sinai and Khanin (1984), Collet and Eckmann (1980)).

6.5. The Period-Doubling Operator for One-Dimensional Mappings. We consider a mapping $x \mapsto f(x)$ of the interval into itself, whose graph has the form shown in Fig. 31a. The graph of its second iterate $x \mapsto f(f(x))$ is shown in Fig. 31b. The shape of this graph, modulo a rescaling and reversal of axes, recalls that of the original graph. This observation motivates:

[9] An elliptic periodic trajectory of a Hamiltonian system is a cycle with nonreal multipliers whose moduli equal one; a hyperbolic trajectory of a Hamiltonian system is a cycle with multipliers whose moduli are not equal to one.

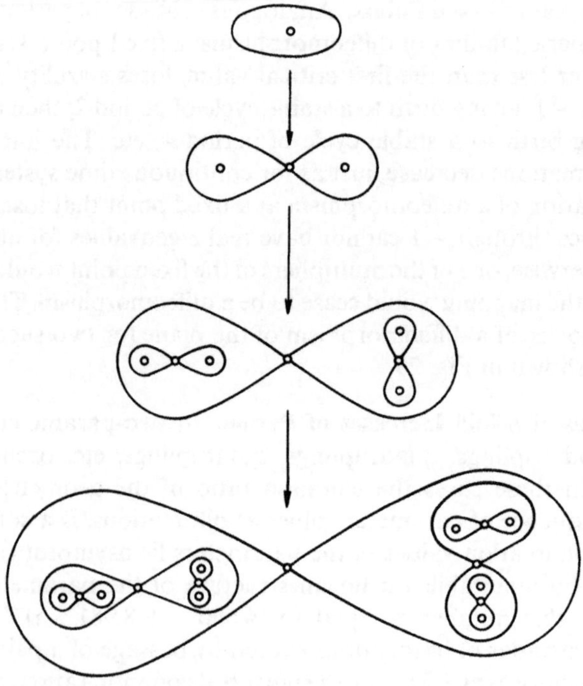

Fig. 30. Three successive period-doubling bifurcations in a generic family of area-preserving maps (the Hamiltonian case)

Definition. A mapping of the interval $I = [-1, 1]$ into itself is called *autoquadratic* if it is conjugate to the restriction of its square to a smaller interval and, moreover, the conjugating diffeomorphism is linear.

The last requirement can generally be fulfilled by a coordinate change on the line.

Theorem (Lanford (1982), Campanino and Epstein (1981) see ref. 29, 30 in Vul, Sinai, and Khanin (1984))[10]. *There exists an even analytic autoquadratic mapping* $g\colon I \mapsto I$, *for which*

$$g(0) = 1, \quad g(1) < 0, \quad g'(x) > 0 \quad \text{for } x \in [-1, 0), \quad g(g(\alpha^{-1})) < \alpha^{-1} < g(\alpha^{-1}), \tag{25}$$

where $\alpha = -1/g(1)$. *In some neighborhood of* g *in the space of all mappings of the interval, there exists no other autoquadratic mapping satisfing the normalizing requirement* $g(0) = 1$.

An autoquadratic mapping is a fixed point of the "period-doubling operator"

$$Tf = \beta \circ f \circ f \circ \beta^{-1}, \qquad \beta = -1/f(1).$$

[10] For a detailed proof see: K.I. Babenko, V.Yu. Petrovich, "On proofs by computation on a computer". Preprint, the M.V. Keldysh Institute of Appl. Math., Moscow, 1983, 183 pp.

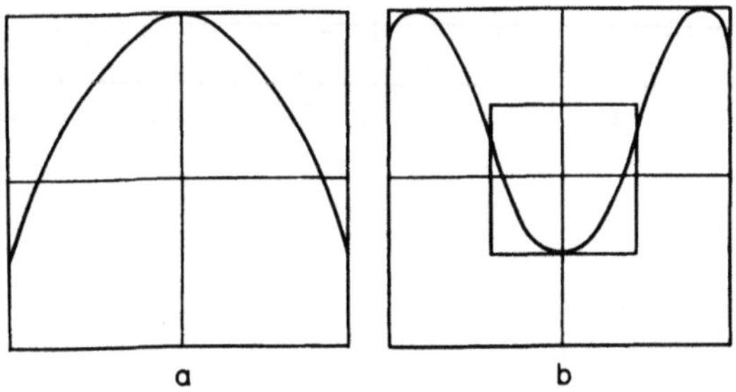

Fig. 31. An almost autoquadratic map of the interval and its square

This operator is defined for all even mappings satisfying the conditions (25), and for all, not necessarily even, mappings close to g for which 0 is mapped to 1.

Remark. The mappings Tf and $f^2 = f \circ f$ are conjugate. Therefore, if Tf has a cycle of period N, then f^2 has a cycle of the same period, and the mapping f has a cycle of double this period.

6.6. The Universal Period-Doubling Mechanism for Diffeomorphisms. We consider the two-dimensional case. Let g be an autoquadratic mapping as in Sect. 6.5. We consider the interval I: $x \in [-1, 1]$, $y = 0$ on the plane, and we construct an extremely degenerate autoquadratic mapping of a neighborhood of the interval I into itself. We assume: $\varphi(x) = g(\sqrt{x})$. Since the function g is even and analytic, the function φ is analytic on the interval $[0, 1]$, and, consequently, it can be analytically continued in some neighborhood of its endpoints. Let $\mathscr{D}_r$, an r-neighborhood of the interval I in the (x, y)-plane, be the union of all disks of radius r with centers on I. For sufficiently small r the mapping

$$G\colon (x, y) \mapsto (\varphi(x^2 - y), 0), \qquad \mathscr{D}_r \to \mathscr{D}_r$$

is well-defined and coincides with g on I. Assume

$$\alpha = -1/g(1) = 2.5029\ldots, \qquad A\colon (x, y) \mapsto (-\alpha x, \alpha^2 y).$$

Consider the doubling operator:

$$T\colon F \mapsto A \circ F \circ F \circ A^{-1}. \tag{26}$$

If the C^1-norm of the difference $F - G$ does not exceed $r/2$, then the mapping TF is well-defined in $\mathscr{D}_r$. It is easy to verify that G is a fixed point of the operator T, and that it is, in this sense, autoquadratic.

Remark. The mapping G can be approximated by a family of diffeomorphisms

$$G_\varepsilon\colon G_0 = G,\ G_\varepsilon(x, y) = (\varphi(x^2 - y), \varepsilon x).$$

The mapping $G_\varepsilon\colon \mathscr{D}_r \to \mathbb{R}^2$ is a diffeomorphism for sufficiently small r and all sufficiently small $\varepsilon \neq 0$.

Consider a neighborhood of the autoquadratic mapping G in a suitable function space of mappings of the domain $\mathscr{D}_r$ into itself. This neighborhood is fibred into the orbits of the action of the affine group of coordinate transformations (more exactly, it is partitioned into classes of affinely equivalent mappings; employing an abuse of language, we shall call these classes "orbits", although they represent only "pieces" of orbits). An orbit of the mapping G, or any mapping near G, is a smooth manifold whose dimension coincides with the dimension of the affine group of the plane $\mathbb{R}^2$. Therefore, a neighborhood of the mapping G is factorized by the action of the affine group; let Π be the projection of this neighborhood onto the corresponding quotient-space. The period-doubling operator respects orbits of the action of the affine group (maps orbits to orbits); therefore it "descends' to an operator acting on the quotient space. The point ΠG is a fixed point of this new operator. It has been proved (see ref. 9 of Vul, Sinai and Khanin (1984)) that this fixed point is hyperbolic and has a one-dimensional unstable manifold W^u, and a stable manifold W^s of codimension 1. In the space of one-parameter families of diffeomorphisms, an open set is formed by families transversally intersecting the manifold $\Pi^{-1}W^s$, having codimension 1 in the space of all mappings of the domain $\mathscr{D}_r$, into itself. In such families a countable number of period-doublings take place; the mechanism of these bifurcations is explained by the hyperbolic properties of the period-doubling operator just as in the one-dimensional case (Bunimovich, Pesin, Sinai and Yakobson (1985), Vul, Sinai and Khanin (1984)).

When the parameter of a family runs through an interval between successive bifurcation values corresponding to period-doublings, one multiplier of the corresponding cycle changes from 1 to -1 along the way, exiting into the complex plane and then returning to the real axis. It is interesting to investigate the asymptotic behavior of the curve followed by the multiplier in $\mathbb{C}$. At the present time, there exists an upper estimate on the radius of the disk with center 0, in which the arc of nonreal values of the multiplier lies. This radius decreases in "geometric progression" with successive bifurcations, that is to say, it decreases like the sequence $\exp(-\alpha 2^n)$.

Let the two-dimensional domain $\mathscr{D}_r$, of the period-doubling operator (26), its fixed point $G: \mathscr{D}_r \to \mathbb{R}$, and its invariant hypersurface $\Pi^{-1}W^s$ be the same as above.

Theorem (M.V. Yakobson (1985), private oral communication). *There exists a neighborhood of the mapping G in function space, having the property that if the one-dimensional family of diffeomorphisms belongs to this neighborhood and intersects the hypersurface* $\Pi^{-1}W^s$ *transversally, then*:

1. *The element* ε_n *of the sequence of bifurcation values of the parameter arising in a period-doubling cascade that corresponds to the exit into* $\mathbb{C}$ *of multipliers of a cycle of period* 2^n *has the form*

$$\varepsilon_n = c\delta^{-n} + O(\delta^{-n}\sigma^n),$$

where δ *is the Feigenbaum constant and* σ *is the maximal contracting eigenvalue of the linearized period-doubling operator at the fixed point G, and c is a constant depending upon the family.*

2. *The corresponding element of the sequence of arcs over which the multiplier of the cycle of period 2^n varies, lies in a circle of radius* $\exp(-a2^n)$ *with center at the origin. Here $a > 0$ is a constant depending on the family.* ▲

A weakened version, $\varepsilon_n = O(\delta^{-n})$, of the first assertion of the theorem easily follows from the theory of Feigenbaum universality. A proof of the first assertion in its full form is beyond the scope of the present survey; we set down a proof of the second assertion.

◄ If the multiplier $\lambda_n(\varepsilon)$ of the fixed point of the diffeomorphism $f_\varepsilon^{2^n}$ of the planar region ($\lambda_n(\varepsilon)$ being also a multiplier of the cycle of period 2^n of the diffeomorphism f_ε) is not real, then the second multiplier is its complex conjugate, and the Jacobian of the diffeomorphism at this point is equal to $|\lambda_n(\varepsilon)|^2$. On the other hand, if the diffeomorphism f_ε is sufficiently close to the mapping G, the image of which is one-dimensional, then everywhere the Jacobian of f_ε is defined, it is less than some constant $\exp(-2\alpha) < 1$. Thus, the Jacobian of the diffeomorphism $f_\varepsilon^{2^n}$ is less than $\exp(-2\alpha 2^n)$ everywhere in $\mathcal{D}_r$. From this it follows that

$$|\lambda_n(\varepsilon)| < \exp(-\alpha 2^n). ►$$

The theorem is also valid for a map of a domain of any dimension (not only two-dimensional). For the proof of conclusion 2 one uses that all maps near to a mapping onto a line decrease two-dimensional volume.

Chapter 3
Nonlocal Bifurcations

In this chapter we describe the bifurcations of systems on the boundary of the set of Morse-Smale systems. We recall that a point P is a *nonwandering point* of a flow $\{f^t\}$ (or a diffeomorphism f) if, for any neighborhood $\mathcal{U}$ containing P, there exist sequences $\{t_i\}$ (or $\{k_i\}$ with $k_i \in \mathbb{Z}$), diverging to ∞ as $i \to \infty$, such that

$$(f^{t_i}\mathcal{U}) \cap \mathcal{U} \neq 0 \; ((f^{k_i}\mathcal{U}) \cap \mathcal{U} \neq 0).$$

A flow (or diffeomorphism) on a compact manifold is called a *Morse-Smale system* if

1. The nonwandering set of the flow or diffeomorphism consists of only a finite number of fixed points and periodic trajectories.
2. All the fixed points and periodic trajectories are hyperbolic.
3. All the stable and unstable manifolds of the fixed points and cycles intersect transversally.

The boundary of the set of Morse-Smale systems can be subdivided into the following parts:

1. Systems with a finite set of nonwandering trajectories, containing either nonhyperbolic fixed points or cycles, or trajectories of nontransversal intersections of the stable and unstable manifolds of fixed points and/or cycles, or some or all of these simultaneously.
2. Systems with an infinite set of nonwandering trajectories.

Bifurcations taking place at crossings of the first part of the boundary of the set of Morse-Smale systems have been studied in a comparatively detailed way, and are described in Sects. 1–6 below. The second part of the boundary, and the bifurcations that correspond to it, have hardly been investigated at all; it has recently been proved that this second part of the boundary is nonempty for systems with a phase space of more than two dimensions. These results are given in Sect. 7.

Some of the bifurcations described in this chapter lead to the birth of strange attractors. There exist various inequivalent definitions of attractors. At the "physics" level of rigor, an attractor is a set of trajectories in phase space corresponding to "steady-state" regimes. Different definitions of attractors are discussed, and some of their bifurcations are described in Sect. 8.

Information on bifurcations in the class of systems with non-trivial nonwandering sets is given in Sects. 5, 6 and others.

§ 1. Degeneracies of Codimension 1. Summary of Results

1.1. Local and Nonlocal Bifurcations. We denote: the Banach space of C^r-smooth vector fields, in the C^r-topology ($r \geqslant 1$), on a C^∞-smooth manifold **M** by $\chi^r(\mathbf{M})$; the set of vector fields generating structurally stable (or rough[11]) dynamical systems by $\Sigma^r(\mathbf{M})$.

Definition. The set $\mathbf{B}^r(\mathbf{M}) = \chi^r(\mathbf{M}) \setminus \Sigma^r(\mathbf{M})$ is called the *bifurcation set*. Let $v(\varepsilon)$, $\varepsilon \in \mathbb{R}^k$, be a k-parameter, continuous family of vector fields.

Definition. The v values of ε for which $v(\varepsilon) \in \mathbf{B}^r(\mathbf{M})$ are called *bifurcation values*, and a change in the topological structure of the subdivision of phase space into

[11] We recall (see Andronov and Pontryagin (1937) or Lefschetz (1957)) that the original definition of structural stability differed from that of roughness by the absence of the requirement of nearness to the identity of the homeomorphism realizing the topological equivalence between the original and the perturbed systems. The set of vector fields generating structurally stable systems is open. This follows immediately from their definition, in contrast to the case for rough systems. On the other hand, we do not know of any structurally stable systems that are also not rough. Therefore, at the present time "structural stability" is often used as a synonym for "roughness". Translator's Note: More explicitly, $f \in C^1$ is structurally stable (according to Lefschetz's definition) if for any g that is sufficiently close to f there is a homeomorphism h such that $g \circ h = h \circ f$. According to Andronov and Pontryagin, an $f \in C^1$ is rough if for any $\varepsilon > 0$ there is a $\delta(\varepsilon) > 0$ such that, for any g, $\mathrm{dist}_{C^1}(f, g) < \delta(\varepsilon)$, there exists a homeomorphism h, $\mathrm{dist}_{C^0}(h, \mathrm{Id}) < \varepsilon$ such that $g \circ h = h \circ f$. This translation uses "structurally stable" as a synonym for "rough", corresponding to the standard English usage.

trajectories of the dynamical system generated by the vector field $v(\varepsilon)$ as ε passes through a bifurcation value is called *a bifurcation.*

Bifurcations for discrete time dynamical systems – diffeomorphisms – are defined analogously.

It is evident that a bifurcation set contains vector fields having nonhyperbolic singular points or nonhyperbolic cycles, as well as vector fields having hyperbolic singular points and/or cycles whose stable and unstable manifolds intersect nontransversally.

Definition. A phase curve of a vector field is called a *homoclinic trajectory* of a singular point (or cycle) if it converges to this point (winds onto the cycle) both for $t \to \infty$ and $t \to -\infty$. In other words, if its α- and ω-limit sets coincide with the singular point (cycle). A phase curve is called a *heteroclinic trajectory* if its α- and ω-limit sets are distinct singular points or cycles.

Definitions. Bifurcations taking place in a small, fixed neighborhood of an equilibrium (or a cycle) and connected with the destruction of its hyperbolicity are called *local.* Bifurcations taking place in a small, fixed neighborhood of a finite number of homoclinic or heteroclinic trajectories are called *semilocal.* All the rest (nonlocal and non-semilocal) are called *global.*

We note that these definitions relate primarily to the setting of a problem: local bifurcations may be accompanied by semilocal ones, and semilocal ones may be accompanied by global ones.

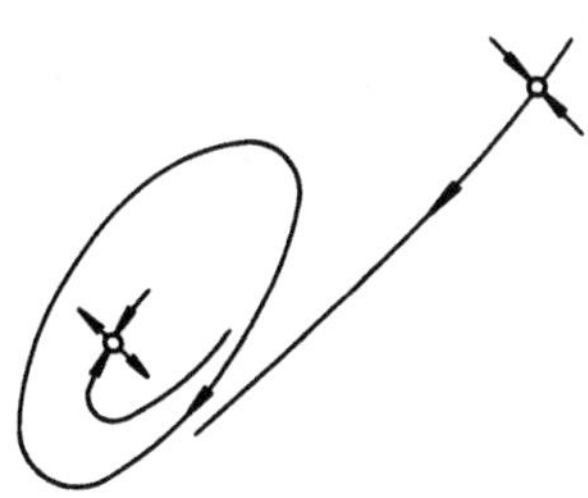

Fig. 32. Phase curves of a vector field on the plane, a one-parameter deformation of which has a countable set of bifurcation values

Example. The system depicted in Fig. 32 has a semi-stable limit cycle at $\varepsilon = \varepsilon_0$: inside the cycle one stable separatrix of the saddle winds from the cycle, and outside the cycle one unstable separatrix of another saddle winds onto the cycle. After the disappearance of the cycle, say, for $\varepsilon > \varepsilon_0$, the separatrices of these saddles join when the parameter ε varies through a sequence of values $\{\varepsilon_i\}$, $\varepsilon_i > \varepsilon_0$, $\varepsilon_i \to \varepsilon_0$. The local bifurcation here is the merging of stable and unstable cycles into a semistable cycle for $\varepsilon = \varepsilon_0$ and its disappearance for $\varepsilon > \varepsilon_0$. It is accompanied by a countable set of semilocal bifurcations, the closure of the separatrices for $\varepsilon = \varepsilon_i$.

Now we describe the degeneracies of codimension 1 that are related to the breaking of the requirement that the system be Morse-Smale.

1.2. Nonhyperbolic Singular Points. There are systems with nonhyperbolic points (cycles) on the boundary of the set of Morse-Smale systems. The local bifurcations of such points and cycles were described in Chaps. 1 and 2. However, degeneracies of a nonlocal character are connected with nonhyperbolic points and cycles and lead to semilocal bifurcations. We shall describe trajectories homoclinic to nonhyperbolic singular points.

Definition. The union of all positive (negative) semi-trajectories of a vector field that converge to a nonhyperbolic point is called an *unstable* (*stable*) *set*.

Stable and unstable sets of a nonhyperbolic cycle or nonhyperbolic fixed point of a diffeomorphism are defined analogously.

Remark. The total dimension of the stable and unstable sets of a nonhyperbolic singular point with a one-dimensional center manifold is equal to $n + 1$ (n being the dimension of phase space). Therefore, in the class of vector fields with such singular points, the presence of homoclinic trajectories at these points is generic.

The stable, unstable and center manifolds of points and cycles were defined by Hirsch, Pugh, and Shub (1977) and are denoted W^s, W^u and W^c (or W_O^s, W_O^u, W_O^c, W_L^s, W_L^u, W_L^c, where O and L are the corresponding point and cycle, respectively). The stable and unstable sets of points and cycles are denoted S^s and S^u (or S_O^s, S_O^u; S_L^s, S_L^u, where O and L are the corresponding point and cycle, respectively).

If all the eigenvalues of the matrix of the linear part of a vector field at a singular point that do not lie on the imaginary axis are found in the right (left) half-plane, then we say that the singular point is an unstable (stable) *node in its hyperbolic variables.* Otherwise, the singular point is called *a saddle in its hyperbolic variables.*

Example 1. Consider a nonhyperbolic singular point O of a vector field with a one-dimensional center manifold, on which the field reduces to the form $(ax^2 + \cdots)\partial/\partial x$, $a \neq 0$. If this singular point is a node in its hyperbolic variables, then the germ at the point O of one of the sets S^s and S^u is diffeomorphic to the germ of a ray at its vertex, and the germ of the other set is diffeomorphic to a germ of a halfspace at a boundary point. If the singular point O is a saddle in its hyperbolic variables, then the germs of the sets S^s and S^u are diffeomorphic to germs of halfspaces of dimensions greater than one at a boundary point;

$$\dim S^s = \dim W^s + 1 \quad \text{and} \quad \dim S^u = \dim W^u + 1.$$

Example 2. Consider a nonhyperbolic singular point of a vector field with a pair of purely imaginary eigenvalues and a two-dimensional center manifold. Then restricting the field to its center manifold, one obtains a normalized three-jet given by an equation of the form

$$\dot{z} = i\omega z + az|z|^2, \qquad \operatorname{Re} a \neq 0$$

(see Chap. 1, Sect. 2). Suppose, for definiteness, that Re $a < 0$. Then the germs of the sets S^s and S^u are germs of manifolds of dimensions

$$\dim S^s = \dim W^s + 2 \qquad \text{and} \qquad \dim S^u = \dim W^u,$$

respectively, the sum of these dimensions being n.

Remark. In the class of vector fields with singular points with a pair of purely imaginary eigenvalues, generic fields do not have a homoclinic trajectory of a singular point.

1.3. Nonhyperbolic Cycles. We investigate homoclinic trajectories of nonhyperbolic cycles. Nonhyperbolic cycles having one multiplier $+1$ or -1, or a pair of nonreal multipliers $e^{\pm i\varphi}$ can be found in generic one-parameter families. If the rest of the multipliers lie inside (outside) the unit circle, then we shall say that such *a cycle is of stable (unstable) nodal type in its hyperbolic variables.* Otherwise, we shall say it is *a cycle of saddle type in its hyperbolic variables.* Analogous definitions are given for fixed or periodic points of a diffeomorphism. We now describe stable and unstable sets of nonhyperbolic cycles, assuming that they satisfy the condition of genericity from Chapter 2, Sect. 1.

Example 1. For a vector field on $\mathbb{R}^n$ having a cycle L with multiplier $+1$, a fixed point of the monodromy transformation of the transversal D in a neighborhood of L has a one-dimensional center manifold, and the germs of the sets $S_L^s \cap D$, $S_L^u \cap D$ at the fixed point are the same as the germs S_O^s, S_O^u of a vector field on $\mathbb{R}^{n-1}$ at a singular point with a one-dimensional center manifold (see Example 1 of Sect. 1.2). The germ of the set S_L^s (S_L^u) on L is diffeomorphic to the germ on $\{0\} \times S^1$ of the product, or skew product, of an s-dimensional (u-dimensional) half-space (with the origin on its boundary) and the circle S^1. Here $s = \dim W_L^s$ and $u = \dim W_L^u$. In particular, if L is a stable node in its hyperbolic variables, then the germ of S_L^u on L is diffeomorphic to the germ on $\{0\} \times S^1$ of the product of a ray having vertex the origin, and the circle S^1.

Remark. Since $\dim S_L^s + \dim S_L^u = n + 2$, the presence of homoclinic trajectories and even one-parameter families of such trajectories is generic in the class of vector fields with a nonhyperbolic cycle having multiplier $+1$.

Example 2. Consider a vector field on $\mathbb{R}^n$ having a cycle with multiplier -1. A fixed point of the monodromy transformation of the transversal corresponding to the cycle has a one-dimensional center manifold on which the monodromy transformation can be written as:

$$x \mapsto -x + ax^2 + bx^3 + \cdots.$$

The square of this transformation is written as:

$$x \mapsto x - 2(a^2 + b)x^3 + \cdots,$$

from which it is clear that for $a^2 + b > 0$ (<0), the fixed point on the center manifold is stable (unstable). The same is true for a cycle. Therefore, since

$$u + s = n, \qquad s = \dim W_L^s, \qquad \text{and} \qquad u = \dim W_L^u,$$

for $a^2 + b > 0$ we have

$$\dim S^s = s + 1 \qquad \text{and} \qquad \dim S^u = u,$$

but for $a^2 + b < 0$ we have

$$\dim S^s = s \qquad \text{and} \qquad \dim S^u = u + 1.$$

Consequently, the presence of isolated homoclinic trajectories in the class of vector fields with such a cycle is generic for $s \geqslant 2$, $u \geqslant 2$.

Example 3. Assume that a vector field in $\mathbb{R}^n$ has a cycle with a pair of nonreal multipliers $e^{\pm i\varphi}$, $\varphi \notin \{\pi/2; 2\pi/3\}$. The monodromy transformation has a two-dimensional center manifold on which (in the coordinates $x + iy = z$) it can be written in the form $z \mapsto vz + az|z|^2 + \cdots$, $v = e^{i\varphi}$. From this it is straightforward to conclude that for $\operatorname{Re} a < 0$ ($\operatorname{Re} a > 0$), the fixed point of this transformation is stable (unstable) on the center manifold. The same is true for a cycle. It is not difficult to convince oneself that $u + s = n - 1$, where $u = \dim W^u$, $s = \dim W^s$, and therefore for $\operatorname{Re} a < 0$,

$$\dim S^s = s + 2 \qquad \text{and} \qquad \dim S^u = u,$$

but for $\operatorname{Re} a > 0$,

$$\dim S^s = s \qquad \text{and} \qquad \dim S^u = u + 2.$$

Since $\dim S^s + \dim S^u = n + 1$, the presence of isolated homoclinic trajectories in the class of vector fields with such a cycle is generic.

Lemma. (V.S. Afrajmovich (1985)). *If a vector field satisfying the conditions in Example 2 or 3 has a homoclinic trajectory of a cycle, along which the sets S^u and S^s intersect transversally, then all the vector fields from some neighborhood of this field in $\chi^r(\mathbf{M})$ have an infinite set of nonwandering trajectories and, consequently, none of these fields belongs to the boundary of the set of Morse-Smale vector fields.*

Since in this survey only bifurcations in a neighborhood of the boundary of the set of Morse-Smale systems are considered, homoclinic trajectories of a non-hyperbolic cycle are investigated below only if one of the multipliers is equal to 1.

1.4. Nontransversal Intersections of Manifolds

Definition. Two smooth submanifolds **A** and **B** of an n-dimensional manifold have a *simple tangency* at a point P if the sum of their dimensions is not less than n, and, moreover:

1) the direct sum of their tangent planes at P is an $(n-1)$-dimensional submanifold: $\dim (T_P\mathbf{A} + T_P\mathbf{B}) = n - 1$;

2) if f is a smooth function with a noncritical point P, vanishing on A, and having a critical point P on $\mathbf{B}$, then the second differential (Hessian) at P of the restriction of f to $\mathbf{B}$ is a quadratic form on $T_P\mathbf{B}$. It is required that the restriction of this form to $T_P\mathbf{A} \cap T_P\mathbf{B}$ be non-degenerate.

Remark. The simplicity of the tangency depends neither on the choice of the function, nor on which of the two submanifolds the function reduces to zero.

Definition. Two smooth submanifolds $\mathbf{A}$ and $\mathbf{B}$ of an n-dimensional manifold M^n have a *quasi-transversal intersection* at a point P if $\dim \mathbf{A} + \dim \mathbf{B} = n - 1$ and there exist a neighborhood $\mathscr{U}$ of P, and an $(n-1)$-dimensional smooth submanifold $\mathbf{M}^{n-1}$ of $\mathbf{M}^n$ such that the manifolds $\mathbf{A} \cap \mathscr{U}$ and $\mathbf{B} \cap \mathscr{U}$ belong to $\mathbf{M}^{n-1}$ and, as submanifolds of $\mathbf{M}^{n-1}$, intersect transversally at P.

Lemma (Newhouse, Palis and Takens (1983)). *Two smooth submanifolds* $\mathbf{A}$ *and* $\mathbf{B}$ *of an n-dimensional manifold* M^n *have a simple tangency at a point P if and only if there exists a system of coordinates* $\{(x_1, \ldots, x_n)\}$ *in some neighborhood* $\mathscr{U}$ *of P such that the intersections* $\mathbf{A} \cap \mathscr{U}$ *and* $\mathbf{B} \cap \mathscr{U}$ *are given by the equations*:

$$\mathbf{A} \cap \mathscr{U} = \{x_k = 0, a + 1 \leqslant k \leqslant n\};$$

$$\mathbf{B} \cap \mathscr{U} = \left\{x_k = 0, 1 \leqslant k \leqslant n - (b + 1), x_n = \sum_{i=n-b}^{a} \varepsilon_i x_i^2, \varepsilon_i = \pm 1\right\}.$$

Here $a = \dim \mathbf{A}$, $b = \dim \mathbf{B}$; *if* $n = b + 1$, *then the linear equations of the second system are absent*; $x(P) = 0$.

Definition. Two invariant manifolds of a vector field have *a trajectory of simple tangency* (*a quasi-transversal intersection*) if they intersect in a phase curve that is not a single point, and at some (and hence, every) point of this curve their intersection with a hypersurface transversal to the field has a simple tangency (quasi-transversal intersection).

1.5. Contours. Nonlocal bifurcations connected with simple tangencies and quasi-transversal intersections split into two classes having essentially different properties, depending upon the existence or nonexistence of so-called contours. (In Western mathematical literature the term "cycle" is used instead of "contour." We introduce the term "contour" in order to avoid confusion between "cycles" = closed phase curves and "cycles" = "contours.")

Definition. A sequence $Q_0, \ldots, Q_k$, where each Q_i is either an equilibrium point or a limit cycle, $k \geq 2$, $Q_0 = Q_k$, forms a *contour*[12] if

$$S^s_{Q_{i+1}} \cap S^u_{Q_i} \neq \varnothing, \qquad i \in \{0, \ldots, k-1\}.$$

[12] In the literature, a term "cycle" is also used, and it is not always assumed that $k \geqslant 2$.

Theorem (S. Smale (1967)). *If all Q_i are hyperbolic, and all intersections are transversal, then the vector field (and all nearby fields) has a countable set of limit cycles.* ▲

Therefore, for a generic vector field having a contour and lying on the boundary of the set of Morse-Smale vector fields, either:

Case 1) all Q_i are hyperbolic and there exists trajectory of simple tangency (or a quasi-transversal intersection), or

Case 2) all intersections are transversal, except that one (and only one) of the Q_i is nonhyperbolic (and belongs to the class described in Sects. 1.2 and 1.3).

Remark. Suppose a contour exists for a vector field on a two-dimensional surface. If Q_j is an equilibrium point, then it is either a saddle, or a saddle-node, or, if it is a cycle, then a cycle with multiplier $+1$. (Of course, we assume that all singular points and cycles belong to the class described in Sects. 1.2 and 1.3). If more than one equilibrium point or more than one cycle lies on the contour, then the vector field belongs to a set of codimension at least two in the space of vector fields. Actually, if i cycles lie on such a contour, $i \in \{0; 1; 2\}$, then there exist no less than $2 - i$ separatrices which join neighboring saddles or saddle-nodes.

In this way, we conclude that the unique possibility (in codimension 1) is for a contour to consist of a cycle with multiplier $+1$ and a saddle. A vector field with such a contour may arise on a surface of positive genus (but not on a sphere or a projective plane); see Fig. 33. The field in this case is quasi-general, but not of the first degree of structural instability (or "nonroughness"); see Sect. 2 below.

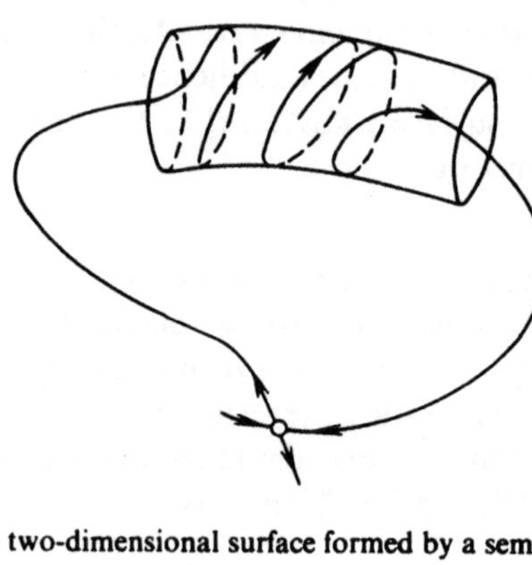

Fig. 33. A contour on a two-dimensional surface formed by a semi-stable cycle and a saddle

Lemma. *Suppose we are in case* 2) *and Q_j is nonhyperbolic. Then, either*

a) *Q_j is an equilibrium point with a one-dimensional center manifold, a saddle in its hyperbolic variables, and a countable set of homoclinic trajectories, or*

b) *Q_j is a cycle with multiplier $+1$, having a homoclinic trajectory.*

Proof. Using the transversality of the intersections of stable and unstable sets, and the λ-lemma (see deMelo and Palis (1982)), analogously to Smale (1967), we can show that Q_j has a homoclinic trajectory along which the stable and unstable sets intersect transversally. Consequently, if Q_j is an equilibrium point, then by virtue of Sect. 1.2, this equilibrium cannot have a two-dimensional center manifold. If, further, $\dim W^c_{Q_j} = 1$, then Q_j cannot be a node in its hyperbolic variables (for then it would be the case that either

$$\dim S^s_{Q_j} = 1 \quad \text{and} \quad S^s_{Q_j} \subset S^u_{Q_j}, \quad \text{or} \quad \dim S^u_{Q_j} = 1 \quad \text{and} \quad S^u_{Q_j} \subset S^s_{Q_j},$$

that is, $k = 1$, which is impossible), and finally a) follows from (Afrajmovich and Shil'nikov (1974)).

Now suppose Q_j is a cycle. If it has multiplier -1 or a pair of multipliers $e^{\pm i\varphi}$, then, by the lemma of Sect. 1.3, the vector field does not belong to the boundary of the set of Morse-Smale vector fields.

Case 2a will be considered in Sect. 3, Case 2b in Sect. 4, and Case 1 in Sect. 6.

1.6. Bifurcation Surfaces. Consider the set $\tilde{\mathscr{B}}_1$ of all vector fields on **M** having either a nonhyperbolic singular point, a nonhyperbolic limit cycle, or a trajectory belonging to a nontransversal intersection of the stable and unstable manifolds of two hyperbolic singular points or cycles or a point and a cycle.

Theorem (Sotomayor (1973a, 1973b, 1974)). *There exists an open everywhere dense subset $\mathscr{B}_1 \subset \tilde{\mathscr{B}}_1$ which is a smooth hypersurface of codimension 1 in a neighborhood of each singular point in $\chi^r(\mathbf{M})$. The vector fields in $\mathscr{B}_1$ have singular points, or cycles, or trajectories of a simple tangency, or of a quasi-transversal intersection of stable and unstable manifolds, listed in Sect.* 1.2–1.4.

We shall call the components of the set $\mathscr{B}_1$, corresponding to vector fields with degeneracies listed in Sect. 1.2–1.4, *bifurcation surfaces.* The smoothness of bifurcation surfaces can be proved by constructing smooth functionals, whose nondegenerate level surfaces coincide with these bifurcation surfaces.

As an example, we exhibit a functional for a nonhyperbolic singular point with a one-dimensional center manifold. Suppose a vector field $v_0 \in \mathscr{B}_1$ has a nonhyperbolic singular point O with a one-dimensional center manifold. We introduce a system of coordinates $\{(x, y_1, \ldots, y_{n-1})\}$ such that the x-axis is tangent to the center manifold at O, and for $y = (y_1, \ldots, y_{n-1})$ we choose a chart in the complementary (to x) hyperplane. Then any vector field v, C^2-close to v_0, can be written in the form $\dot{x} = f(x, y)$, $\dot{y} = g(x, y)$, with $\det(\partial g/\partial y)(0) \neq 0$. Therefore, the equation $g = 0$ has a unique solution $y = \varphi(x)$. The value of f at an extremum of the function $f(x, \varphi(x))$ is defined to be the value of the functional κ on v.

As is obvious, to construct the functional we consider the vector field on a zero isocline of the hyperbolic variables and project it onto the axis of the nonhyperbolic variable; for the value of the functional we take the value of this projection at the point where it takes its extremum. By the hypotheses on v_0, the isocline, its projection and its point of the extremum each depend smoothly on v.

Example. Consider the family $v(\varepsilon)$ of equations on $\mathbb{R}^1$:

$$\dot{x} = \alpha(\varepsilon) + \beta(\varepsilon)x + \gamma(\varepsilon)x^2 = f(x, \varepsilon), \qquad \alpha(0) + \beta(0) = 0, \qquad \gamma(0) \neq 0.$$

By definition, $\kappa(v(\varepsilon)) = f(x_0(\varepsilon), \varepsilon) = -(\beta^2 - 4\alpha\gamma)/(4\gamma)$. If α, β and γ are smooth functions and $\partial f(x_0(\varepsilon), \varepsilon)/\partial\varepsilon \neq 0$ at $\varepsilon = 0$, then $v(\varepsilon)$ is transversal to $\mathscr{B}_1$ at $v(0)$.

From the example it is evident that the values of functionals defining bifurcation surfaces permit the construction of a one-parameter family of vector fields transversal to the bifurcation surfaces.

1.7. Characteristics of Bifurcations. It is convenient to classify bifurcations by the following characteristics of bifurcation surfaces:

a) Accessibility or nonaccessibility of a bifurcation surface from the domain of structurally stable systems (a bifurcation surface lies on the boundary of this domain[13]). Obviously, a surface may be accessible from one side, both sides, or neither side. As an example, the bifurcation surface described in Sect. 1.1 is not accessible from the side $\varepsilon > \varepsilon_0$, because of the existence of a closed separatrix at each $\varepsilon = \varepsilon_i$.

b) For bifurcation surfaces belonging to the boundary of the set of Morse-Smale systems, we introduce the following:

Definition. A bifurcation is said to be *nonderivable* from the class of Morse-Smale systems in $\chi^r(\mathbf{M})$ if the class of Morse-Smale systems is everywhere dense from both sides of the corresponding bifurcation surface in a ball in $\chi^r(\mathbf{M})$ of sufficiently small radius with center on this surface.

A bifurcation surface can separate Morse-Smale systems from systems with infinite nonwandering sets. Upon passage through such a surface, for example, a strange attractor, a nontrivial hyperbolic set (for its definition see Smale (1967)), or a complicated limit set, containing infinitely many trajectories, may be born.

1.8. Summary of Results. In order to present the properties of bifurcations of codimension 1 investigated up to the present, it is convenient to create a table, see Table 1 below, with the following structure. It consists of seven columns. In the first column under "class", the type of singularity is given. In the second column under "subclass", the presence or absence of homoclinic trajectories, or some other characteristic of the bifurcation, is described. In the third column under "accessible, nonaccessible", the symbols "+ +" denote accessibility from both sides of a bifurcation surface, the symbols "+ −" denote accessibility from one side, and the symbols "− −" denote inaccessibility from both sides. In the fourth column under "nonderivable, derivable from the class of Morse-Smale systems", a "+" denotes that they are derivable, and a "−" denotes that they are not. In the fifth column under "a cycle is born", if p cycles are born a "p"

[13] We recall that a boundary point v_0 of an open set $\mathscr{U}$ is called *accessible* if there exists a path (a homeomorphic image of a closed interval) all of whose points, except for the boundary point that coincides with v_0, lie in $\mathscr{U}$.

Table 1

1	2	3	4	5	6	7
Class	Subclass	Accessible, Nonaccessible	Non-derivable/derivable from Morse-Smale systems	No. of cycles	Non-trivial limit set is born	A strange attractor is born
Nonhyperbolic singular point	No homoclinic trajectories	+ +	+	0	–	–
	One homoclinic trajectory	+ +	+	1	–	–
	At least two homoclinic trajectories	+ +	+	∞	Ω	–
Nonhyperbolic cycle with multiplier +1	No homoclinic trajectories	+ + or + –	+	0 or 2	–	–
	The union of homoclinic trajectories and the cycle is compact	+ –	+ or –	f or ∞	$\mathbb{T}^2$ or $\mathbb{K}^2$ or Ω or ?	+ or –
	The union of homoclinic trajectories and the cycle is not compact	+ + or + –	–	∞	Ω	+ or –
Hyperbolic singular point with a homoclinic trajectory		+ +	+	1 or ∞	– or +?	– or +?
A simple tangency or quasi-transversal intersection of stable and unstable manifolds	Neither contours nor homoclinic trajectories	+ + or + – or – –	+	0	–	–
	Contours or homoclinic trajectories exist	+ – or – – or ?	–	∞	Ω	–

appears, and if there exist infinitely many cycles the symbol "∞" appears. In the sixth and seventh columns under "a nontrivial limit set is born" and "a strange attractor is born", respectively, the sign "+" means the object is born, the sign "−" means the object is not born.

In the sixth column, $\mathbb{T}^2$ and $\mathbb{K}^2$ denote a two-dimensional torus, and a two-dimensional Klein bottle, respectively. In the fifth column, the symbol "f" denotes the possibility of the birth of a finite set of limit cycles. In the sixth column, the sign "Ω" denotes the presence of a nontrivial hyperbolic set. The symbol "?" indicates that the question is open. The answers separated by "or" all correspond to open subsets of the space of vector fields. The answers "+?" in the seventh row mean an unproved conjecture; the sign + here means that only examples exist that have the required properties (Sect. 5.8).

§ 2. Nonlocal Bifurcations of Flows on Two-Dimensional Surfaces

The simplest example of a nonlocal bifurcation on a two-dimensional surface is the appearance of a "saddle connection" when, as a parameter changes and reaches a certain value, an outgoing separatrix from one saddle intersects an incoming separatrix of another saddle (and, consequently, combines with it at this value of the parameter). As the parameter passes through this bifurcation value, the separatrices of both saddles "change places". This bifurcation occurs generically in one-dimensional families of vector fields.

We note that a saddle connection is the unique possible realization of a nontransversal intersection of stable and unstable manifolds of hyperbolic equilibria and/or cycles in the twodimensional phase space.

Our goal in this section is to describe, as far as is possible, bifurcations in generic one-parameter families of vector fields on closed surfaces, and also to describe the structure of the bifurcation set in the function space of vector fields.

Remark. We recall the following result:

Theorem (Andronov, Leontovich, Gordon, and Majer (1966, 1967), de Melo and Palis (1982)). 1. *Any structurally stable system on a closed surface is a Morse-Smale system.* 2. *The set of all structurally stable vector fields on a closed orientable two-dimensional surface* **M** *is open and dense in* $\chi^r(\mathbf{M})$, *for any natural number* r.

Thus, we see that any structurally unstable system lies on the boundary of the set of Morse-Smale systems.

2.1. Semilocal Bifurcations of Flows on Surfaces

Theorem (Andronov, Leontovich, Gordon, and Majer (1966, 1967)). *In generic one-parameter families of vector fields on the plane only the following semilocal bifurcations are possible*:

a. The birth of a cycle from a homoclinic trajectory of a saddle-node.
b. The appearance and disintegration of a saddle-connection.
c. The birth of a limit cycle from a separatrix loop at a nondegenerate saddle. If the saddle number (the trace of the linearization of the vector field at the saddle) is negative, then the cycle that is born of the separatrix loop is stable, and if the saddle number is positive, then the cycle is unstable.

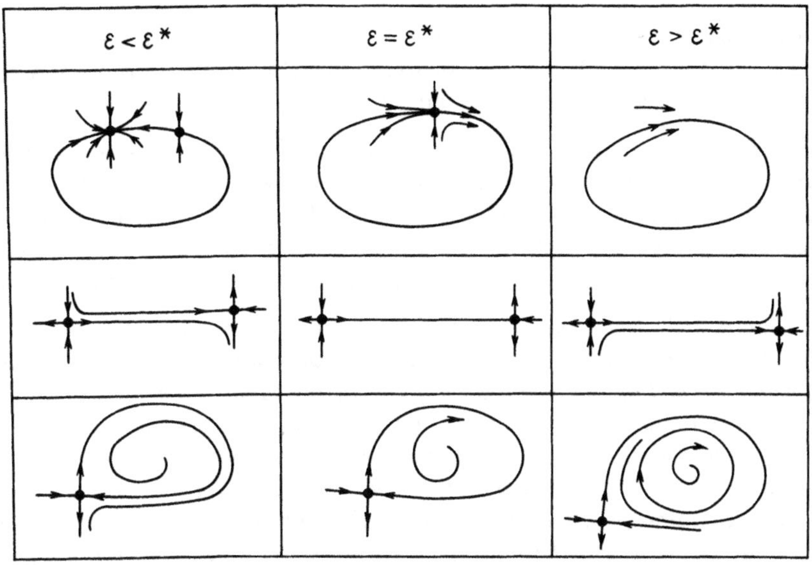

Fig. 34. Semi-local bifurcations of codimension 1 on surfaces

These bifurcations are illustrated in Fig. 34. Their higher-dimensional analogs are investigated in Sects. 3, 5, and 6, respectively.

Vector fields with separatrix loops at saddles, that have zero saddle numbers, occur in generic families with at least two parameters. The bifurcations of such fields in generic two-parameter families are described in Sect. 2.6. Bifurcations of separatrix loops in generic multi-parameter families are investigated in the work of Leontovich (1951) and also in a paper by R. Roussarie (1986). See also the survey by Il'yashenko and Yatovenko [3*].

2.2. Nonlocal Bifurcations on a Sphere: The One-Parameter Case. We begin with some definitions. Let **M** be a two-dimensional, smooth, closed surface, and let $\Psi^{k,r}(\mathbf{M})$ be the set of C^k-smooth families of C^r-smooth vector fields on M. This set consists of the C^k-mappings of the interval $\mathbf{I} = [0, 1]$ ($\varepsilon \in \mathbf{I}$) into the space $\chi^r(\mathbf{M})$. A family is *generic* if it belongs to a set of second Baire category[14] in $\Psi^{k,r}(\mathbf{M})$.

[14] A set of second Baire category is an intersection of countably many open, everywhere dense sets (otherwise known as "residual sets" or "thick sets").

Definition (Malta and Palis (1981)). Two *families* $\{v_\varepsilon\}$, $\{w_\varepsilon\} \subset \Psi^{k,r}(\mathbf{M})$ are *topologically equivalent* if there exists a homeomorphism

$$H = (h, \eta)\colon \mathbf{M} \times \mathbf{I} \to \mathbf{M} \times \mathbf{I}$$

(where η is an orientation-preserving homeomorphism of $\mathbf{I}$) such that for any $\varepsilon \in \mathbf{I}$ the homeomorphism

$$(h, \eta(\varepsilon)) \equiv h_\varepsilon\colon \mathbf{M} \to \mathbf{M}$$

is a topological equivalence between v_ε and w_ε. A *family* is called *structurally stable* if it is topologically equivalent to any nearby family; in other words, if it belongs to the interior of its equivalence class.

The definition of *weak topological equivalence* of families is obtained from the previous definition if one assumes that the mapping $h_\varepsilon = \mathbf{H}(\cdot, \eta)\colon \mathbf{M} \to \mathbf{M}$ is not necessarily continuous in ε. A family is said to be *weakly structurally stable* if it is weakly topologically equivalent to any nearby family. It is easy to see that any family containing only structurally stable vector fields is structurally stable.

"Theorem". 1. *In a generic one-parameter family of vector fields on S^2 ($r \geqslant 2$, $k \geqslant 1$), there is an at most countable set of bifurcation values of the parameter (in a neighborhood of these values the fields of the family change their topology). The field is structurally stable at the remaining values of the parameter.*

2. *For isolated bifurcation values of the parameter, only those nonlocal bifurcations are possible that are listed in the theorem in Sect.* 2.1.

3. *Accumulation points of bifurcation values of the parameter are one-sided limits, and can be only of the following two types*: a) *at the moment of bifurcation corresponding to the accumulation point, the vector field has a separatrix loop of a saddle which is the limit of either stable or unstable separatrices of another saddle* (see Fig. 35); b) *the field has a cycle, with multiplier* $+1$, *which is the limit of stable and unstable separatrices of two distinct saddles* (see Fig. 32). *Bifurcation values corresponding to vector fields having saddle connections accumulate at such points.*

4. *A one-parameter deformation of the field, corresponding to the bifurcation value of the parameter in a generic family for values of the parameter close to the bifurcation value, is topologically versal and structurally stable*: *any other deformation is topologically equivalent to a deformation induced from the given one, and any nearby one-parameter deformation is topologically equivalent to the given one.*

5. *In the large, the family is structurally stable.* ▲

A complete proof of this "theorem" has not been published. Conclusion 2 is proved by Andronov, Vitt, Gordon and Majer (1966, 1967), from whose proof, moreover, it is possible to derive Conclusion 4 above. See also Guckenheimer (1973). Separate results are contained in the work of Malta and Palis (1981) and Sotomayor (1974). Generic families are described in more detail below, and the parts of the theorem that are unproved are made more precise.

2.3. Generic Families of Vector Fields. A generic family of vector fields is an arc in function space, transversally intersecting a bifurcation surface at a "generic

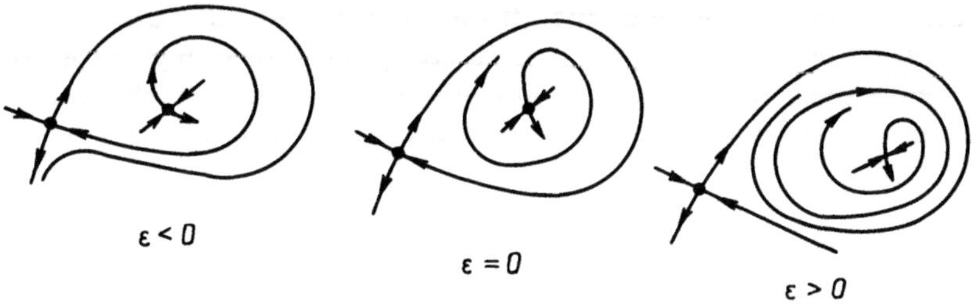

Fig. 35. Bifurcation of a separatrix loop that is the limit of a separatrix of another saddle. For $\varepsilon < 0$ the moment of birth of a saddle connection is shown.

point". To strictly define these points it is necessary to separate the class of "generic systems" in the set of all structurally unstable systems.

Definition (Sotomayor (1974)). A vector field on a two-dimensional surface is called *quasi-generic* if the nonwandering set of the dynamical system generated by the vector field consists of a finite number of equilibria and cycles, and if, in addition, one of the following two conditions is fulfilled:

1) all equilibria and cycles are hyperbolic, and there is a unique saddle connection, a separatrix going from saddle to saddle;

2) all equilibria and cycles are hyperbolic, except one; the set of eigenvalues of the nonhyperbolic singular point or multipliers of the nonhyperbolic cycle is degenerate; the corresponding degeneracy has codimension 1, and is described above (see Chap. 1, Sect. 2 and Chap. 2, Sect. 1). Moreover, there exist neither any saddle connections nor any separatrices joining a saddle-node and a saddle.[15]

A *dynamical system* generated by a quasi-generic vector field is called *quasi-generic*.

Theorem (Sotomayor (1974)). *If $r \geqslant 4$ and* **M** *is either a closed, orientable surface or a closed nonorientable surface of genus $g \leqslant 3$,[16] then the set of quasi-generic vector fields of class C^r on* **M** *: 1) is a C^{r-1}-smooth submanifold of the space $\chi^r(\mathbf{M})$, immersed in $\chi^r(\mathbf{M})$; 2) is everwhere dense in the bifurcation set.*

Let **B** be an arbitrary connected subset of a topological space χ. The *neighborhood of a point $x \in$* **B** *in the inner topology* is defined as a connected component containing the point x of the intersection of **B** and some neighborhood of x in the ambient space x. This definition provides an "intrinsic" topology for the set of quasi-generic vector fields (in the ambient space $\chi^r(\mathbf{M})$). In the case where quasi-generic systems are dense in the bifurcation set in the sense of the intrinsic topology, a generic one-parameter family contains only structurally stable and

---

[15] A separatrix of a saddle-node equilibrium is understood here to be the part of a center manifold not belonging to a two-dimensional stable or unstable set, in other words, the common boundary of two hyperbolic sectors.

[16] A condition on the topological type of the surface appears here because, for surfaces not included in the statement of the theorem, the closing lemma is unproved in the C^r-topology for $r \geqslant 2$.

quasi-generic systems (see Fig. 36a). If quasi-generic systems are dense in the bifurcation set only in the topology induced by the embedding in the space of vector fields, then, in a way nonremovable by a small perturbation in one-parameter families, one finds structurally unstable and nonquasi-generic fields (see Fig 36b).

Fig. 36. Possible arrangements of bifurcation surfaces: "1" – structurally unstable vector fields, "2" – a quasi-generic field. (a) Dense in the intrinsic topology. (b) Dense in the topology of the ambient space. The curve that transversally intersects the bifurcation surfaces represents a generic one-parameter family.

Theorem. *The set of quasi-generic vector fields on a two-dimensional sphere or the projective plane is dense in the set of all structurally unstable vector fields in the intrinsic topology.*

This theorem follows from classical results of Andronov, Leontovich, Gordon and Majer (1966, 1967).

We introduce the class $\Phi^{k,r} \subset \Psi^{k,r}$ of one-parameter families of vector fields on the sphere satisfying the following conditions: 1) each vector field in any family is either structurally stable or quasi-generic, 2) each family intersects the bifurcation set transversally, 3) if a family contains a quasi-generic vector field, corresponding to the situation shown in Fig. 32, then the conditions of genericity, formulated in the next section, are satisfied.

Apparently, Theorem 2.2 holds for families belonging to $\Phi^{k,r}(S^2)$.

2.4. Conditions for Genericity. We assume that a family $\{v_\varepsilon\}$ contains a vector field corresponding to the situation shown in Fig. 32. The first return map f_0 of the vector field v_0 is defined on a transversal l to a cycle with multiplier $+1$. Let x be a local coordinate on l such that: 1) the cycle corresponds to $x = 0$; 2) $P_1, \ldots, P_k$ are points on separatrices of distinct saddles that are ω-asymptotic to the cycle, and $Q_1, \ldots, Q_m$ are points on the likely separatrices α-asymptotic to the cycle, that are such that

$$x(P_1) < \cdots < x(P_k) < x(f_0 P_1) < 0 < x(Q_1) < \cdots < x(Q_m) < x(f_0 Q_1).$$

If $k \leq 1$, or $m \leq 1$, then no further conditions, except 1) and 2) (in the previous section), are imposed on the family. Suppose $k > 1$ and $m > 1$. As is shown by Malta and Palis (1981) and Nitecki (1971), f_0 can be embedded in a uniquely defined flow $\{g^t\}$ on l, so that $f_0(P) = g^1(P)$. Set

$$P_2 = g^{t_1} P_1, \ldots, P_k = g^{t_{k-1}} P_1, \qquad Q_2 = g^{\bar{t}_1} Q_1, \ldots, Q_m = g^{\bar{t}_{m-1}} Q_1.$$

The genericity conditions given by Malta and Palis (1981) are:

$$|t_i - t_j| \neq |\bar{t}_\alpha - \bar{t}_\beta|, \quad 1 - |t_i - t_j| \neq |\bar{t}_\alpha - \bar{t}_\beta| \quad \text{for } i \neq j \text{ and } \alpha \neq \beta,$$

$i, j \in \{1, \ldots, k-1\}$, and $\alpha, \beta \in \{1, \ldots, m-1\}$.

Malta and Palis (1981) sketched a proof of the structural stability of such families (more exactly, a sketch for proving versality of deformations for the vector field v_0) if the conditions formulated above are fulfilled. If these conditions are not fulfilled, then there exist families, arbitrarily close to the original one, containing vector fields with two (or more) saddle connections (Yu.S. Il'yashenko and S.Yu. Yakovenko [4*]).

2.5. One-Parameter Families on Surfaces different from the Sphere. It is clear that for any surface **M**, one can identify the class of one-parameter families of vector fields $\Phi^{k,r}(\mathbf{M})$ analogously to the class $\Phi^{k,r}(\mathbf{S}^2)$, that is, the class of arcs in function space intersecting the bifurcation set only at points of the set of quasi-generic vector fields. This was done by Malta and Palis (1981), where a sketch for proving the openness of this class in the set of all one-parameter families is given. Isolated bifurcation values in families of this class correspond to systems of the first degree of structural instability, which we now define.

Definition (Andronov and Leontovich (1965)). A dynamical system is called a system of the *first degree of structural instability* (*or nonroughness*) if it is not structurally stable and there exists a neighborhood of it such that each dynamical system from this neighborhood is either structurally stable, or is orbitally topologically equivalent to the original system; and, moreover, the conjugating homeomorphism is close to the identity. Vector fields generating systems of the first degree of structural instability are called vector fields of the *first degree of structural instability*.

Theorem (Andronov and Leontovich (1965), Andronov, Leontovich, Gordon, and Majer (1966), (1967), Aranson (1968, 1970)). *Suppose that* **M** *is a closed surface that is either orientable, or nonorientable and of genus* $g \leqslant 3$. *Then a smooth dynamical system on* **M** *is a system of the first degree of structural instability provided it has the following properties*: 1) *it is quasigeneric*; 2) *it does not have separatrices of saddles that contain separatrix loops of other saddles* (*or of the same saddle*) *in the set of their limit points*; 3) *it does not have a separatrix of a saddle whose* α*-limit* (ω*-limit*) *points contain a nonhyperbolic cycle also contained in the set of* ω*-limit* (α*-limit*) *points of some separatrix of another saddle, or of the same saddle, and, in particular, it has no contours*; 4) *it does not have homoclinic trajectories of nonhyperbolic cycles.*

Corollary. A bifurcation of a system of the first degree of structural instability on a surface admitted by the theorem is semi-local. (One can always find a set of trajectories in a neighborhood of which only nonwandering trajectories and separatrices are born or disappear.) In fact, these are bifurcations of semi-stable cycles, saddle connections and separatrix loops (see Fig. 34).

Accumulation points of bifurcation values in families in $\Phi^{k,r}(\mathbf{M})$ and bifurcations in neighborhoods of these points can be looked at analogously to the corresponding bifurcations in families in $\Phi^{k,r}(\mathbf{S}^2)$, at least if the surface under consideration is orientable (Malta and Palis (1981)). However, for surfaces on which a system may have nontrivial (that is, different from an equilibrium or a cycle) Poisson-stable trajectories (that is, for all surfaces except a sphere $\mathbf{S}^2$, a projective plane Π^2 and a Klein bottle $\mathbb{K}^2$), generically there are vector fields with infinite nonwandering sets in a generic one-parameter family. Bifurcations in such families are completely undescribed, except for bifurcations of systems with a global transverse section on a two-dimensional torus (see the following subsection). However, it is known that there exist generic one-parameter families on surfaces, other than $\mathbf{S}^2$, Π^2 and $\mathbb{K}^2$, which contain non-structurally stable vector fields of infinite degree of structural instability (Aranson (1986)). For systems on the sphere $\mathbf{S}^2$ the following result holds.

Theorem (Andronov and Leontovich (1965, 1970), Andronov, Leontovich, Gordon, and Majer (1966), (1967)). *The set of systems of the first degree of structural instability is open and dense in the set of all nonstructurally stable systems on* $\mathbf{S}^2$.

For vector fields on a two-dimensional torus a weaker result has been established.

Theorem (Aranson, 1986). *The set of vector fields of the first degree of structural instability on a torus is open and dense in the space of non-structurally stable vector fields with no singular points in the topology induced from* $\chi^r(\mathbb{T}^2)$. *The conclusion also holds for* Π^2 *and* $\mathbb{K}^2$.

In both of these theorems it is assumed that the vector fields are at least C^2-smooth.

2.6. Gobal Bifurcations of Systems with a Global Transversal Section on a Torus. The investigation of flows on a torus with a global transversal section leads to the investigation of diffeomorphisms of the circle (appearing as first-return maps). Here the basic characteristic determining the topological structure is the Poincaré rotation number. It characterizes the global bifurcations that occur as a parameter is varied.

Majer (1939) noted that the dependence of the rotation number on the parameter may be described by a Cantor function.

Definition. A continuous function $\omega(\varepsilon)$: $[a, b] \to \mathbb{R}^1$ is a *Cantor function* if 1) a set C is given on $[a, b]$ that is homeomorphic to a perfect Cantor set; 2) the function ω is constant on all contiguous intervals (connected components of the difference set $[a, b] \setminus C$) and is not equal to a constant on $[a, b]$.

For a C^r-vector field v_ε ($r \leqslant 2$) that depends continuously on ε the intervals of constant rotation number may correspond to both rational and irrational values of ω, moreover, some rational values may not correspond to intervals of constancy.[17]

[17] We recall that a vector field with a cycle corresponds to a rational rotation number.

The following theorem, proved by Aranson et al. (1984), using the methods of Arnol'd (1961) (see Anosov (1977)), establishes sufficient conditions for "genericity" of one-parameter families of mappings of the circle.

Theorem. *Let $f_\varepsilon\colon \mathbf{S}^1 \to \mathbf{S}^1$, $\varepsilon \in [a, b]$, be a family of analytic diffeomorphisms, analytically depending on* ε, *(or analytic homeomorphisms that are not diffeomorphisms) such that:* 1) *the covering maps $\bar{f}_\varepsilon\colon \mathbb{R}^1 \to \mathbb{R}^1$ have the form: $\bar{f}_\varepsilon(x) = x + h(x, \varepsilon)$, where h is an analytic function in x, ε having period* 1 *in x, and such that for all $x \in \mathbb{R}^1$, and $\varepsilon \in [a, b]$, either $\partial h/\partial\varepsilon < 0$ or $\partial h/\partial\varepsilon > 0$;* 2) *$\bar{f}_\varepsilon(z)$ is an entire function of one complex variable, and on the complex z-plane there is at least one root of the equation $d\bar{f}_\varepsilon(z)/dz = 0$. Then:* 1) *the rotation number $\omega(\varepsilon)$ of the diffeomorphism f_ε is a Cantor function, non-decreasing for $\partial h/\partial\varepsilon > 0$ and nonincreasing for $\partial h/\partial e < 0$;* 2) *the function ω takes each of its rational values on intervals;* 3) *the function ω strictly increases for $\partial h/\partial\varepsilon > 0$ (strictly decreases for $\partial h/\partial\varepsilon < 0$) on the set of those values of ε which correspond to irrational values of ω.*

Example. The mapping

$$\bar{f}_\varepsilon(x) = x + \varepsilon - (1/2\pi)\sin 2\pi x$$

satisfies all the conditions of the theorem. The graph of $\omega(\varepsilon)$ is presented in Fig. 37.

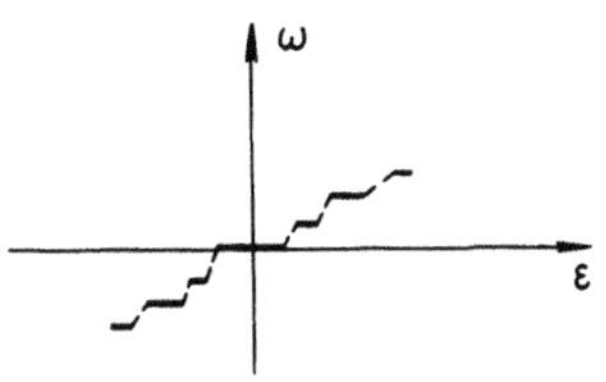

Fig. 37. The graph of the dependence of the rotation number ω on the parameter ε

Remark. Knowledge of the dependence of the rotation number on ε allows one to find all bifurcations taking place as ε changes, with the exception, perhaps, of bifurcations that occur at constant rational rotation numbers, that is, bifurcations of merging and disappearing (or birth) of cycles under the condition that some other cycles are preserved as this takes place (see also Sect. 7.1).

2.7. Some Global Bifurcations on a Klein Bottle. Until recently there remained an unsolved problem: does there exist on a compact manifold a one-parameter family of vector fields $\{v_\varepsilon\}$ with base [0, 1] having for $\varepsilon < 1$ a limit cycle whose length grows unboundedly as $\varepsilon \to 1$, a cycle that is situated a positive distance from 0, that is bounded uniformly in ε away from singular points of the vector field v_ε, and which disappears for $\varepsilon = 1$. The name "blue-sky catastrophe" was given to such a bifurcation of a cycle (Palis and Pugh (1975)).

Medvedev (1980) constructed a one-parameter families of vector fields $\{v_\varepsilon\}$ on a Klein bottle, and on a two-dimensional torus, in which a blue-sky catastrophe takes place; moreover, on the Klein bottle the family is generic and the vector field v_1 is quasi-generic: it has a double limit cycle L, and the remaining trajectories are doubly asymptotic to the cycle (for $\varepsilon = 1$ on the Klein bottle there is no global transverse section). For $\varepsilon < 1$ this cycle disappears, and two cycles L_ε^i ($i \in \{1, 2\}$) arise that are not homotopic to L, one of which is stable, the other unstable, and all other trajectories are wandering trajectories. For all $\varepsilon \in [0, 1)$, the field v_ε is structurally stable. From this it follows that the bifurcation surface is accessible at v_1 from the domain of structurally stable systems.

For the vector field $\tilde{v}_\varepsilon$ that lifts v_ε in the two-sheeted covering of the Klein bottle with the torus, for $\varepsilon \neq 1$ there exist two limit cycles $\tilde{L}_\varepsilon^1$, $\tilde{L}_\varepsilon^2$ that are the preimages of L_ε^1, L_ε^2, respectively. As $\varepsilon \to 1$, each cycle behaves as follows: it winds many times clockwise around the torus into a narrow ring K_1, and then it winds out the same number of times counter-clockwise into another ring K_2; $K_1 \cap K_2 = \varnothing$ and the boundaries of K_1 and K_2 are homotopic to each other and to a circle; see Fig. 38.

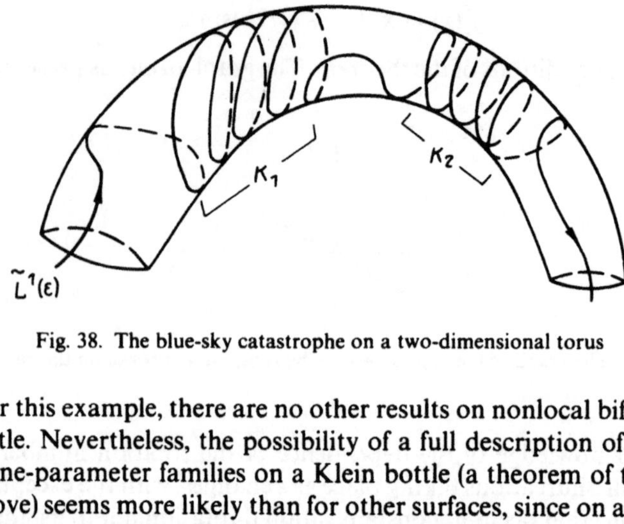

Fig. 38. The blue-sky catastrophe on a two-dimensional torus

Except for this example, there are no other results on nonlocal bifurcations on a Klein bottle. Nevertheless, the possibility of a full description of bifurcations in generic one-parameter families on a Klein bottle (a theorem of type given in Sect. 2.2 above) seems more likely than for other surfaces, since on a Klein bottle there cannot exist nontrivial Poisson-stable trajectories; see Aranson (1970) and Markley (1969).

2.8. Bifurcations on a Two-Dimensional Sphere: The Multi-Parameter Case. Although even local bifurcations in high codimensions (at least three) on a disc are not fully investigated, it is natural to discuss nonlocal bifurcations in multi-parameter families of vector fields on a two-dimensional sphere. For their description, it is necessary to single out the set of trajectories defining perestroikas in these families.

Definitions and Examples (V.I. Arnol'd, 1985)

Definition 1. A finite subset of phase space is said to *support a bifurcation* if there exists an arbitrarily small neighborhood of this subset and a neighborhood of the bifurcation value of the parameter (depending on it) such that, outside this neighborhood of the subset, the deformation (at values of the parameter from the second neighborhood) is topologically trivial.

Example 1. Any point of a saddle connection (including both saddles) supports a bifurcation, even if one adds to it any other points. In a system with two saddle connections an interior point on one connection supports a bifurcation only with a point on the other connection.

Definition 2. The *bifurcation support* of a bifurcation is the union of all minimal sets supporting a bifurcation ("minimal" means not containing a proper subset that supports a bifurcation).

Example 2. In a system with one saddle connection (bifurcating in a standard way), the support coincides with the saddle connection, including its endpoints, the saddles.

Definition 3. Two deformations of vector fields with bifurcation supports Σ_1 and Σ_2 are said to be *equivalent on their supports* or *weakly equivalent on their supports* if there exist arbitrarily small neighborhoods of the supports, and neighborhoods of the bifurcation values of the parameters depending on them, such that the restrictions of the families to these neighborhoods of the supports are topologically equivalent, or weakly equivalent[18], over these neighborhoods of bifurcation values.

Example 3. All deformations of vector fields with a simple saddle connection are equivalent to each other, independent of the number of hyperbolic equilibria or cycles in the system as a whole.

Example 4. Four-parameter deformations of a vector field close to a cycle of multiplicity four are weakly topologically equivalent, but, generally, not equivalent: the classification of such deformations with respect to topological equivalence involves functional invariants (see Chap. 2 Sect. 5.11)

Conjectures (V.I. Arnol'd, 1985). For a generic l-parameter family of vector fields on $\mathbf{S}^2$:

1) On their supports, all deformations are equivalent to a finite number of deformations (the number depends only upon l).

2) Any bifurcation diagram is (locally) homeomorphic to one of a finite number (depending only upon l) of generic examples.

3) There exist versal and weakly structurally stable deformations.

4) The family is globally weakly structurally stable.

5) The bifurcation supports consist of a finite number (depending only upon l) of (singular) trajectories.

6) The number of points in a minimal supporting set is bounded by a constant depending only on l.

---

[18] The definitions of topological equivalence and weak equivalence of families and their structural stability are analogous to those presented in Sect. 2.2. It is only necessary to replace the interval I by a neighborhood of the bifurcation value.

Certainly proofs or counterexamples to the above conjectures are necessary for investigating nonlocal bifurcations in generic l-parameter families. *Remark added in proof:* Recently A. Kotova and V. Stanzo found a counterexample to Conjecture 2. Little is now known: even for families of structurally stable and quasi-generic vector fields, Conjectures 3 and 4 (the only nontrivial ones in this case) are unproved. As far as we know, for $l = 2$ only two nonlocal bifurcations have been investigated in detail.

Theorem 1 (Nozdracheva (1970). *In a generic two-parameter family of C^r-vector fields ($r \geqslant 3$) there occur only fields with separatrix loops of a saddle (having zero saddle number), whose bifurcations are shown in Fig. 39.*

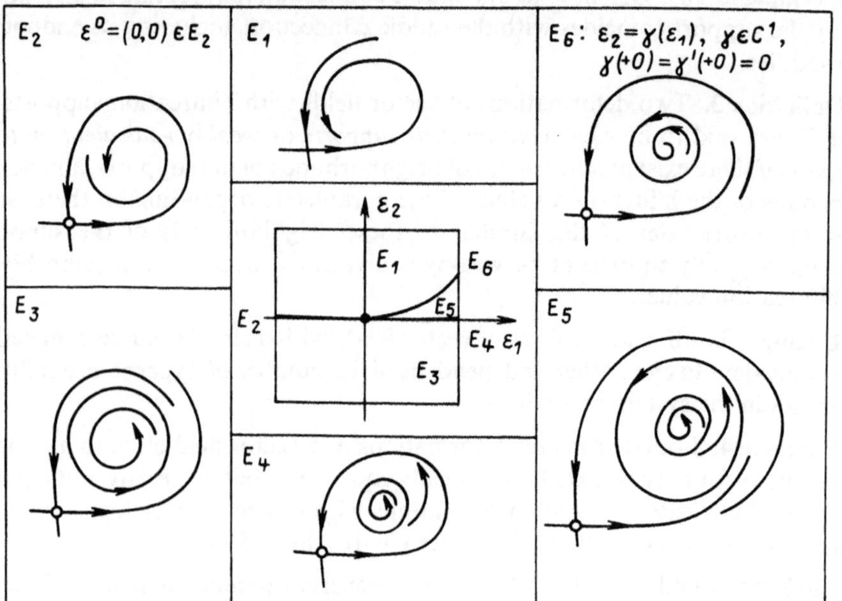

Fig. 39. The bifurcation diagram and perestroikas of phase portraits for generic two-parameter deformations of a vector field with a separatrix loop. The bifurcation curve, corresponding to a semi-stable cycle, has a tangency of infinite order with the ε_1-axis at the origin.

Theorem 2. *Suppose a vector field $v_0 \in \chi^r(\mathbf{M})$, $r \geqslant 6$, has a contour Γ consisting of two saddles O_1 and O_2 and two separatrices Γ_1 and Γ_2 such that $\alpha(\Gamma_1) = \omega(\Gamma_2) = \{O_1\}$, $\alpha(\Gamma_1) = \omega(\Gamma_1) = \{O_2\}$. Let λ_{ij} be the eigenvalues of the linear part of the vector fields v_0 at O_i ($i, j \in \{1, 2\}$), $\sigma_i = \lambda_{i1} + \lambda_{i2}$, and $\Delta = \lambda_{11}\lambda_{22} - \lambda_{12}\lambda_{21}$. Assume that Γ has a neighborhood $\mathcal{U}$, homeomorphic to $\mathbb{R} \times \mathbf{S}^1$, such that one of the components of $\mathcal{U} \setminus \Gamma$ does not intersect the stable and unstable manifolds of O_1 and O_2. Set $k = 1$ ($k = 2$) if $\sigma_1\sigma_2 < 0$ and $\Delta > 0$ ($\Delta < 0$), and set $k = 3$ ($k = 4$) if $\sigma_i < 0$ ($\sigma_i > 0$) ($i = 1, 2$). Then a generic two-parameter deformation $\{v_\varepsilon\} \subset \chi^r(\mathbf{M})$, $r \geqslant 6$, with support on Γ has a bifurcation diagram, shown in* Fig. 40a (Fig. 40b) *for $k = 1$ ($k = 3$), and for $k = 2$ ($k = 4$) it has a diagram obtained from the diagram in*

Fig. 40a (Fig. 40b) *by reversing time. The perestroikas of the phase portraits in Theorem* 2 *are also depicted in* Figs. 40a,b.

Case a) of this theorem was investigated by V.Sh. Rojtenberg (1985); Case b) by V.P. Nozdracheva (1981; see Ref. Zh. Mat. 1981, 8B233); the birth of cycles at bifurcation of the contours on the plane was investigated by J.W. Reyn (1980).

2.9. Some Open Questions. Besides those introduced previously in this section, we point out some more open questions.

1. What sort of structure does a component of the bifurcation set of a system with an infinite nonwandering set have? In particular, can these components contain submanifolds of codimension 1?

2. What can be said about bifurcations of systems on a nonorientable surface of genus $g > 3$?

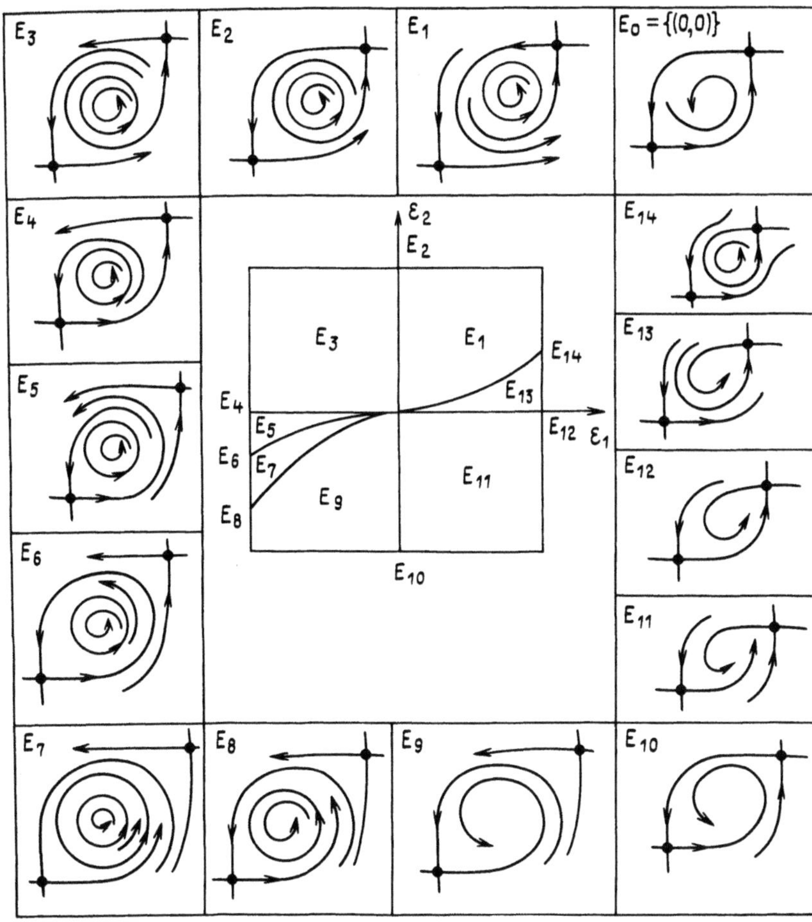

Fig. 40a. Bifurcation diagrams and perestroikas of phase portraits for generic two-parameter deformations of a vector field with a contour from two saddles. Case (a) saddle numbers with different signs.

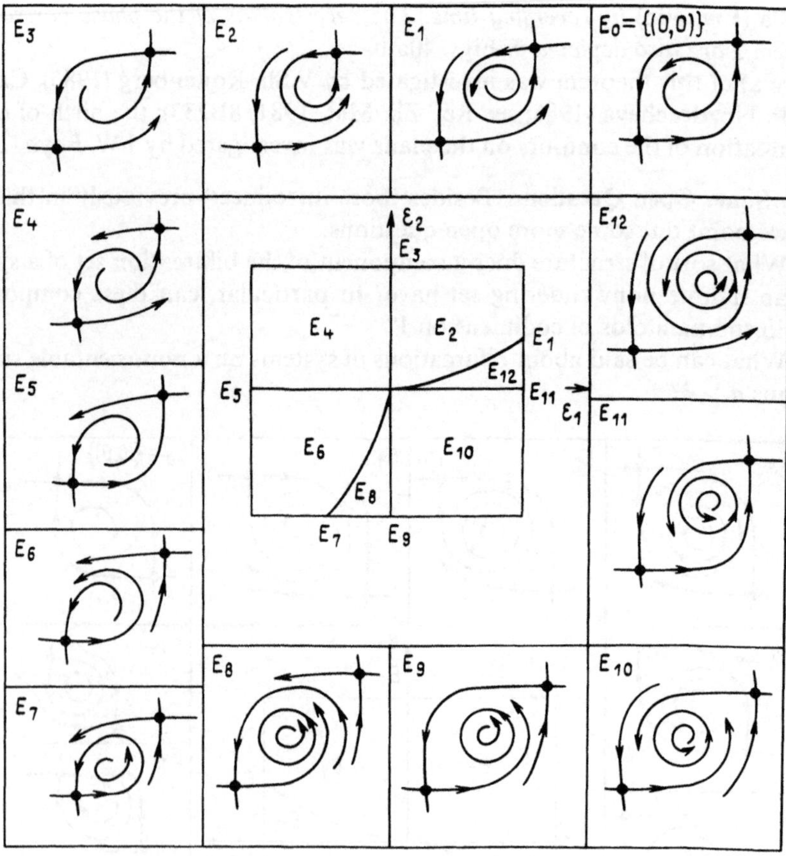

Fig. 40b. Case (b), saddle numbers with the same signs

3. How are one-parameter deformations of quasi-generic systems described if they are not systems of the first degree of structural instability? In particular, how does one describe the bifurcations that produce the appearance and disappearance of nontrivial Poisson-stable trajectories? (Here, one probably needs to use symbolic dynamics such as the theory of kneading sequences; see Collet and Eckmann (1981) and Jonker and Rand (1981).)

§ 3. Bifurcations of Trajectories Homoclinic to a Nonhyperbolic Singular Point

The bifurcations described in this Section occur in generic one-parameter families and lead to the birth of either a structurally stable limit cycle, or a nontrivial hyperbolic set.

3.1. A Node in its Hyperbolic Variables

Theorem (Shil'nikov (1963)). *Suppose that in a generic one-parameter family there is a vector field v_0 with a degenerate singular point O, having one eigenvalue* 0, *which is a node in its hyperbolic variables and which has a homoclinic trajectory Γ. Then all the noncritical vector fields in the family that are sufficiently close to v_0 either have two singular points near to O (depending on which side of O the parameter lies) or have a stable (completely unstable*[19]*) limit cycle. This cycle tends to $\Gamma \cup O$ as the parameter tends to* 0.

Requirements of genericity. 1. The same requirements of genericity are imposed upon the germ of a family at the point (0, 0) (in the product of phase space and parameter space) as those in the Sect. 2.1 of Chap. 1.

2. The following nonlocal requirement is imposed on the field v_O: $\Gamma \cap W^s = \varnothing$. In other words, the homoclinic trajectory enters the interior without crossing the boundary of the stable set.

3. A local family transversally intersects a hypersurface of vector fields with a degenerate singular point.

It is possible to formulate the previous result in the language of spaces of vector fields.

Theorem. *Suppose a field v_0 satisfies all the requirements above. Then, in the space $C^2(U)$ of vector fields on some neighborhood U of the curve $\Gamma \cup O$ with the C^2-topology, there exists a neighborhood W of the vector field v_0 having the following property. The neighborhood W is divided into two domains by a hypersurface* **B** *passing through v_0; all fields lying on one side of* **B** *have two singular points close to O, and all vector fields lying on the other side of* **B** *have a stable or completely unstable limit cycle. All fields on* **B** *are topologically equivalent to v_0 in the domain U.*

Remark. All theorems on bifurcations with degeneracies of codimension 1 have dual formulations: one in the language of one-parameter families and the other in the language of hypersurfaces in a function space. We shall mostly formulate the theorems given below in the language of one-parameter families.

3.2. A Saddle in its Hyperbolic Variables: One Homoclinic Trajectory. A vector field with a degenerate singular point that is a saddle in its hyperbolic variables may have an arbitrary finite number of homoclinic trajectories at the singular point; such fields occur generically in generic one-parameter families. We denote the number of homoclinic trajectories at a degenerate singular point O by p. The cases $p = 1$ and $p > 1$ are distinctly different from each other.

Theorem (Shil'nikov (1966)). *Suppose the zero critical value of the parameter in a generic one-parameter family corresponds to a vector field v_0 with a degenerate*

[19] A cycle is called completely unstable if it becomes stable upon reversing time.

singular point O, O being a saddle in its hyperbolic variables, but with one eigenvalue 0, *and having exactly one homoclinic trajectory. Then the conclusions of the first theorem in Sect.* 3.1 *hold, but the cycle born is of saddle type (that is, hyperbolic but neither stable nor completely unstable).*

The requirements of genericity on the vector field v_0, and on the family, are the same as in Sect. 3.1 and, additionally, it is required that the stable and unstable sets intersect transversally.

If several homoclinic trajectories bifurcate, fields are obtained that are described with the aid of the topological Bernoulli automorphism.

3.3. The Topological Bernoulli Automorphism. Let Ω be the space of doubly infinite sequences on p symbols $\{1, \ldots, p\}$ with the metric

$$\rho(\omega, \omega') = \sum_{k=-\infty}^{\infty} |\alpha_k - \alpha'_k|/2^{|k|},$$

$$\omega = (\ldots, \alpha_{-1}, \alpha_0, \alpha_1, \ldots), \qquad \omega' = (\ldots, \alpha'_{-1}, \alpha'_0, \alpha'_1, \ldots).$$

We denote by $\sigma: \Omega \to \Omega$ the homeomorphism that shifts each element to the right by one place:

$$\sigma\omega = \omega', \qquad \omega = \{\alpha_k\}, \qquad \omega' = \{\beta_k\}, \qquad \beta_{k-1} = \alpha_k.$$

The pair (σ, Ω) is called the *topological Bernoulli automorphism*, the *topological Bernoulli shift*, or the *topological two-sided shift*. A *suspension over the topological Bernoulli shift* is a periodic vector field x_σ whose monodromy transformation is conjugate to σ. This field is obtained from the standard vector field $\partial/\partial t$ on the product $\mathbf{I} \times \Omega$, $\mathbf{I} = \{t \in [0, 1]\}$, after identifying the points $(0, \sigma\omega)$ and $(1, \omega)$ by the gluing map κ. A phase flow on a subset Σ of Euclidean space is topologically equivalent to a suspension over the topological Bernoulli shift if there exists a homeomorphism from $\Sigma \to \mathbf{I} \times \Omega|\kappa$ transforming the original vector field into x_σ.

Remark. The subset Σ is similar to the product of a Cantor set and a circle.

Example. Let K_1 and K_2 be two unit squares on the plane with sides parallel to the coordinate axes and centers (1, 0) and (3, 0). Consider the function $f: K_1 \cup K_2 \to \mathbb{R}^2$, where the mapping

$$f|K_i: (x, y) \mapsto A((x, y) + a_i)$$

is the composition of translation by the vector a_i and the affine transformation $A: \mathbb{R}^2 \to \mathbb{R}^2$ that maps $(x, y) \to (10x, 0.1y)$ (see Fig. 41) and where $a_1 = (-1, 1)$, $a_2 = (-3, 3)$. The planar set on which all (positive or negative) iterations of the function f are defined is mapped homeomorphically onto the space of sequences of two symbols in the following way: a point P corresponds to the sequence $\{\alpha_k(P)\}$ where $\alpha_k(P) = i$ if and only if $f^k(P) \in K_i$ $(i = 1, 2)$. It is not difficult to prove that this mapping is a homeomorphism. Clearly, it conjugates the mapping f with the shift σ.

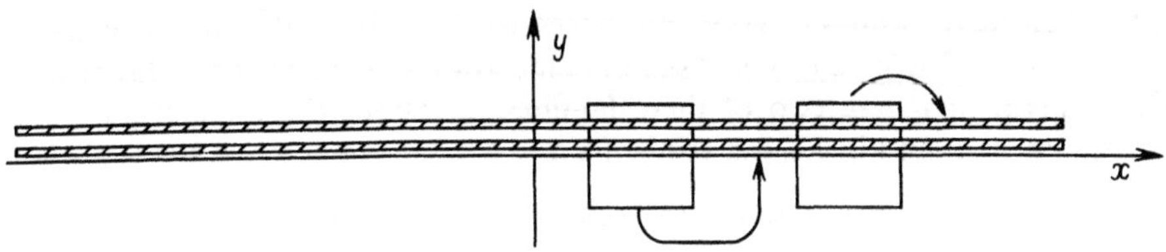

Fig. 41. A model map for the problem of bifurcations of two homoclinic trajectories at a saddle-node

3.4. A Saddle in its Hyperbolic Variables: Several Homoclinic Trajectories

Theorem (Shil'nikov (1969)). *In a generic one-parameter family of vector fields there are vector fields with a degenerate singular point O, having one eigenvalue* 0, *which is a saddle in its hyperbolic variables, and which has p homoclinic trajectories Γ_i $(p > 1)$. Then, for all vector fields v_ε corresponding to values of the parameter ε lying sufficiently close to and on one side of the critical value* 0, *the following assertion is true. For some neighborhood U of $O \cup \Gamma_i$, the restriction of the flow of the field v_ε to the set of nonwandering trajectories is topologically equivalent to a suspension over a topological Bernoulli shift on p symbols.*

The requirements of genericity upon the family are the same as in Sect. 3.2. The mechanism by which the invariant set arises is illustrated for $p = 2$ by the example in Sect. 3.3.

We now suppose that

$$S_O^u \cap S_O^s = O \cup \left(\bigcup_{i=1}^{p} \Gamma_i \right),$$

and, moreover, that the stable and unstable sets intersect transversally along Γ_i $(i = 1, \ldots, p)$. We also assume that the field v_0 lies on the boundary of the set of Morse-Smale vector fields, that its nonwandering set is finite and hyperbolic, except for O, and that the stable and unstable manifolds of the hyperbolic nonwandering trajectories transversally intersect each other and also the manifolds S_O^u, S_O^s, W_O^u, and W_O^s. The accessibility of the bifurcation surface from both sides may be obtained from the following theorem.

Theorem. *Under the conditions formulated above on the vector field v_0, there exists a neighborhood U of v_0 in $\chi^r(\mathbf{M})$ such that for any system $v \in U$ not having an equilibrium in a neighborhood of the point O, Axiom A and Smale's strong transversality condition hold.*

We recall that a vector field satisfies Axiom A if its set of nonwandering trajectories is hyperbolic and the periodic trajectories of the field are dense in it. The strong transversality condition consists of the following: the stable and unstable manifolds of all nonwandering trajectories intersect transversally. For details of the hyperbolic theory see EMS, Dynamical Systems 9.

3.5. Principal Families. To begin, we construct principal families which are normal forms of deformations of vector fields from Sect. 3.2 in three-dimensional phase space; there are two of these families. Consider the cube K_0: $|x_1| \leqslant 1$, $|x_2| \leqslant 1$, $|z| \leqslant 1$, and in K_0 the vector field:

$$v_\varepsilon^0 = -x_1(\partial/\partial x_1) + x_2(\partial/\partial x_2) + (z^2 + \varepsilon)(\partial/\partial z).$$

We glue together the boundary faces $z = 1$ and $z = -1$ of K_0 in two ways. Set

1) $f_0^+ : (x_1, x_2, 1) \to (x_1, x_2, -1)$; 2) $f_0^- : (x_1, x_2, 1) \to (-x_1, -x_2, -1)$.

We obtain two three-dimensional manifolds K^+ and K^- homeomorphic to each other (and to the product of a two-dimensional disk by S^1), and two vector fields v_ε^+ and v_ε^- defined on K^+ and K^-, respectively. It is easy to verify that: 1) for $\varepsilon < 0$, $v_\varepsilon^\pm$ in $K^\pm$ have two hyperbolic equilibria O_1 and O_2 with dim $W_{O_1}^s = 2$, dim $W_{O_2}^u = 2$, and, moreover, $W_{O_1}^s$ and $W_{O_2}^u$ intersect transversally along two trajectories Γ_1 and Γ_2; 2) as $\varepsilon \to 0$, O_1 and O_2 tend along the trajectory Γ_1 to O and coincide for $\varepsilon = 0$; Γ_2 becomes a homoclinic trajectory Γ; 3) for $\varepsilon > 0$, v_ε has a limit cycle L_ε of saddle type ($x_1 = x_2 = 0$) which is the unique nonwandering trajectory of the fields $v_\varepsilon^\pm$ in $K^\pm$; moreover, for the field v_ε^+ the stable and unstable manifolds of the cycle are cylinders, and for v_ε^- they are Möbius bands.

Theorem (on versality). *A germ of a generic one-parameter family of vector fields $\{w_\varepsilon\}$ on a homoclinic trajectory of a nonhyperbolic singular point in $\mathbb{R}^3$, which is of saddle type in its hyperbolic variables, is topologically equivalent to the germ of one of the principal families $\{v_\varepsilon^+\}$ or $\{v_\varepsilon^-\}$ on the homoclinic trajectories of the fields $\{v_\varepsilon^+\}$ or $\{v_\varepsilon^-\}$.*

There is an analog of this theorem for arbitrary n: the principal family is obtained by a suspension of the hyperbolic equilibrium over $\{v_\varepsilon^+\}$ or $\{v_\varepsilon^-\}$.

Principal deformations of the equations described in Sect. 3.1 are constructed analogously, and a theorem on their versality can also be formulated. For each n the principal deformation is unique.

§4. Bifurcations of Trajectories Homoclinic to a Nonhyperbolic Cycle

The bifurcations described here lead to the birth of invariant tori, Klein bottles, complex invariant sets with countably many cycles, and strange attractors. Some cases are not yet fully studied: some open questions are formulated in Sect. 4.11. At the end of this section the structural stability of one-parameter families of diffeomorphisms is investigated.

4.1. The Structure of a Family of Homoclinic Trajectories. As was mentioned in Sect. 1 of this Chapter, a field with homoclinic trajectories of a nonhyperbolic cycle corresponds to a generic point on the boundary of the set of Morse-Smale systems only if one of the multipliers of this cycle is equal to 1. The compactness or noncompactness of the union of a cycle and the set of its homoclinic trajectories has a substantial influence on the bifurcations of such fields.

We consider the compact case first; the noncompact case is considered in Sect. 4.7, below.

Lemma (Afrajmovich and Shil'nikov (1972, 1982). *Suppose that in a generic one-parameter family there is a vector field with a nonhyperbolic cycle, with multiplier* 1, *for which the union of the cycle together with all of its homoclinic trajectories is compact. Then this union consists of a finite number (say,* p) *of continuous two-dimensional manifolds, each of which is homeomorphic to a torus or a Klein bottle. If the cycle is of nodal type in its hyperbolic variables, then* $p = 1$ *and the union coincides with* S^u *(resp.,* S^s*) for stable (resp., unstable) nodes.*

Another important property defining the character of bifurcations (and also the smoothness of the manifolds described in the above lemma) is the so-called criticality of a cycle, which is considered in the next subsection.

4.2. Critical and Noncritical Cycles. Suppose a smooth vector field has a limit cycle, with multiplier 1, which is a stable node in its hyperbolic variables, i.e., all other multipliers have moduli less than 1. Then some neighborhood of the cycle is endowed with a smooth foliation, with leaves of codimension 1, invariant relative to the flow and *strongly stable*: each leaf contracts exponentially upon a shift along trajectories of the field corresponding to an increase in time (Hirsch, Pugh, and Shub (1977), Newhouse, Palis and Takens (1983)). One of the fibers coincides with the stable manifold of the cycle. The *strongly unstable foliation* that arises in the case of a node which is unstable in its hyperbolic variables is described analogously.

Suppose a cycle of a vector field has multiplier 1, and is of saddle type in its hyperbolic variables. Then the restriction of the field to its center-stable (center-unstable) manifold W^{sc} (W^{uc}) has a stable (unstable) cycle that is of nodal type in its hyperbolic variables. One may define, as above, *strongly stable* and *strongly unstable foliations* on W^{sc} and W^{uc}, which are denoted by $\mathcal{F}^{ss}$ and $\mathcal{F}^{uu}$, respectively.

Definition (Newhouse, Palis and Takens (1983)). A limit cycle of a vector field with multiplier 1 is called *s-critical* if either, there exists a hyperbolic equilibrium or hyperbolic cycle, whose stable or unstable manifold is tangent to one of the leaves of $\mathcal{F}^{ss}$ on S^s, or the unstable set of the cycle is tangent to one of these leaves. In the latter case the union of the homoclinic trajectories of the cycle is called *s-critical.* The concepts of a *u-critical* cycle and a *u-critical* union of its homoclinic trajectories are defined analogously by appropriately interchanging the superscripts *u* and *s*. A cycle and the union of its homoclinic trajectories are called *critical* if they are either *s* or *u*-critical and *noncritical* otherwise; see Fig. 42.

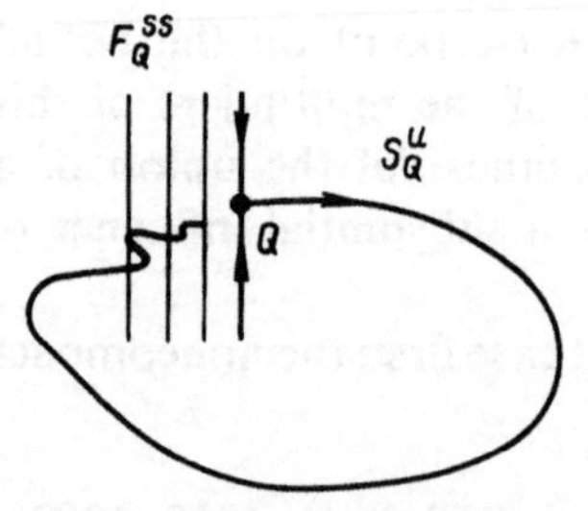

Fig. 42. A transversal section of a set of homoclinic trajectories of an s-critical cycle (the compact case)

Remark. The tori and Klein bottles in the lemma of Sect. 4.1 are smooth if the union of homoclinic trajectories of the cycle is noncritical, otherwise, there are nonsmooth ones among them.

4.3. Creation of a Smooth Two-Dimensional Attractor. We use the definition of an attractor presented in Arnol'd and Il'yashenko (1985, p. 42), which is reproduced in Sect. 8.3 below. The results of this and the following subsections are parallel to the results in Sect. 3; only, instead of nonhyperbolic singular points with eigenvalue zero, nonhyperbolic cycles with a multiplier 1 undergo bifurcations. As a result, instead of hyperbolic equilibria, hyperbolic cycles are born, and, instead of cycles, tori and Klein bottles are born, etc.

Theorem (Afrajmovich and Shil'nikov (1974)). *In a generic one-parameter family, vector fields having the following properties may occur*:

1. *The vector field has a nonhyperbolic cycle L with multiplier* 1.
2. *The union of the cycle and its homoclinic trajectories are noncritical and compact.*
3. *The cycle L is of stable nodal type in its hyperbolic variables.*

Suppose that such a field corresponds to the zero value of the parameter ε *of the family. Then*:

a. *All fields of the family corresponding to values of* ε *on one side of and sufficiently close to* 0 *have smooth two-dimensional attractors* $\mathbf{M}_\varepsilon^2$, *diffeomorphic to a torus or to a Klein bottle. As* $\varepsilon \to 0$, *the attractor* $\mathbf{M}_\varepsilon^2$ *converges to the union* $S_L^u \cup L$, *to which it is homeomorphic.*

b. *All fields of the family corresponding to values of* ε *on the other side of* 0 *have two structurally stable limit cycles, and no other nonwandering trajectories, in some neighborhood of the union of L and all of its homoclinic trajectories.* ▲

The case of an unstable node in its hyperbolic variables leads to the previous case by reversing time: a smooth *repeller*[20] is born, diffeomorphic to a torus or a Klein bottle.

---

[20] A repeller is an invariant set of a dynamical system that turns into an attractor upon reversal of time.

4.4. Creation of Complex Invariant Sets (The Noncritical Case).

Theorem. *In a generic one-parameter family there may be a vector field having properties* 1 *and* 2 *of the theorem in Sect.* 4.3, *and also the property*:

3′. *The cycle L is of saddle type in its hyperbolic variables, and the union of its homoclinic trajectories is connected.*

Suppose that such a vector field corresponds to the parameter value $\varepsilon = 0$ *of the family. Then, for such families, the conclusions a and b of the theorem in Sect.* 4.3 *hold if one changes "attractor* $\mathbf{M}_\varepsilon^2$*" in conclusion a to "invariant submanifold* $\mathbf{M}_\varepsilon^2$*"; it is neither an attractor nor a repeller.* ▲

Upon bifurcation of a cycle, such that the union of its homoclinic trajectories is noncritical and consists of p tori and Klein bottles ($p > 1$), an invariant set is born that contains a countable number of two-dimensional invariant manifolds.

Theorem. *In a generic one-parameter family there may be a vector field having properties* 1 *and* 2 *of the theorem in Sect.* 4.3, *and also the property*:

3″. *The cycle L is of saddle type in its hyperbolic variables, and the union of its homoclinic trajectories consists of p connected components.*

Suppose that such a field corresponds to the parameter value $\varepsilon = 0$ *of the family. Then*:

a. *All fields of the family corresponding to values of* ε *on one side of, and sufficiently close to,* 0 *have invariant sets* Ω_ε.

b. *All of the path-connected components of the space* Ω_ε *are two-dimensional. There exists a one-to-one mapping of the set of these components onto the set of trajectories of a topological Bernoulli shift on p symbols. The path-connected components are compact if and only if the corresponding trajectories are periodic.*

c. *The conclusion b of the theorem in Sect.* 4.3 *holds for such a family.* ▲

The results of this subsection were announced by Afrajmovich and Shil'nikov (1982) for $n = 4$.

4.5. The Critical Case. In cases where the the union of homoclinic trajectories of a cycle with multiplier 1 is compact and critical, strange attractors may be born upon bifurcations in the corresponding fields.

"Theorem" (Afrajmovich and Shil'nikov (1974), Newhouse, Palis and Takens (1983)). *In generic one-parameter families there may be a vector field (say,* v_0*) having properties* 1 *and* 3 *of the theorem in Sect.* 4.3, *and also having the property*:

2′. *The union of the cycle L and its homoclinic trajectories is compact and critical; the set* S_L^u *is tangent to some leaves of the strongly stable foliation* $\mathscr{F}_L^{ss}$.

Suppose such a field v_0 *corresponds to the parameter value* $\varepsilon = 0$ *of the family. Then*:

a. *On one side of* $\varepsilon = 0$ *there is an open set with limit point* 0, *consisting of a countable union of intervals, such that to each* ε *in this set there corresponds a vector field* v_ε *in the family that has a strange attractor* $\mathbf{M}_\varepsilon$. *This attractor contains a countable set of periodic trajectories; it converges to* $S_L^u \cup L$ *as* $\varepsilon \to 0$.

b. *Conclusion b of the theorem in Sect.* 4.3 *holds.* ▲

This theorem, in somewhat different terms, was formulated by Newhouse, Palis and Takens (1983), where an outline of its proof was given.[21] A complete proof of the theorem was obtained by Afrajmovich and Shil'nikov (1974), with an additional hypothesis on the field (which does not raise the codimension of the degeneracy, but which shrinks the domain of degenerate fields under consideration in function space). We formulate this hypothesis, and at the same time we clarify the mechanism by which the strange attractor arises. For simplicity, assume that the phase space is three-dimensional.

We assume for simplicity that the monodromy transformation of the cycle L (as a function of the initial conditions and the parameter) may be extended into a neighborhood of the intersection of the plane, transversal to the vector field and the union of homoclinic trajectories of the cycle. On this plane a fixed point Q of the diffeomorphism f_0, corresponding to the vector field v_0, corresponds to the cycle. One multiplier of this fixed point is equal to 1, and the rest are less than 1 in modulus. The union of homoclinic trajectories traces out a curve S_Q^u on the transversal plane, which is a closed curve if one adds the point Q to it (see Fig. 42). The strongly stable foliation corresponding to the field v_0 gives rise to a strongly stable foliation $\mathscr{F}_Q^{ss}$ of the diffeomorphism f_0 on the transversal plane. The curve S_Q^u is tangent to some leaves of this foliation.

Before formulating the additional condition on the vector field v_0, we give a rough argument supporting the existence of an attractor. Since the diffeomorphism f_0 is contracting in its hyperbolic variables, in some neighborhood of the point Q there exists some neighborhood $\mathscr{U}$ of the "homoclinic curve" $S_Q^u \cup Q$ whose closure $\bar{\mathscr{U}}$ is compact and which is mapped into $\mathscr{U}$ under the action of f_0. Thus, for all sufficiently small ε, $f_\varepsilon \bar{\mathscr{U}} \subset \mathscr{U}$. The intersection

$$A_\varepsilon = \bigcap_{k=1}^{\infty} f_\varepsilon^k \mathscr{U}$$

will be the maximal attractor of the diffeomorphism f_ε. In the remainder of this subsection we omit the adjective "maximal".

Assume that for small $\varepsilon > 0$ the point Q disappears, and for $\varepsilon < 0$, Q splits into two nondegenerate points. Suppose w is a neighborhood of Q in which the projection $\pi: w \to W_Q^c$ along the leaves of the strongly stable foliation $\mathscr{F}_Q^{ss}$ of the diffeomorphism f_0 onto its center manifold is defined. The neighborhood w is divided by the manifold W_Q^s into two parts w^+ and w^-, defined by the conditions $\pi f_0 w^- \subset w$, $\pi f_0^{-1} w^+ \subset w$. Since all points on S_Q^u are homoclinic, for any arc $\Gamma \subset w^+$, there exists a k such that $f_0^k \Gamma \subset w^-$.

The additional condition on f_0 is the following. There exists an arc $\Gamma \subset S_Q^u \cap w^-$ having the following properties:

---

[21] The analog of this theorem for the case of a saddle in its hyperbolic variables (in which instead of a strange attractor a complicated invariant set is born) is announced in Afrajmovich and Shil'nikov (1982) We note that a complete proof of this theorem has not been published up to this time, and, probably one has not been obtained. Some progress has been made by F. Przytycki, "Chaos after bifurcation of a Morse-Smale diffeomorphism through a one-cycle saddle-node and iterations of an interval and a cycle," Preprint 347, *Inst. of Math Polish Acad. Sci.*, 1985, 62 pp.

1. The first point of Γ is mapped into its last point by f_0.
2. There exists an integer k and a leaf $F \subset w^-$ of the foliation $\mathscr{F}^{ss}$ such that the domain included in the neighborhood w between the leaves F and f_0F cuts two arcs transversal to the foliation $\mathscr{F}^{ss}$ out of the curve $\Gamma' = f_0^k\Gamma$ (see Fig. 43).

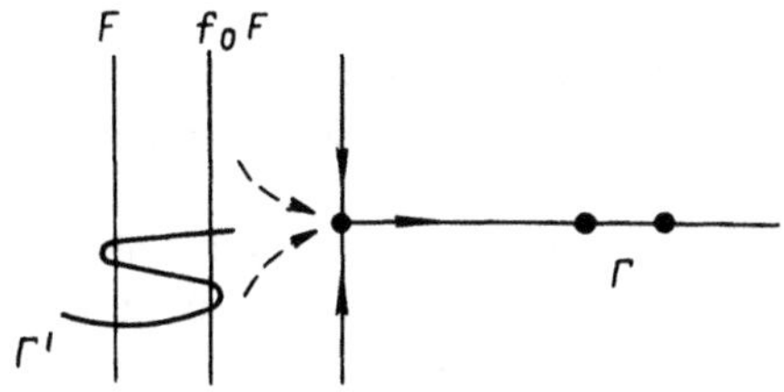

Fig. 43. Intersection of the unstable set of an s-critical cycle with the connected component of a fundamental domain of the monodromy transformation

We now explain why the attractor A_ε is strange for sufficiently small ε. Consider a neighborhood V of the arc Γ, the image of which $V' = f_0^k V$ is a neighborhood of the arc Γ' and belongs entirely to w^-. For each $\varepsilon_0 > 0$ there exists a positive $\varepsilon < \varepsilon_0$ and a natural number $N(\varepsilon)$ such that:

1. The image $V'' = f_\varepsilon^{N(\varepsilon)} V$ is a horseshoe, strongly contracted in the hyperbolic variables (as $\varepsilon \to 0$ the exponent $N(\varepsilon) \to \infty$), and not strongly distorted in directions parallel to the tangent to W_Q^c at the origin (the last distortion may be estimated uniformly in ε).
2. There exists a sequence of intervals in the interval $(0, \varepsilon_0)$, converging to zero such that for values of ε from any of these intervals the horseshoe V'' intersects the domain V in two connected components, each of whose images under the projection π contains Γ. Although the mapping $\varphi_\varepsilon = f_\varepsilon^{k+N(\varepsilon)}$: $V \to V''$ is not a real Smale horseshoe[22] (there is contraction in one direction but not expansion in the other), the existence of a countable number of cycles of the diffeomorphism φ_ε can be proved, and hence the same holds for f_ε. Thus, the attractor A_ε is not a manifold of dimension 1.

On the other hand, for sufficiently small ε, some power of the diffeomorphism f_ε decreases two-dimensional volume. Consequently, the attractor A_ε is not a manifold of dimension higher than 1, and therefore A_ε is strange.

4.6. A Two-Step Transition from Stability to Turbulence. It is possible to imagine a one-parameter family of vector fields in which, to values of the parameter less than some first critical value, there correspond fields with a globally stable critical point. As the parameter passes through the first critical

[22] See EMS, Dynamical Systems 2, pp. 115–118 for a description of a Smale horseshoe, or Guckenheimer and Holmes (1983, Sect. 51).

value, a stable limit cycle is born; as the parameter passes through a second critical value, this cycle disappears, as was described in Sect. 4.5. Moreover, a strange attractor is born and chaos sets in.

Here only bifurcations are considered that are noticeable by "physical observation," which "sees" only the perestroikas of stable (steady-state) regimes (and those approaching them).

4.7. A Noncompact Set of Homoclinic Trajectories. Everywhere in this subsection the cycle L is a node in its hyperbolic variables, and, for definiteness, stable.

Let us assume that a vector field, having a cycle with multiplier 1, and with a noncompact set of homoclinic trajectories including L, satisfies the following genericity conditions: its nonwandering set consists of a finite number of hyperbolic equilibria and hyperbolic cycles, besides L, whose stable and unstable manifolds intersect transversally with each other and with S_L^u, S_L^u, W_L^s, and W_L^u. The last four manifolds (two of them have boundaries) intersect transversally at each point of intersection not belonging to L.

The following lemma is proved analogously (see Smale, (1967)).

Lemma. *Under the assumptions formulated above, a vector field v_0 has a contour $Q_0, Q_1, \ldots, Q_k$, containing $L = Q_j$, and such that the stable and unstable sets of elements of the contour intersect transversally* (*Case* 2, Sect. 1.5).

We rename the elements of the contour so that $L = Q_0 (= Q_k)$. It is simple to derive the following corollary from the transversality of the manifolds and the λ-lemma.

Corollary. For any family of vector fields $\{v_\varepsilon\}$, intersecting the bifurcation set at the point v_0, and not having limit cycles in a neighborhood of L for $\varepsilon > 0$, there exist $(k-1)$ sequences

$$\{\varepsilon_i^s\} \quad (i \in \mathbb{N}, s \in \{1, \ldots, k-1\}, \varepsilon_i^s \to 0 \text{ as } i \to \infty)$$

such that for $\varepsilon = \varepsilon_i^s$ the vector field v_ε has a homoclinic trajectory of an equilibrium or a cycle Q_s.

We shall say that Case B holds if $k = 2$ and Q_1 is an equilibrium of saddle type, either with a leading stable direction corresponding to a real eigenvalue, or in the opposite case, with a negative saddle number (see Sect. 5.1 below for the definitions of saddle number and leading direction). In all other cases we shall say that Case A holds.

From the previous Corollary we have:

Theorem (Afrajmovich (1974)). *If Case A holds, then in an interval* $(0, \varepsilon_0)$, ε_0 *sufficiently small, there exist $(k-1)$ sequences of intervals* $\{\delta_i^s\}$ $(i \in \mathbb{N}, s \in \{1, \ldots, k-1\})$, *contracting to zero as $i \to \infty$, such that for $\varepsilon \in \delta_i^s$ each vector field v_ε has a countable set of limit cycles of saddle type.*

Corollary. Let us assume that in addition to the conditions of the theorem the following condition is satisfied: for any equilibrium or cycle Q such that

$S^u_L \cap W^s_Q \neq \varnothing$, the following inclusion occurs: $W^u_Q \backslash Q \subset S^s_L$. Then for $\varepsilon \in \delta^s_i$, each vector field v_ε has a strange attractor in a neighborhood of the closure of S^u_L and converging to it as $\varepsilon \to 0$.

Case B, as far as we know, has not been investigated.

4.8. Intermittency. Let us assume that either the conditions of the previous corollary are satisfied, or the conditions of the theorem in 4.5 are satisfied, that is, the vector field v_ε has a strange attractor for $\varepsilon > 0$. Consider an arbitrary continuous function $\psi(x)$, mapping phase space into $\mathbb{R}^1$. Suppose $x = x(t)$ is a trajectory belonging to the strange attractor. Then the graph of the function $\psi(x(t))$ has the following form in general: a long sequence of nearly periodic oscillations (on this interval of time $x(t)$ lies in a small neighborhood of the disappearing cycle), then a burst of "turbulence", then an interval of periodicity, etc. Such a regime has been called *intermittent* by Manneville and Pomeau (1980a). Intermittency accompanies a bifurcation in which a strange attractor arises upon the disappearance of a semi-stable cycle and is often found in models of real processes (see, for example, Gapanov-Grekhov and Rabinovich (1984), Manneville and Pomeau (1980b)).

Intermittency can, in addition to the cases listed above, accompany the disappearance of a cycle with multiplier 1, which is of nodal type in its hyperbolic variables and which has a homoclinic trajectory belonging to W^s (in this case the vector field actually does not lie on the boundary of the set of Morse-Smale systems (see Luk'yanov Shil'nikov (1978))).

4.9. Accessibility and Nonaccessibility. Let v_0 be a generic vector field (that is, satisfying conditions analogous to those formulated at the beginning of Sect. 4.7 above) that lies on the boundary of the set of Morse-Smale systems $\mathscr{B}_1$, and has a nonhyperbolic cycle L. Let us assume that one of the following possibilities holds: (1) L is a cycle with multiplier -1; (2) L is a cycle with a pair of nonreal multipliers (we recall, see Sect. 1.4 and 1. 6, that in Cases 1 and 2, L is not a part of a contour, and there is no trajectory doubly asymptotic to L other than L itself); (3) L is a cycle with multiplier $+1$ and either: (3a) $S^u_L \cap S^s_L = L$ (there are no homoclinic trajectories of the cycle L), (3b) $S^u_L \cap S^s_L$ is a Klein bottle, smoothly embedded in phase space, or (3c) $S^u_L \cap S^s_L$ is a smooth torus.

Lemma. *If the conditions formulated above hold in a neighborhood of* v_0 *in* $\chi^r(\mathbf{M})$, *then Morse-Smale systems are everywhere dense in a neighborhood of* v_0 *in* $\chi^r(\mathbf{M})$.

This lemma follows from the Kupka-Smale Theorem (de Melo and Palis (1982)) and the fact that Morse-Smale systems on a torus or a Klein bottle are everywhere dense. The question of accessibility or inaccessibility of the bifurcation surface and, in the case of inaccessibility, to identify the bifurcations that accompany it remains to be answered.

Proposition. *In Case* 1, *the intersection* $\mathscr{B}_1 \cap \mathbb{A}$ *is connected, where* $\mathbb{A} \subset \chi^r(\mathbf{M})$ *is a ball of sufficiently small diameter and with center* v_0, *and all vector fields in* $\mathbb{A} \setminus \mathscr{B}_1$ *are Morse-Smale vector fields.*

As a consequence we obtain the accessibility from both sides of $\mathscr{B}_1$ at v_0. The proposition follows easily from a variant of the theorem on continuous dependence of invariant manifolds on parameters in, for example, Hirsch, Pugh and Shub (1977).

In Case 2, after birth of a torus for "almost" any one-parameter family of vector fields the rotation number changes as the parameter varies; consequently, an infinite number of bifurcations takes place. However, there are families for which the rotation number on the torus does not change with the parameter so that the bifurcation surface may be accessible.

In Case 3, information on accessibility is collected in Table 2 below, which presents details of part of Table 1 in Sect. 1.8 of this chapter.

Table 2

	Subclass	Accessibility
(3a) $S_L^s \cap S_L^u = L$	s-critical cycle and $S_L^u \cap W_1^s \neq \varnothing$, dim $W_1^s < n$ OR u-critical cycle and $S_L^s \cap W_2^u \neq \varnothing$, dim $W_2^u < n$	+ −
	The remaining cases	+ −
(3b)	Noncritical	+ +
$S_L^s \cap S_L^u = \mathbb{K}^2$	Critical	+ −
(3c) $S_L^s \cap S_L^u = \mathbb{T}^2$		+ −

Here W_1^s and W_2^u denote the stable and unstable manifolds of hyperbolic equilibria or cycles. We explain why inaccessibility may arise in Case 3a in Fig. 44, where a diffeomorphism of a two-dimensional disc is pictured, having a fixed point Q with multiplier 1 and two saddles Q_1, Q_2 at $\varepsilon = 0$; moreover, S_Q^u intersects $W_{Q_2}^s$ transversally, and $W_{Q_1}^u$ contains a point P of simple tangency with a leaf of $\mathscr{F}_Q^{ss}$. For $\varepsilon > 0$ a neighborhood of P maps diffeomorphically into a neighborhood of a point on $W_{Q_2}^s$, and, for a suitable choice of ε, $W_{Q_2}^u$ and $W_{Q_1}^s$ have a point of simple tangency.

In Case 3c, inaccessibility is connected with a change in the rotation number on the torus that arises, and in Case 3b, with the birth of points of simple tangency of the stable and unstable manifolds of hyperbolic cycles on the Klein bottle and the "distantly located" equilibria or cycles (critical case), and with the "blue sky catastrophe" (noncritical case, Li Weigu and Zhan Zhifen).

4.10. Stability of Families of Diffeomorphisms. In the papers of Newhouse and Palis (1976) and Newhouse, Palis and Takens (1976, 1983), general properties of one-parameter families of diffeomorphisms are studied. Various definitions of stability were formulated, and necessary and (or) sufficient conditions for various

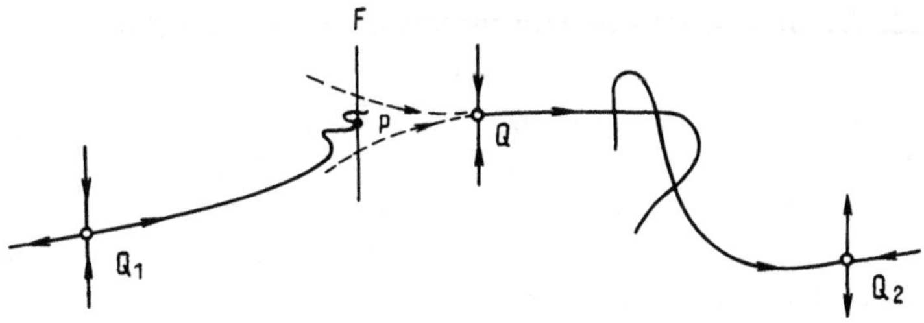

Fig. 44. Fixed points and invariant curves of a diffeomorphism of a disk, belonging to the inaccessible part of a bifurcation surface

types of stability were set up, some of which were proved. The account given here follows that of Newhouse, Palis and Takens (1983).

Suppose: **M** is a compact, C^∞-smooth manifold without boundary, Diff(**M**) is the set of C^∞-diffeomorphisms of **M**, **MS** is the set of Morse-Smale systems on **M**, and $\mathscr{P}(\mathbf{M})$ is the set of C^∞-arcs of diffeomorphisms of **M**. That is, if **I** is the unit interval, then $\mathscr{P}(\mathbf{M})$ consists of the C^∞-mappings Φ: $\mathbf{M} \times \mathbf{I} \to \mathbf{M} \times \mathbf{I}$ such that $\varphi(m, \varepsilon) = (\varphi_\varepsilon(m), \varepsilon)$, where $m \to \varphi_\varepsilon(m)$ is a C^∞-diffeomorphism for each $\varepsilon \in \mathbf{I}$. The elements of $\mathscr{P}(\mathbf{M})$ will be called *one-parameter families of diffeomorphisms* or *arcs of diffeomorphisms.*

For each arc $\{\varphi_\varepsilon\} \subset \mathscr{P}(\mathbf{M})$ with $\varphi_0 \in \mathbf{MS}$, let

$$b(\varphi) = \inf\{\varepsilon \in \mathbf{I},\ \varphi_\varepsilon \notin \mathbf{MS}\}.$$

We assume that $b(\varphi) < 1$. Consider the arcs $\{\varphi_\varepsilon\}$, $\{\varphi'_\varepsilon\} \subset \mathscr{P}(\mathbf{M})$, then we say that $(h, \{H_\varepsilon\})$ is a conjugacy between them if h: $[0, 1] \to [0, 1]$ is a homeomorphism such that $h(b(\varphi)) = b(\varphi')$, H_ε: $\mathbf{M} \to \mathbf{M}$ is a homeomorphism, conjugating φ_ε and $\varphi'_{h(\varepsilon)}$ for all ε in some neighborhood of $[0, b(\varphi)]$, and H_ε is continuous in ε. If the homeomorphism H_ε conjugates φ_ε and φ'_ε only for $\varepsilon \leqslant b(\varphi)$, but is not necessarily continuous in ε, then we say that $(h, \{H_\varepsilon\})$ is a left-conjugacy for $\{\varphi_\varepsilon\}$, $\{\varphi'_\varepsilon\}$. Conjugacy and left-conjugacy each define an equivalence relation on the set of all arcs in $\mathscr{P}(\mathbf{M})$ originating with Morse-Smale diffeomorphisms. An arc is called *stable* or *left-stable* if it is an interior point of the corresponding equivalence class.

We denote by $v(\varphi)$ a vector field generating a flow that is a suspension over the diffeomorphism φ. We denote by **R** the set of arcs $\{\varphi_\varepsilon\}$ in the space of diffeomorphisms such that, $v(\varphi_b) \in \mathscr{B}_1$, where $v(\varphi_\varepsilon)$ transversely intersects $\mathscr{B}_1$ at the point $v(\varphi_b)$; $v(\varphi_b)$ satisfies conditions of genericity, the principal one of which consists of the following. The nonwandering set of $v(\varphi_b)$ consists of a finite set of cycles; moreover, if one of them is not hyperbolic, then its stable and unstable sets and manifolds transversally intersect each other and the manifolds of the other cycles. Moreover, if all the cycles are hyperbolic, then their invariant manifolds transversally intersect each other along all trajectories, except for one.

Newhouse, Palis, and Takens (1983) imposed some additional conditions on the local behavior of trajectories in the neighborhood of hyperbolic points, conditions which do not destroy genericity, but which do reduce the class of arcs considered. We do not reformulate these conditions here, but we do consider them to be satisfied.

Theorem 1. 1) *The arc* $\{\varphi_\varepsilon\} \in \mathbf{R}$, *is left-stable if and only if* $v(\varphi_b)$ *has a nonhyperbolic limit cycle.* 2) *The arc* $\{\varphi_\varepsilon\}$ *is stable if and only if*: a) $\{\varphi_\varepsilon\}$ *is left-stable,* b) $v(\varphi_b)$ *does not have a cycle with a pair of nonreal multipliers,* c) *if* $v(\varphi_b)$ *has a cycle with multiplier* 1, *then that cycle is noncritical, is not a part of a contour, and does not have homoclinic trajectories.*

Theorem 2. *Suppose* $\{\varphi_\varepsilon\}$, $\varepsilon \in [0, 1]$, *is an arc of diffeomorphisms such that the limit set of each diffeomorphism* φ_ε *consists of only a finite set of trajectories. Then* $\{\varphi_\varepsilon\}$ *is stable if and only if there exists only a finite set of bifurcation values on* $[0, 1]$, *say* $b_1, \ldots, b_k$, *and for each* $i \in \{1, \ldots, k\}$ *the following assertions hold*:

a) $v(\varphi_{b_i}) \in \mathscr{B}_1$, *and does not have a cycle with a pair of nonreal multipliers;*

b) $v(\varphi_\varepsilon)$ *transversally intersects* $\mathscr{B}_1$ *at the point* $v(\varphi_{b_i})$*;*

c) *If* $v(\varphi_{b_i})$ *has a cycle with multiplier* 1, *then that cycle is noncritical, is not a part of a contour, and does not have homoclinic trajectories.*

These restrictive conditions on stability are connected with the existence of numerical invariants of topological equivalence, namely, moduli, that arise upon nontransversal intersections of stable and unstable manifolds (see Sect. 6 below).

4.11. Some Open Questions. We list some problems on codimension 1 bifurcations of Morse-Smale vector fields, connected with the violation of hyperbolicity of cycles.

1. Investigate the bifurcations of vector fields, having a contour, which contain only cycles with multiplier 1, and an equilibrium of saddle type, either with a real stable leading direction, or with a complex one but with a negative saddle number (Case B of Sect. 4.7 above).

2. Give, if possible, a complete description of bifurcations of vector fields having a critical cycle of nodal type in its hyperbolic variables, with multiplier 1 and with a compact set of homoclinic trajectories. For the one-dimensional analog of this problem some results were found by Newhouse, Palis, and Takens (1983), where the language of kneading sequences and rotation sets is used.

3. Investigate the bifurcations of vector fields of saddle type in their hyperbolic variables, having critical cycles with multiplier 1, at least in the case of a compact set of homoclinic trajectories.

Remark. All the bifurcations in Sect. 4 are global; *a priori* we do not know a finite set of trajectories, in the neighborhood of which bifurcation phenomena take place.

§ 5. Hyperbolic Singular Points with Homoclinic Trajectories

In this section we describe bifurcations that take place as one passes through a hypersurface in a function space consisting of vector fields with a homoclinic trajectory of a hyperbolic singular point. We investigate the neighborhood of generic points on this hypersurface, whether or not they belong to the boundary of the set of Morse-Smale systems.

5.1. Preliminary Notions: Leading Directions and Saddle Numbers. Consider a germ $v(x) = Ax + \cdots$ of a smooth vector field at a hyperbolic singular point O of saddle type, $\dim W_O^s = s > 0$, $\dim W_O^u = u > 0$. We order the eigenvalues $\{\lambda_j, \mu_k\}$ of the operator A so that

$$\operatorname{Re} \lambda_s \leqslant \cdots \leqslant \operatorname{Re} \lambda_1 < 0 < \operatorname{Re} \mu_1 \leqslant \cdots \leqslant \operatorname{Re} \mu_u.$$

The sum $\sigma = \operatorname{Re} \lambda_1 + \operatorname{Re} \mu_1$ is called the *saddle number* of the germ (and the corresponding singular point O).

If $\operatorname{Re} \lambda_1 = \cdots = \operatorname{Re} \lambda_k > \operatorname{Re} \lambda_{k+1}$, then the invariant subspace of the operator A corresponding to the eigenvalues $\lambda_1, \ldots, \lambda_k$ is called the *leading stable direction* of the germ at the singular point; the *leading unstable direction* is analogously defined. This notation is explained by the fact that almost all phase curves of the equation $\dot{x} = v(x)$ beginning on the stable manifold of the singular point O arrive at this point tangentially to the leading stable direction; the exceptions are curves, filling a submanifold of dimension less than $\dim W_O^s$. For a linear equation this is clear; for nonlinear equations this was proved by Petrowskii (1934).

5.2. Bifurcations of Homoclinic Trajectories of a Saddle that Take Place on the Boundary of the Set of Morse-Smale Systems. In generic one-parameter families there are vector fields with a homoclinic trajectory of a hyperbolic saddle that is not removed by small perturbations of the corresponding family. We shall assume that in one-parameter families such fields correspond to the zero value of the parameter, which is also called the critical value.

Theorem (Shil'nikov (1963, 1968)). *Suppose that in a generic one-parameter family, there corresponds to the zero value of the parameter a differential equation with a homoclinic trajectory of a hyperbolic saddle, satisfying one of the following conditions:*

1. *The saddle number is negative, and the leading unstable direction is one-dimensional.*

2. *The saddle number is positive, and the leading stable direction is one-dimensional.*

Then all noncritical vector fields of the family sufficiently close to the critical field correspond to Morse-Smale systems in some neighborhood of the homoclinic trajectory with no more than two nonwandering trajectories, one of which is the singular point of the vector field. The vector fields of the family corresponding to parameter values on one side of zero have no other nonwandering trajectories; the vector fields of the family corresponding to parameter values on the other side of zero have a limit cycle. The dimension of the stable manifold of the cycle is one more than the dimension of the stable manifold of the saddle, or coincides with it, depending upon whether the saddle number σ is negative or positive. ▲

Remarks. 1. The assertion of the theorem for $\sigma > 0$ is obtained from the assertion for $\sigma < 0$ simply by reversing time.

2. In the generic case, the leading direction is either one-dimensional (one real eigenvalue), or two-dimensional (there is a pair of complex conjugate eigenvalues in this case). We shall say that in the first case the leading direction is *real*, and in the second case that it is *complex*.

From the theorem one may conclude that, if its hypotheses are satisfied, the bifurcation surface $\mathscr{B}_1$ is accessible from both sides at a point in general position.

5.3. Requirements for Genericity. In order for the conclusions of the previous theorem to be satisfied by a one-parameter family of vector fields, this family must satisfy the following requirements of genericity. The first three are conditions upon the vector field at the critical value of the parameter.

1. The leading direction, stable in the first case ($\sigma < 0$) and unstable in the second case ($\sigma > 0$), is either real and one-dimensional, or complex and two-dimensional.

2. A homoclinic trajectory reaches the singular point at $t \to +\infty$ and $t \to -\infty$, tending to the leading direction.

In order to formulate the third condition, some information on the variational equations along a homoclinic trajectory of a saddle is needed. Suppose the equation

$$\dot{x} = v(x), \qquad v(0) = 0, \qquad v_*(0) = A,$$

corresponds to the critical value of the parameter, $\varphi(t)$ is a homoclinic trajectory of the saddle O, $\varphi(0) = x$, and suppose X is an operator-valued solution of the initial-value problem for the variational equations:

$$\dot{X} = (v_* \circ \varphi(t))X(t), \qquad X(0) = E.$$

Proposition. *For each nonzero vector $\xi \in T_x\mathbb{R}^n$ the limits*

$$\lambda_\pm(\xi) = \lim_{t \to \pm\infty} \frac{\ln|X(t)\xi|}{t}$$

exist. Each of these limits is equal to the real part of one of the eigenvalues of the operator A (and they are called the Lyapunov exponents of the variational equation). The set of vectors ξ given by the inequalities $\lambda_+(\xi) \leqslant \lambda$ or $\lambda_-(\xi) \leqslant \lambda$ is a plane without the origin. The dimension of this plane is what it would be if $\exp(At)$ *were used instead of $X(t)$ in the definition of the Lyapunov exponents.*

Remarks. 1. This proposition is obvious if the germ of the vector field at O is smoothly equivalent to its linear part, and it has been proved for an arbitrary germ; see Bylov et al. (1966).

2. Suppose the unstable leading direction of v at O is real and one-dimensional and $\sigma < 0$. Then the planes

$$L_-(x) = \{\xi \in T_x\mathbb{R}^n | \lambda_+(\xi) \leqslant \lambda_1\} \quad \text{and} \quad L_+(x) = \{\xi \in T_x\mathbb{R}^n | \lambda_-(\xi) \leqslant -\lambda_1\}$$

have dimensions s and $u + 1$, respectively. We observe that

$$L_-(x) = T_x W_O^s \quad \text{and} \quad L_+(x) \supset T_x W_O^u.$$

The third condition of genericity on the vector field v for $\sigma < 0$ is:

3. Suppose x is a point on the homoclinic trajectory. It is required that the planes $L_{\pm}(x)$ intersect transversally (that is, along the straight line generated by the vector $v(x)$).

For $\sigma > 0$ the third condition on the vector field is obtained from the condition above by reversing time.

A fourth condition is placed upon the family of vector fields $\{v_\varepsilon\}$; $v_0 = v$ below.

4. Consider a point x on the homoclinic trajectory, and a germ of the $(n-1)$-dimensional plane Π at this point, transversal to the field v_ε for small ε. The stable manifold W_ε^s and the unstable manifold W_ε^u of the field v_ε at the singular point O intersect Π in two submanifolds of total dimension $n-2$. For $\varepsilon = 0$ these submanifolds intersect at the point x. The fourth condition of genericity is: for $\varepsilon \neq 0$ the distance between these manifolds is of order ε.

Remark. Condition 4 may be weakened and the theorem in Sect. 5.2 still holds. This follows from the theorem in Sect. 5.5 below.

5.4. Principal Families in $\mathbb{R}^3$ and their Properties. In this subsection we construct "topological normal forms of families in the neighborhood of a trajectory homoclinic to a saddle in $\mathbb{R}^3$". The corresponding versality theorem is formulated in Sect. 5.5. The principal families are constructed with the help of a collage created by gluing linear and standard vector fields together as we describe below. We shall suppose that the stable manifold W^s of the linear field is two-dimensional; the case $\dim W^s = 1$ is reduced to the two-dimensional case by reversing time. There are four principal families: they are distinguished from each other by the signs of their saddle numbers, and the topology of the invariant manifolds obtained by extending the manifolds W^s.

We denote two copies of the cube $|x| \leqslant 1$, $|y| \leqslant 1$, $|z| \leqslant 1$ by K_1 and K_2. In the cube K_1 we consider the vector fields v^- and v^+:

$$v^- = -4y\partial/\partial y - 3x\partial/\partial x + 2z\partial/\partial z, \qquad \sigma = -1,$$
$$v^+ = -4y\partial/\partial y - x\partial/\partial x + 2z\partial/\partial z, \qquad \sigma = 1.$$

In the cube K_2 we consider the vector fields $v_\varepsilon = -\partial/\partial x + \frac{1}{2}\varepsilon\partial/\partial z$.

We consider the following gluing mappings:

$$f\colon (-1, y, z) \mapsto (1, y, z), \qquad f^{\pm}\colon (x, y, 1) \mapsto (1, \pm y, \pm x);$$

the domains where f, f^+, f^- are defined are called $\mathrm{dom}(f)$, $\mathrm{dom}(f^+)$, $\mathrm{dom}(f^-)$, respectively. We glue together pairs of points $P \in K_2 \cap \mathrm{dom}(f)$ and $f(P) \in K_1$, and also pairs $Q \in K_1 \cap \mathrm{dom}(f^{\pm})$ and $f^{\pm}(Q) \in K_2$ (see Fig. 45). On the sets of points inside each of these spaces, one can provide the structure of a smooth manifold so that the resulting vector fields are smooth. We denote these manifolds by M^+ and M^- ($M^{\pm}$ is obtained with the help of $f^{\pm}$).

We define V^{++} to be the family of vector fields v_ε^{++} on M^+ corresponding to (v^+, v_ε), V^{-+} the family of vector fields v_ε^{-+} on M^+ corresponding to (v^-, v_ε), V^{+-} the family of vector fields v_ε^{+-} on M^- corresponding to (v^+, v_ε), and V^{--} the family of vector fields v_ε^{--} on M^- corresponding to (v^-, v_ε). The manifolds M^+

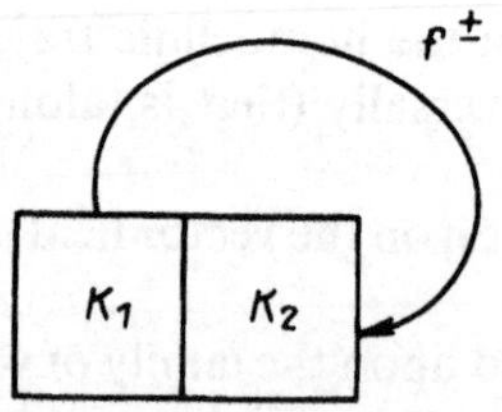

Fig. 45. Construction of the manifolds $M^{\pm}$

and M^-, together with the vector fields that have been just defined, can be smoothly embedded in $\mathbb{R}^3$. The four families $V^{++}, \ldots, V^{--}$ are called principal families. Fields in the principal families corresponding to $\varepsilon = 0$ have a homoclinic trajectory formed from pieces of the coordinate axes $0x$ and $0z$. For sufficiently small ε and h, first-return (Poincaré) maps of each of the principal families are defined on the two-dimensional transversal

$$\mathscr{D}_h = \{(x, y, z) | x = 1, |y| \leqslant 1, 0 < z < h\} \subset K_1.$$

A point $P \in \mathscr{D}_h$ maps into the point of first return to the boundary $x = 1$ of the cube K_1 of the positive semi-trajectory with initial point P, of the vector field of the principal family corresponding to ε. We denote the corresponding monodromy transformations by $\Delta_\varepsilon^{++}, \ldots, \Delta_\varepsilon^{--}$. These transformations are now computed.

We denote by

$$\Delta^+ : \mathscr{D}_h \to \{z = 1\}, \qquad \Delta^- : \mathscr{D}_h \to \{z = 1\}$$

the maps that take a point $P \in \mathscr{D}_h$ into the point on the boundary $\{z = 1\}$ through which the positive semi-trajectories of v^+ and v^-, respectively, with initial points P, leave the cube K_1. Let Δ_ε be the map from the boundary $x = 1$ of the cube K_2 into the plane $x = -1$ along a trajectory of the vector field v_ε: $\Delta_\varepsilon(1, y, z) = (-1, y, z + \varepsilon)$. Thus, the map Δ_ε^{++} has the form (see Figs. 45 and 46)

$$\Delta_\varepsilon^{++} = f \circ \Delta_\varepsilon \circ f^+ \circ \Delta^+.$$

We have

$$\Delta^{\pm}(1, y, z) = (z^{\nu^{\pm}}, yz^2, 1), \qquad \nu^- = 3/2, \qquad \nu^+ = \tfrac{1}{2},$$

$$\Delta_\varepsilon^{++}(1, y, z) = (1, yz^2, z^{1/2} + \varepsilon).$$

Analogously, we have

$$\Delta_\varepsilon^{+-}(1, y, z) = (1, -yz^2, -z^{1/2} + \varepsilon), \qquad \Delta_\varepsilon^{-+}(1, y, z) = (1, yz^2, z^{3/2} + \varepsilon),$$

and

$$\Delta_\varepsilon^{--}(1, y, z) = (1, -yz^2, -z^{3/2} + \varepsilon).$$

For sufficiently small h the maps Δ_ε^{-+} and Δ_ε^{--} are contracting. The maps Δ_ε^{++} and Δ_ε^{+-} are hyperbolic; they are expanding in the z-direction and contracting

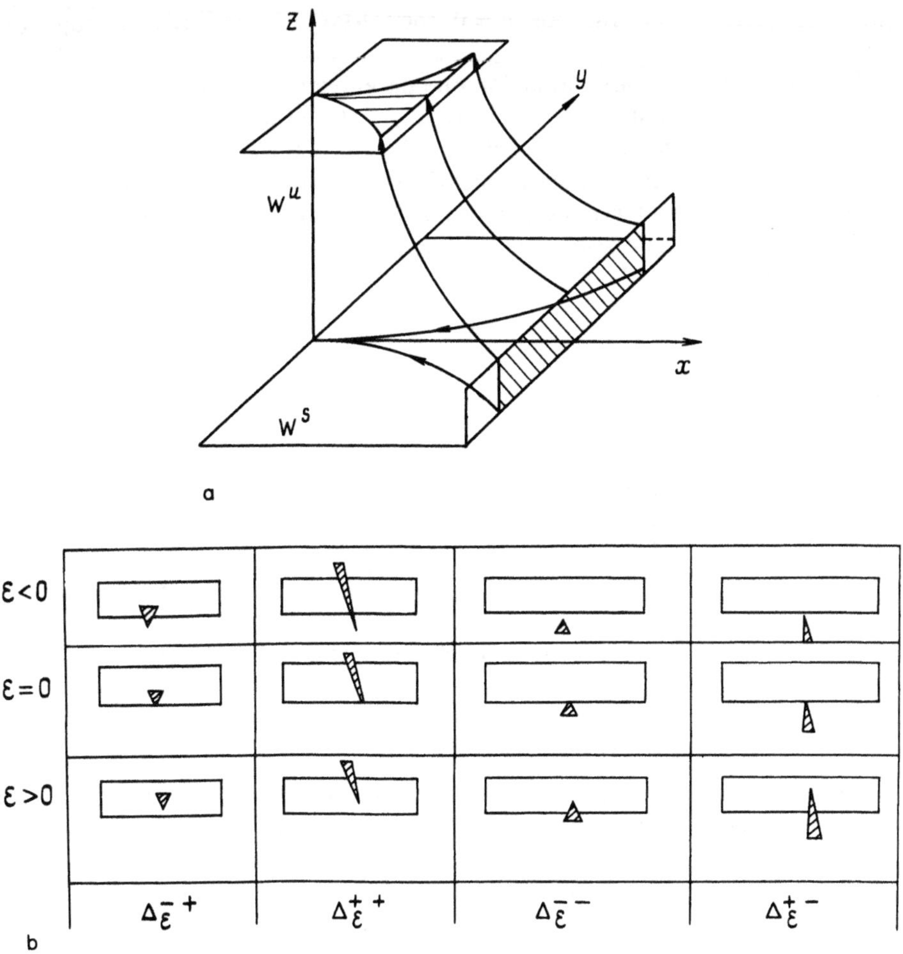

Fig. 46. (a) A correspondence map for a hyperbolic saddle. (b) The image and preimage of the first return map corresponding to a homoclinic trajectory of a saddle

in the y-direction. From these considerations one can derive the following results:

1. For $\varepsilon > 0$ each of the vector fields V^{-+}, V^{--} has a stable limit cycle $L^-(\varepsilon)$, but for $\varepsilon < 0$ both have none. For $\varepsilon < 0$, the nonwandering set of both V^{-+} and V^{--} consists of the singular point O; for $\varepsilon > 0$ it consists of $O \cup L^-(\varepsilon)$, and for $\varepsilon = 0$ it consists of $O \cup \Gamma$, where Γ is a homoclinic curve.

2. Each of the vector fields V^{++} and V^{+-} has a limit cycle $L^+(\varepsilon)$ of saddle type with a two-dimensional stable manifold and a two-dimensional unstable manifold, for $\varepsilon < 0$ and $\varepsilon > 0$, respectively. Moreover, the stable and unstable manifolds of V^{++} (V^{+-}) are homeomorphic to cylinders (Möbius bands). For $\varepsilon \neq 0$, the vector fields V^{++} and V^{+-} have no nonwandering trajectories except for O

and the cycle $L^+(\varepsilon)$, and for $\varepsilon = 0$ they have no homoclinic trajectories except Γ.

3. Analogous assertions hold for one-parameter families of smooth vector fields that are sufficiently C^1-close in K_i $(i = 1, 2)$ to the principal families.

For the principal families the existence of cycles of the fields v_ε (or alternatively, existence of fixed points of their Poincaré maps) is shown in elementary fashion, since the maps Δ_ε preserve the y-coordinate only for $y = 0$. Consequently, it is sufficient to study the one-dimensional map $\Delta_\varepsilon|_{y=0}$. The graphs of the maps Δ_ε are illustrated in Fig. 47.

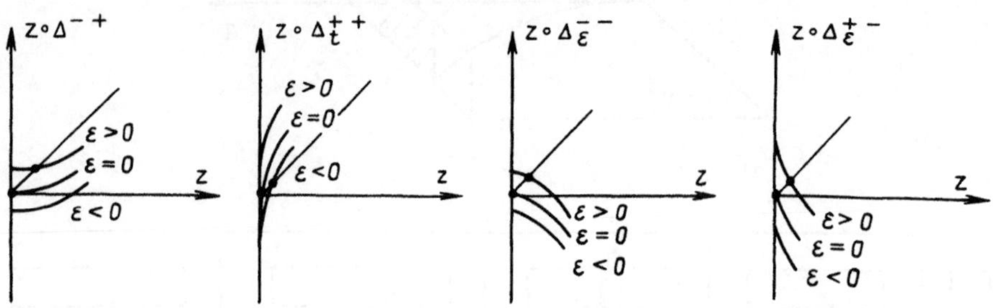

Fig. 47. Graphs of factorized monodromy maps in principal families

5.5. Versality of the Principal Families

Theorem. *A germ of a generic one-parameter family $\{v_\varepsilon\}$ of vector fields on a homoclinic trajectory of a hyperbolic saddle point in $\mathbb{R}^3$ with a real one-dimensional leading direction is topologically equivalent (possibly after reversing time) to a germ of one of the principal families $V^{++}, \ldots, V^{--}$ on a homoclinic trajectory of the corresponding vector field from the list $v_0^{++}, \ldots, v_0^{--}$.*

The principal families of vector fields in $\mathbb{R}^n$ $(n > 3)$ with a hyperbolic saddle for which the leading stable and unstable directions are one-dimensional (and, consequently, real), and which at $\varepsilon = 0$, have a homoclinic trajectory are obtained from those described above for $n = 3$ by a saddle suspension. They are investigated analogously to the case $n = 3$: for arbitrary n the analog of the previous theorem holds.

5.6. A Saddle with Complex Leading Direction in $\mathbb{R}^3$. All families described in Sect. 5.2 have the same nonwandering trajectories. However, the topological equivalence of these families in the case of a complex leading direction is obstructed by the existence of a topological numerical modulus. We describe it for systems in $\mathbb{R}^3$.

Theorem (V.S. Afrajmovich and Yu.S. Il'yashenko, 1985). *Suppose a smooth vector field in $\mathbb{R}^3$ has a homoclinic trajectory of a hyperbolic saddle point with eigenvalues $\alpha \pm i\beta$, and λ, with $\alpha\lambda < 0$. Then α/λ is a topological invariant.*

◀ *Proof*. 1. By rescaling time we can make $\lambda = 1$; we prove that α is a topological invariant. Consider the monodromy map Δ of the homoclinic trajectory γ of the hyperbolic saddle point O. For this purpose, choose a point $P \in \gamma$ ($Q \in \gamma$) sufficiently close to O on its two-dimensional stable manifold W^s (one-dimensional unstable manifold W^u). A condition stating how close is sufficient will be formulated below. The manifold W^s splits a neighborhood of O into two parts: that part into which the trajectory γ enters as $t \to -\infty$ we denote by U^+. We choose two transversal, two-dimensional, smooth disks $\Gamma \ni P$ and $\Gamma' \ni Q$; see Fig. 48a. Let $\Gamma^+ = U^+ \cap \Gamma$. If the region Γ^+ is sufficiently small, then the correspondence map $\Delta_1 : \Gamma^+ \to \Gamma'$ is well defined: this map takes each point $P^+ \in \Gamma^+$ into the endpoint on Γ' of the arc of the phase curve of the vector field under consideration, starting at P^+ and located entirely in U^+; see Fig. 48a. Let $f : (\Gamma', Q) \to (\Gamma, P)$ be the germ of the monodromy transformation (Poincaré map) corresponding to the arc of the homoclinic trajectory γ beginning at Q and ending at P. Obviously, f is a germ of a diffeomorphism. The germ of the monodromy transformation $\Delta : (\Gamma^+, P) \to (\Gamma, P)$ equals the composition of germs $f \circ \Delta_1$. One may assume that a representative of the germ Δ (which we denote with the same symbol) is defined on the region Γ^+, and that its image is contained in a disk $\tilde{\Gamma} \supseteq \Gamma$.

2. We make use of the following theorem of Belitskij (1979).

Theorem. *Suppose a smooth vector field has a hyperbolic saddle O with eigenvalues* $\lambda_1, \dots, \lambda_n$, *and suppose that none of the relations* $\operatorname{Re} \lambda_i = \operatorname{Re} \lambda_j + \operatorname{Re} \lambda_k$ *is fulfilled. Then the germ of the vector field at O is* C^1*-equivalent to its linear part.*

Our vector field satisfies the conditions of Belitskij's theorem, since the real parts of the eigenvalues of the saddle point O are α, α, and 1, with $\alpha < 0$. Consequently, there exists a C^1-smooth chart (x, y, z) in some neighborhood U of the saddle point O that linearizes our field. In this chart W^s is given by the equation $z = 0$, and W^u is given by the equations $x = y = 0$. Suppose $P \in U$, $Q \in U$; this is the condition of nearness of P and Q to the saddle. Stretching the coordinate axes, we obtain the equalities

$$x(P) = 1, \qquad (x, y, z)(Q) = (0, 0, 1).$$

Suppose the disks Γ and Γ' lie in the planes $S_1 : x = 1$ and $S_2 : z = 1$, respectively, with charts (y, z) on S_1 and $(x + iy)$ on S_2. Then in these coordinates

$$\Delta_1(1 + iy, z) = ((1 + iy)z^{-(\alpha + i\beta)}, 1). \tag{1}$$

Actually, the time of transition of the point $(1, y, z)$ to the plane S_2 is equal to $\ln(1/z)$, and the transformation of the phase flow of the linear system

$$(x + iy)^{\cdot} = (\alpha + i\beta)(x + iy), \qquad \dot{z} = z$$

has the form

$$g^t(x, y, z) = (e^{(\alpha + i\beta)t}(x + iy), e^t z).$$

The image $\Delta_1(\Gamma^+)$ of the region Γ^+ on the disk Γ' is a "thick" spiral with center 0;

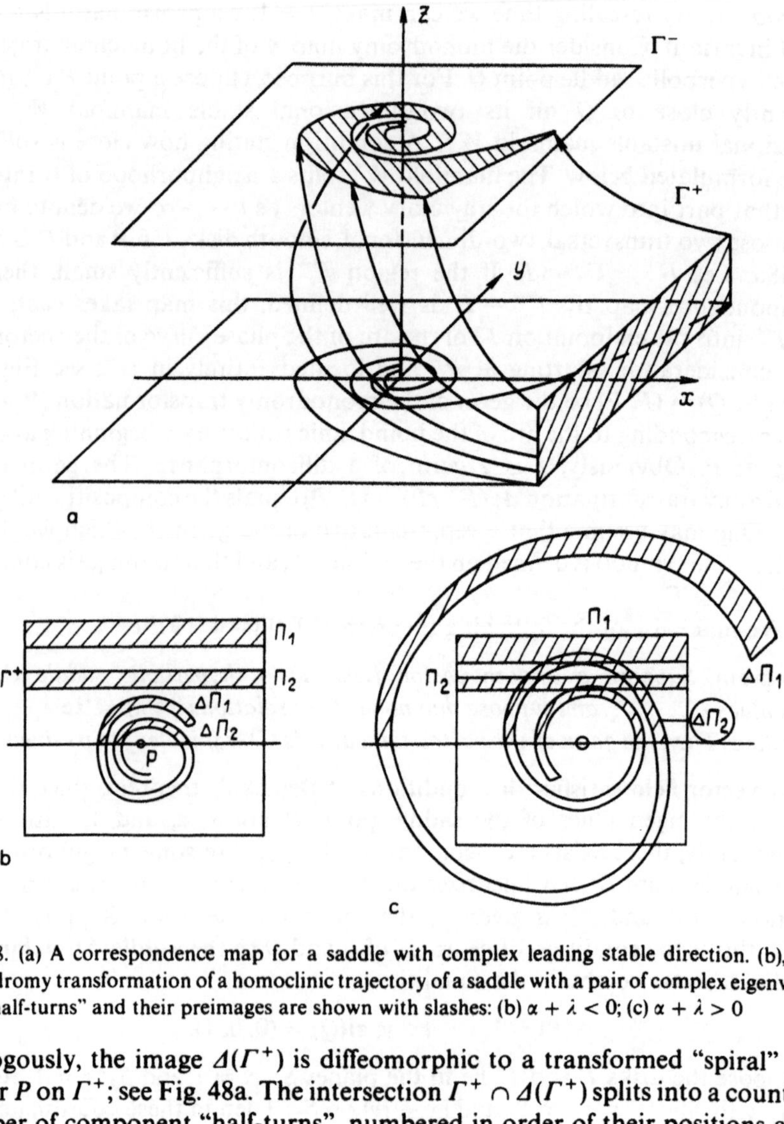

Fig. 48. (a) A correspondence map for a saddle with complex leading stable direction. (b), (c) A monodromy transformation of a homoclinic trajectory of a saddle with a pair of complex eigenvalues. The "half-turns" and their preimages are shown with slashes: (b) $\alpha + \lambda < 0$; (c) $\alpha + \lambda > 0$

analogously, the image $\Delta(\Gamma^+)$ is diffeomorphic to a transformed "spiral" with center P on Γ^+; see Fig. 48a. The intersection $\Gamma^+ \cap \Delta(\Gamma^+)$ splits into a countable number of component "half-turns", numbered in order of their positions along the spiral. Let Π_n be a curvilinear quadrilateral that is the preimage of the nth connected component of

$$\tilde{\Gamma}^+ \cap \Delta(\Gamma^+), \qquad \text{where } \tilde{\Gamma}^+ = U \cap \tilde{\Gamma}.$$

($\tilde{\Gamma}$ is defined at the end of part 1 of the proof.)

Case 1. $\alpha + \lambda \leqslant 0$. We consider the map k of the natural numbers into itself given by the formula

$$k(n) = \min\{k \mid \Pi_k \cap \Delta\Pi_n \neq \varnothing\}$$

(see Fig. 48b).

Case 2. $\alpha + \lambda > 0$ (see Fig. 48c). We set

$$n(k) = \max\{n \mid \Pi_k \cap \Delta\Pi_n \neq \varnothing\}.$$

Remark. The functions k and n are also defined for $\alpha + \lambda > 0$ and $\alpha + \lambda \leqslant 0$, respectively. But in this case, $k(n) \equiv n$ ($n(k) \equiv k$). This may be derived from the proof of the following lemma.

3. **Lemma.** $\lim_{n\to\infty} (k(n)/n) = -\alpha, \qquad \lim_{k\to\infty} (n(k)/k) = -1/\alpha$

for $\alpha + \lambda < 0$ and $\alpha + \lambda > 0$, *respectively.*

◂ We define $\arg \Delta(y, z)$ to be a continuous function on Γ^+ such that

$$\arg \Delta(y, z)|_{\Pi_1} \in [0, \pi].$$

Then $|\arg \Delta(y, z)|_{\Pi_n} \in [2\pi(n-1), \pi(2n-1)]$.

To be definite, suppose the map f preserves orientation. Then the polar angle changes under the action of f by a bounded amount. By formula (1)

$$\arg \Delta_1(y, z) = -\beta \ln z + \arg(1 + iy).$$

Thus,

$$\arg \Delta(y, z) = -\beta \ln z + O(1)$$

as $z \to 0$. Consequently,

$$\max|\ln z||_{\Delta\Pi_n} = |2\pi n\alpha/\beta| + O(1).$$

But

$$|\ln z||_{\Pi_n} = |2\pi k/\beta| + O(1).$$

The intersection $\Pi_k \cap \Delta\Pi_n$ is certainly nonempty if

$$\max|\ln z||_{\Delta\Pi_k} \leqslant \max|\ln z||_{\Pi_n},$$

and is empty if

$$\min|\ln z||_{\Pi_k} \geqslant \max|\ln z||_{\Delta\Pi_n},$$

Consequently,

$$k(n) = n|\alpha| + O(1) \qquad \text{for } |\alpha| \geqslant 1.$$

and

$$n(k) = k/|\alpha| + O(1) \qquad \text{for } |\alpha| < 1,$$

which proves the lemma. ▸

4. The theorem follows easily from the lemma. Indeed, let two vector fields v_1 and v_2, satisfying the conditions of the theorem, be orbitally topologically equivalent. Then the functions k and n coincide up to $O(1)$ for v_1 and v_2. Indeed, suppose Γ_1^+ and Γ_1' are transversal planes for v_1, that Γ_2^+ and Γ_2' are the analogous planes for v_2, and suppose H is a homeomorphism transforming the phase curves of v_1 into those of v_2. The images $H\Gamma^+$ and $H\Gamma'$ are not smooth, but they intersect

each phase curve of v_2, situated in some neighborhood of these disk-images, in one point since H is a homeomorphism. Decreasing the size of Γ^+ and Γ' if necessary, and mapping the disks $H\Gamma^+$ and $H\Gamma'$ by the projection π onto Γ_1^+ and Γ_1' along the curves of v_2, we obtain disks $\pi H\Gamma_1^+$ and $\pi H\Gamma_1'$ belonging to Γ_1^+ and Γ_1', respectively. It is clear that decreasing the size of Γ^+ and Γ' changes the functions k and n only by $O(1)$. ►

5.7. An Addition: Bifurcations of Homoclinic Loops Outside the Boundary of a Set of Morse-Smale Systems

Theorem. *Suppose that in the theorem of Sect. 5.2 both conditions 1 and 2 are violated, that is, for $\sigma < 0$ ($\sigma > 0$) the leading unstable (respectively, stable) direction is complex (and two-dimensional). Then all vector fields of the family $\{v_\varepsilon\}$, sufficiently close to the critical field, have hyperbolic invariant sets; for $\varepsilon \neq 0$, the monodromy transformation of v_ε has a finite number of Smale horseshoes. The number grows unboundedly as $\varepsilon \to 0$ and is equal to infinity for the vector field v_0. For sufficiently small ε each field v_ε has a countable set of hyperbolic limit cycles, the stable manifolds of which each have the same dimension as the stable manifold of the hyperbolic saddle.*

A more exact description of the structure of the hyperbolic subsets for $v_\varepsilon \neq 0$ is given by the following theorem.

Theorem (Shil'nikov (1967, 1970)). *Suppose $\Omega(\rho)$, $\rho > 1$, is a subset of a topological Bernoulli shift on an infinite number of symbols, defined in the following way: $(\ldots m_{-1}, m_0, \ldots, m_i, \ldots) \in \Omega(\rho)$ if and only if $m_{j+1} < \rho m_j$ for all $j \in \mathbb{Z}$. Then, for $\sigma < 0$ the field v_0 has a hyperbolic subset whose trajectories are in a 1–1 correspondence with $\Omega(\rho)$, where ρ is not greater than $-\mathrm{Re}\,\lambda_1/\mathrm{Re}\,\mu_1$. This correspondence preserves asymptotic properties.*[23]

Remark. The limiting value ρ coincides with the modulus given in Sect. 5.6.

For three-dimensional systems, bifurcations appearing with changes of a parameter depend not only upon σ, but also upon a new saddle number $\sigma_1 = 2\,\mathrm{Re}\,\lambda_1 + \mu_1$.

Theorem (Belyakov (1980), Gaspard (1984)). *If $\sigma < 0$, then: 1) for $\sigma_1 < 0$ and for values of ε in a countable set of intervals, each field v_ε has a limit cycle that changes its stability as it undergoes a period-doubling bifurcation, 2) for $\sigma_1 > 0$, there exists a countable set of intervals such that for values of ε in these intervals each field v_ε has an unstable cycle (stable for $t \to -\infty$).* ▲

We now clarify the mechanism though which a countable number of periodic trajectories arise for $n = 3$. In this case the return map, corresponding to a homoclinic trajectory for $\varepsilon = 0$, has already been studied in Sect. 5.4, its image and preimage are illustrated in Fig. 48c. The restriction of the return map to the

---

[23] That is, periodic trajectories correspond to periodic trajectories, and asymptotic trajectories correspond to asymptotic trajectories, etc.

curvilinear rectangle Π_k, for sufficiently large k, is a Smale horseshoe; the number of such horseshoes is countable. For any natural number N, for values of ε sufficiently close to 0 the return map has no less than N Smale horseshoes. A countable set of periodic trajectories corresponds to each horseshoe.

To end this section we present Table 3 in which the conclusions of this section are summarized.

Table 3

$\sigma < 0$			$\sigma > 0$		
Stable \ Unstable	$\mathbb{R}$	$\mathbb{C}$	Stable \ Unstable	$\mathbb{R}$	$\mathbb{C}$
$\mathbb{R}$	$\dim W_L^s = \dim W_O^s + 1$	Ω	$\mathbb{R}$	$\dim W_L^s = \dim W_O^s$	$\dim W_L^s = \dim W_O^s$
$\mathbb{C}$	$\dim W_L^s = \dim W_O^s + 1$	Ω	$\mathbb{C}$	Ω	Ω

Here the symbols $\mathbb{R}$ and $\mathbb{C}$ denote the real and complex leading directions, and the symbol Ω denotes the case in which a nontrivial hyperbolic set exists.

5.8. An Addition: Creation of a Strange Attractor upon Bifurcation of a Trajectory Homoclinic to a Saddle. In generic one-parameter families of smooth vector fields in $\mathbb{R}^n$ ($n \geqslant 3$), perturbed, one can find vector fields V having the following properties:

1. The field V has a saddle with a negative saddle number and a one-dimensional unstable manifold. One of the separatrices emanating from this saddle forms a homoclinic trajectory.

2. The second separatrix coming from this saddle winds onto the first; more precisely, its ω-limit set consists of the union of the homoclinic trajectory and the saddle.

Denote the saddle by O and the "whole" unstable manifold of the point O by W^u, that is, W^u is the curve consisting of the union of the singular point O and the separatrices emanating from it. The set W^u consists of the image of a straight line under an immersion in $\mathbb{R}^n$, but is not a submanifold of $\mathbb{R}^n$ (Fig. 48d); it is closed.

Proposition 1. *Suppose a vector field V satisfies properties* 1) *and* 2) *above. Then, in any neighborhood of the curve W^u, there exists an attracting domain of v (homeomorphic to the interior of a sphere with two handles for $n = 3$); that is, a domain with a smooth boundary, on which the field points strictly into the interior of the domain.*

The proof is elementary; it is based upon the negativity of the saddle number.

Conjecture. *Suppose that in a one-parameter family of vector fields, the field corresponding to the zero value of the parameter ε satisfies the conditions* 1) *and* 2)*. Then for any neighborhood U of the curve W^u in phase space, and any neighbor-*

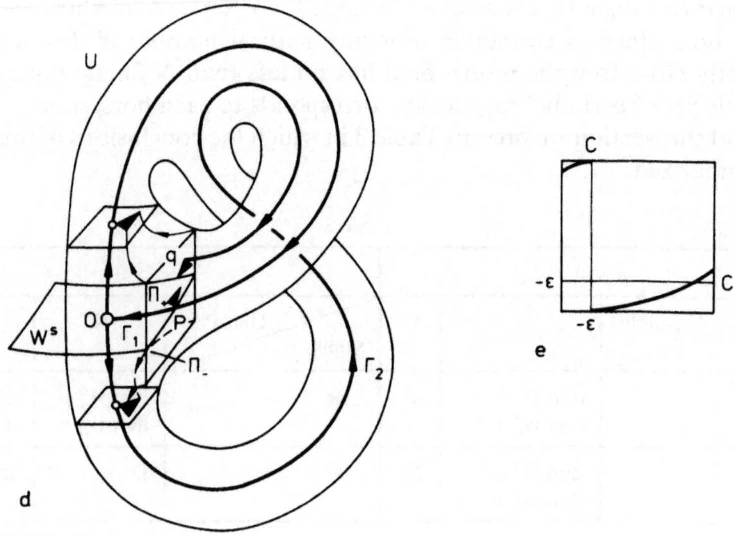

Fig. 48. (d) The set W^u; (e) A family of vector fields satisfying the conditions 1) and 2) and the conclusion of the Conjecture

hood N of zero in the one-dimensional space of ε, there exists a value ε ∈ N such that the corresponding field of the family has a strange attractor lying in the neighborhood U.

At the time of writing the Conjecture is open. However, an example of a family of vector fields has been constructed for which the vector field corresponding to $\varepsilon = 0$ satisfies conditions 1) and 2) and the conclusion of the Conjecture.

We describe this example. To begin, consider the family of mappings of an interval with one jump-point of discontinuity, at which the one-sided derivatives both exist, which are C^1 away from the point of discontinuity and which have the following properties (see Fig. 48e):

$$f_\varepsilon\colon [-\varepsilon, c] \to [-\varepsilon, c],\ 0 \leqslant f_\varepsilon'(x) \leqslant q < 1 \qquad \text{for} \qquad x \neq 0,$$

$$\lim_{x\to 0^+} f_\varepsilon(x) = \lim_{x\to 0^-} f_\varepsilon(x) = 0; \qquad \lim_{x\to 0^+} f_\varepsilon(x) = -\varepsilon, \qquad f(-\varepsilon) > f(c).$$

The results of J.P. Keener (1980) imply the following proposition.

Proposition 2. *For the family $\{f_\varepsilon\}$ described above there exists a positive value of the parameter ε, arbitrarily close to zero, for which the nonwandering set of the mapping f_ε is homeomorphic to a Cantor set consisting of recurrent trajectories, and is the ω-limit set of the trajectories of all points of the interval.*

We return again to vector fields. Let V be a vector field satisfying the conditions 1) and 2) stated at the beginning of this section. We define the monodromy transformation of the field V (and nearby fields). Choose an arbitrary point p on the homoclinic trajectory of the saddle O, and an $(n-1)$-dimensional plane Π passing through p and lying on a hyperplane transversal to the field V at the point

p. The stable manifold W^s of the point O splits the plane Π into two parts. Denote by Π^+ that part for which the germ of the monodromy transformation Δ^+ is defined, namely, $\Delta^+ : (\Pi^+, p) \to \Pi$ along the homoclinic trajectory of the saddle. Let Π^- denote the other half of the plane Π. The exiting separatrix of the saddle O, winding onto the homoclinic trajectory, intersects Π at a point q belonging to the domain in which a representative of the germ Δ^+ is defined. This follows from the properties 1) and 2) of the field V. Let Γ_2 be the arc of the exiting separatrix from the point O to the point q; let Γ_1 be the arc of the homoclinic trajectory from the point p to the point O. There exists a germ of the monodromy transformation $\Delta^- : (\Pi^-, p) \to (\Pi, q)$ along the phase curves of the field V near the union of the phase curves $\Gamma_1 \cup O \cup \Gamma_2$ (see Fig. 48c). We call the discontinuous mapping $\Delta : \Pi \to \Pi$, coinciding on Π^+ with Δ^+, and on Π^- with Δ^-, the monodromy transformation of the field. For each field V_ε close to V, a monodromy transformation Δ_ε is defined which is discontinuous on $\Pi \cap W^s$, and is smooth inside Π^+ and Π^-.

Performing an identification like that of Sect. 5.4, one can construct a family of vector fields $\{V_\varepsilon\}$ such that the field V_0 has properties 1) and 2), and the monodromy transformation Δ_ε, corresponding to V_ε, has a smooth invariant foliation, one of whose leaves is the discontinuous surface $\Pi \cap W^s$. The family $\{\Delta_\varepsilon\}$ induces a mapping of the one-dimensional quotient-space of the plane Π factored by the invariant foliation. The family $\{V_\varepsilon\}$ can be constructed in such a way that the corresponding factor-family will coincide with the family $\{f_\varepsilon\}$ of Proposition 2. The construction of the family $\{V_\varepsilon\}$ is complete.

From Propositions 1 and 2 we deduce:

Proposition 3. *For the family of vector fields $\{V_\varepsilon\}$ constructed above, the conclusion of the above conjecture holds. The corresponding strange attractor, close to any of its points different from O, is locally diffeomorphic to the product of a Cantor set and a disk.*

(This section was added for the English translation by V.S. Afrajmovich and M.I. Malkin. D.V. Turaev communicated to us that the proofs are given in his Ph.D. Thesis, 1991.)

§ 6. Bifurcations Related to Nontransversal Intersections

In this section we consider bifurcations of a vector field lying on the boundary of the set of Morse-Smale systems whose nonwandering set consists of a finite number of hyperbolic equilibria and hyperbolic cycles, with stable and unstable manifolds intersecting transversally along all trajectories, with one exception – a simple tangency or a quasi-transversal intersection.

6.1. Vector Fields with No Contours and No Homoclinic Trajectories. A simple consequence of the Kupka-Smale theorem is:

Proposition. *If a vector field v_0, as described in the beginning of this section, has no contours and no homoclinic trajectories, then Morse-Smale vector fields are everywhere dense in a neighborhood of v_0 in $\chi^r(\mathbf{M})$.*

(We assume below that $r \geqslant 2$ if v_0 has neither equilibria with pure imaginary eigenvalues nor cycles with multipliers $e^{\pm i\varphi}$, otherwise $r \geqslant 3$.)

Nevertheless, perturbations of v_0 may undergo bifurcations.

Definition. Two trajectories Γ_1 and Γ_2 of a dynamical system are called *internally equivalent* if there exists a homeomorphism of phase space[24] onto itself that maps trajectories into trajectories, preserving their orientations and transforming Γ_1 into Γ_2.

Obviously, the partition into classes of internally equivalent trajectories is a topological invariant of a dynamical system. Bifurcations may occur without creation or destruction of nonwandering trajectories, and can be related to changes of the classes of internal equivalence.

Definition (Afrajmovich and Shil'nikov (1972)). A trajectory is called *special* if there exists an $\varepsilon > 0$ such that this trajectory remains invariant under each homeomorphism of phase space onto itself that is within an ε-neighborhood of the identity homeomorphism, and which maps trajectories into trajectories preserving their orientations.

Obviously, a special trajectory belongs to an internal equivalence class containing no more than a countable number of trajectories. Equilibria, limit cycles, and heteroclinic trajectories in $W_1^s \cap W_2^u$ with $\dim W_1^s + \dim W_2^u - n = 1$ are special.

6.2. A Theorem on Inaccessibility. Suppose L_1 and L_2 are cycles of a vector field v_0 such that the intersection $W_{L_1}^s \cap W_{L_2}^u$ contains a trajectory of simple tangency or quasi-transversal intersection.

Theorem. *If $W_{L_1}^s$ ($W_{L_2}^s$) contains a special trajectory not coinciding with $L_1(L_2)$, then the bifurcation surface $\mathscr{B}_1$ is inaccessible at the point v_0 even from one side.*

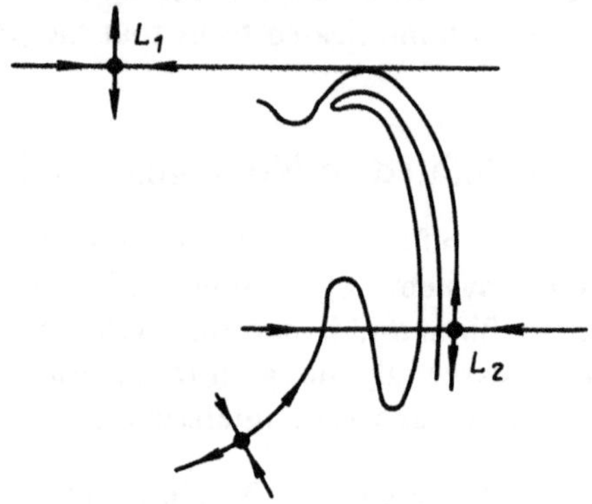

Fig. 49. Fixed points and invariant curves of a diffeomorphism of the plane belonging to the inaccessible part of a bifurcation surface

---

[24] It is required here to be compact.

If the conditions of this theorem are fulfilled, $W^s_{L_1}$ ($W^u_{L_2}$) is a "smooth" limit of manifolds (of the same dimension) of other equilibria or cycles both for v_0 and for nearby vector fields v. Therefore, for any family $\{v_\varepsilon\}$ of vector fields, one finds arbitrarily many values of ε close to zero for which $W^u_{L_2}(\varepsilon)$ ($W^s_{L_1}(\varepsilon)$) will have a trajectory of a nontransversal intersection (see Fig. 49). Here $W^u_{L_2}(\varepsilon)$ is the unstable manifold of a hyperbolic cycle of a field v_ε lying in a neighborhood of L_2; $W^s_{L_1}(\varepsilon)$ is defined analogously.

6.3. Moduli. J. Palis (1978) found that a topological conjugacy of diffeomorphisms with the "same" geometric arrangement of their stable and unstable manifolds implies some condition of equality on the multipliers of their periodic trajectories. More precisely, suppose f (f') is a diffeomorphism of a closed manifold with hyperbolic fixed points p, q (p', q') of saddle type. Suppose $\lambda_1(\lambda'_1)$ is the eigenvalue of largest modulus among the eigenvalues of $Df(p)$ ($Df'(p')$) with modulus less than 1, and suppose γ_2 (γ'_2) is the eigenvalue of smallest modulus among the eigenvalues of $Df(q)$ ($Df'(q')$) with modulus greater than 1. Assume that $\lambda_1(\lambda'_1)$ and $\gamma_2(\gamma'_2)$ have multiplicity 1. Then (Hirsch, Pugh and Shub (1977)) there exists a smooth invariant manifold $W^{u,1}_p$ ($W^{u,1}_{p'}$) tangent to the sum

$$TW^u_p \oplus R_{\lambda_1}(TW^u_{p'} \oplus R_{\lambda'_1}) \qquad \text{at } p(p'),$$

where TW is the tangent space to W and $R_{\lambda_1}(R_{\lambda'_1})$ is the eigenspace corresponding to $\lambda_1, \bar{\lambda}_1$ ($\lambda'_1, \bar{\lambda}'_1$). [If $\lambda_1 \in \mathbb{R}^1$, then $\dim R_{\lambda_1} = 1$; otherwise, $\dim R_{\lambda_1} = 2$.] There also exists a smooth invariant manifold $W^{s,1}_q(W^{s,1}_{q'})$ tangent to the sum $TW^s_q \oplus R_{\gamma_2}(TW^s_{q'} \oplus R_{\gamma'_2})$ at $q(q')$, where $R_{\gamma_2}(R_{\gamma'_2})$ is the eigenspace corresponding to γ_2, $\bar{\gamma}_2(\gamma'_2, \bar{\gamma}'_2)$.

Definition (de Melo and Palis (1980), and Newhouse, Palis and Takens (1983)). A point r of simple tangency, or a quasi-transversal intersection $W^u_p \cap W^s_q$, is called *a point of regular intersection of codimension* 1 if W^u_p is transverse to $W^{2,1}_q$ and W^s_q is transverse to $W^{u,1}_p$ at r.

Although the manifolds $W^{u,1}_p$ and $W^{s,1}_q$ are not unique, since all the manifolds $W^{u,1}_p$ ($W^{s,1}_q$) are tangent at the point p (q), a point of regular intersection is well defined.

Theorem (de Melo and Palis (1980), and Newhouse, Palis and Takens (1983)). *Let f (f') be a C^2-diffeomorphism having hyperbolic fixed points p (p'), q (q'), and a trajectory Γ consisting of points of regular intersection. Then, if there exists a topological conjugacy between f and f', defined in some neighborhood of the closure of $\bar{\Gamma}$, the following equality holds*:

$$\frac{\log|\lambda_1|}{\log|\gamma_2|} = \frac{\log|\lambda'_1|}{\log|\gamma'_2|}$$

Here $\lambda_1, \lambda'_1, \gamma_2$ and γ'_2 are the same as above.

We illustrate the theorem for $m = 2$ in Fig. 50.

It is not difficult to construct a diffeomorphism having more than one modulus of stability. For this purpose it is sufficient that the unstable (stable) manifold

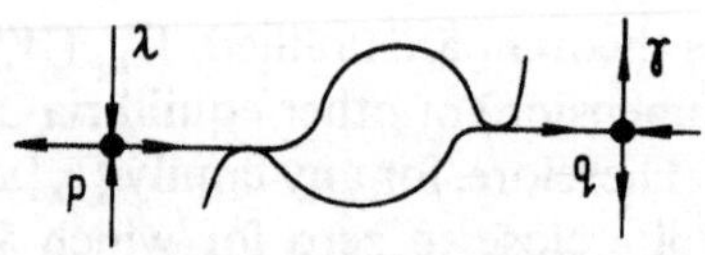

Fig. 50. A diffeomorphism of the plane whose topological invariant is the ratio log γ/log λ

of p (q) be the limit of the unstable (stable) manifolds of other saddle points (as, for example, in the theorem of Sect. 6.2). de Melo, Palis, and van Strien (1981) derived necessary and sufficient conditions for a diffeomorphism lying on the boundary of the set of Morse-Smale systems to have a unique modulus.

6.4. Systems with Contours. We suppose that v_0 has a contour $\{Q_0, \ldots, Q_k\}$ and, moreover, a trajectory of simple tangency or quasi-transversal intersection belonging to $W^s_{Q_{j+1}} \cap W^u_{Q_j}$.

The existence of vector fields with contours on the boundary of the set of Morse-Smale systems was established by Gavrilov (1973). An example of a such a diffeomorphism is given in Fig. 51.

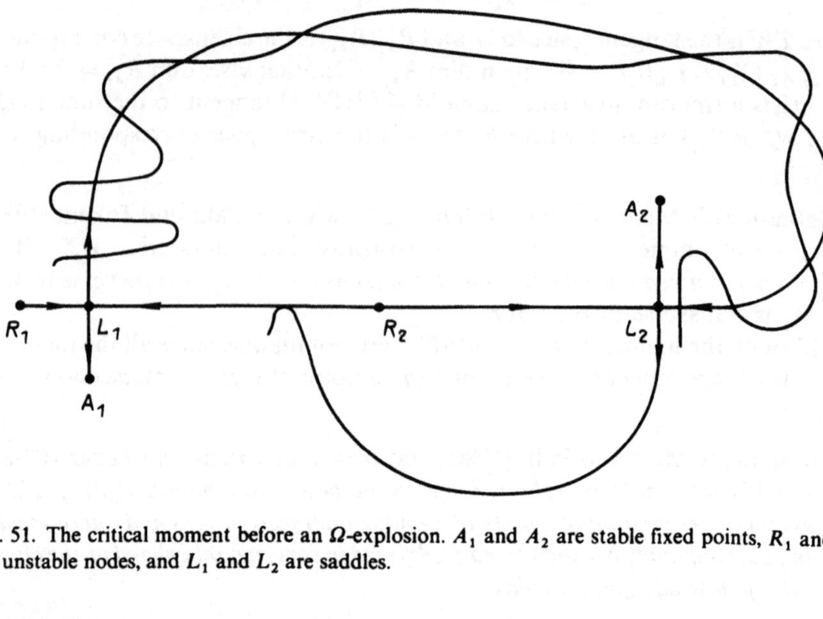

Fig. 51. The critical moment before an Ω-explosion. A_1 and A_2 are stable fixed points, R_1 and R_2 are unstable nodes, and L_1 and L_2 are saddles.

Proposition. *A vector field with a countable set of cycles can be found in any neighborhood of v_0 in χ^r(M).*

The proof consists of establishing, with the aid of the λ-lemma, that a vector field near to v_0 has a homoclinic curve belonging to a transversal intersection of manifolds (see Newhouse and Palis (1976) and Palis (1971)). Such a great enlargement of a nonwandering set is called an Ω-explosion (Palis (1971)).

Remark. If v_0 is a vector field with a homoclinic trajectory of a simple tangency of the stable and unstable manifolds of a cycle, then the Proposition remains true (see Sect. 6.6 below).

6.5. Diffeomorphisms with Nontrivial Basic Sets. The proposition in Sect. 6.4 was strengthened to apply to diffeomorphisms in papers by Newhouse and Palis (1976) and by Newhouse, Palis and Takens (1983): it was shown that in a neighborhood of a point of the bifurcation surface there exist diffeomorphisms satisfying Smale's Axiom A with zero-dimensional non-trivial basic sets. More precisely, suppose $\mathbf{M}$ is a compact, connected C^∞-manifold, $\mathrm{Diff}^r(\mathbf{M})$ is the space of C^r-diffeomorphisms of $\mathbf{M}$ with the uniform C^r-topology, $I = [0, 1]$ and, for k and $r \geqslant 1$, $\Phi^{k,r} = C^k(I, \mathrm{Diff}^r(\mathbf{M}))$ is the space of C^k-mappings of I into $\mathrm{Diff}^r(\mathbf{M})$ furnished with the uniform C^k-topology. An element $\xi \in \Phi^{k,r}$ is a C^k-curve of C^r-diffeomorphisms. Suppose $U^{k,r} \subset \Phi^{k,r}$ is the set of arcs $\xi \in \Phi^{k,r}$ such that $\xi_0 \in \mathbf{MS}$, and if $1 > b = \inf\{\varepsilon\colon \xi_\varepsilon \notin \mathbf{MS}\}$, then $v_0(\xi_b) \in \mathscr{B}_1$ and satisfies the conditions of genericity, where $\mathbf{MS} \subset \mathrm{Diff}^r(\mathbf{M})$ is the set of Morse-Smale diffeomorphisms (see Sect. 4). For $\delta > 0$ suppose $U_\delta = [b_0, b_0 + \delta)$.

Theorem (Newhouse and Palis (1976)). *There exists a set of second category $\mathscr{B} \subset U^{k,r}$, $k \geqslant 1$, $r \geqslant 2$, such that if $\xi \in \mathscr{B}$, then for any $\kappa > 0$ there exists a $\delta > 0$ and an open set $\mathscr{B}_\delta \subset U_\delta$ such that:* (a) *the Lebesgue measure of $\mathscr{B}_\delta$ is less than $\kappa\delta$;* (b) *if ε is in $U_\delta \backslash \mathscr{B}_\delta$, then ξ_ε is a diffeomorphism satisfying Smale's Axiom A (see Sect* 3.4); (c) *there exist ε's in $U_\delta \backslash \mathscr{B}_\delta$ for each of which the nonwandering set is infinite, zero-dimensional and, if the stable manifolds of each Q_i have the same dimension, then this is true for any ε in $\mathscr{B}_\delta$.*

The conclusions of the theorem are most easily understood through examples of vector fields in $\mathbb{R}^3$ for which analogous results hold.

6.6 Vector Fields in $\mathbb{R}^3$ with Trajectories Homoclinic to a Cycle. Suppose a vector field $v_0 \in C^r$ ($r \geqslant 3$) in a three-dimensional Euclidean space has a limit cycle L of saddle type, and a trajectory $\Gamma \subset W_L^s \cap W_L^u$ of a simple tangency of the stable and unstable manifolds of this cycle. Then there exists a neighborhood $\mathbf{U}$ of $L \cup \Gamma$ homeomorphic to a solid torus $\mathbf{U}_0$ with one handle $\mathbf{U}_1$: L lies inside of the solid torus and $\Gamma \cap (\mathbf{U} \backslash \mathbf{U}_0)$ is connected, that is Γ "goes around" the handle just once. For the system v_0 in $\chi^r(\mathbb{R}^3)$, there is a neighborhood $\mathscr{U} = \mathscr{U}_1 \cup \mathscr{U}_0 \cup \mathscr{U}_2$, where (a) $\mathscr{U}_1$ consists of systems without homoclinic trajectories $\tilde{\Gamma}$, with $\tilde{\Gamma} \cap (\mathbf{U} \backslash \mathbf{U}_0)$ connected, (b) $\mathscr{U}_2$ consists of systems with each one having two homoclinic trajectories Γ_1 and Γ_2 such that $\Gamma_i \cap (\mathbf{U} \backslash \mathbf{U}_0)$ is connected ($i = 1, 2$), and each Γ_i belongs to a transversal intersection of the stable and unstable manifolds of a saddle cycle, and (c) $\mathscr{U}_0$ contains systems "similar" to v_0, that is, having a homoclinic trajectory of simple tangency.

Suppose λ and γ are multipliers of a cycle L: $|\lambda| < 1$, $|\gamma| > 1$.

Definition. A cycle L is called *dissipative* if $|\lambda\gamma| < 1$. A *dissipative* fixed point of saddle type of a diffeomorphism of $\mathbf{M}$ is defined analogously.

If $\gamma > 0$, then the manifold W_L^u is homeomorphic to a cylinder, and is divided by the cycle L into two disjoint subsets: W_1^u, W_2^u. Suppose $\Gamma \subset W_1^u$.

Theorem (Gavrilov and Shil'nikov (1972, 1973)). *If*: (1) $\gamma > 0$, (2) $\overline{W_1^u} \cap (W_2^u \setminus L) = \varnothing$, (3) *the cycle L is dissipative, then there exists a neighborhood* $\mathscr{U}$, $\chi^r(\mathbb{R}^3) \supset \mathscr{U} \ni v_0$, *so small that all vector fields from* $\mathscr{U}_1$ *are Morse-Smale vector fields in* $\bar{\mathrm{U}}$.

We clarify this result with an example. Consider a one-parameter family of C^r-diffeomorphisms f_ε: $\mathbb{R}^2 \to \mathbb{R}^2$ which, in a neighborhood U_0 of the fixed point O (the origin), has the form

$$(x, y) \to (\lambda x, \gamma y), \qquad 0 < \lambda < 1 < \gamma, \lambda\gamma < 1.$$

Suppose $P = (0, y^*)$, $Q = (x^*, 0)$, $x^* > 0$, $y^* > 0$, with $f_0^s P = Q$ for some $s \in \mathbb{N}$, are homoclinic points, and for $\varepsilon = 0$ the stable and unstable manifolds of O have a simple tangency at P and Q (see Fig. 52). Suppose

$$\mathrm{U}_0 \supset \Pi_0 = \{(x, y) \,|\, |x - x^*| \leqslant \varepsilon_0, |y| \leqslant \varepsilon_0\},$$
$$\mathrm{U}_0 \supset \Pi_1 = \{(x, y) \,|\, |x| \leqslant \varepsilon_1, |y - y^*| \leqslant \varepsilon_1\},$$

and

$$\Pi_0 \cap f_0(\Pi_0) = \varnothing, \Pi_1 \cap f_0^{-1}(\Pi_1) = \varnothing.$$

We assume that f_0^s in Π_1 can be written in the form

$$x_0 - x^* = b(y_1 - y^*), \qquad y_0 = cx_1 + d(y_1 - y^*)^2 + \varepsilon;$$
$$(x_0, y_0) \in \Pi_0, \qquad (x_1, y_1) \in \Pi_1.$$

This means that the piece of the unstable manifold $x_1 = 0$, $|y_1 - y^*| \leqslant \varepsilon_1$ maps into a piece of the parabola $y_0 = (d/b^2)(x_0 - x^*)^2 + \varepsilon$. Thus, for $d < 0$, $\gamma > 0$, $\varepsilon = 0$, condition (2) of the theorem is satisfied. It is seen that for $d < 0$, $\gamma > 0$, and $\varepsilon < 0$, points from a small neighborhood of P (how small depends upon ε) are mapped into a neighborhood in which y is negative, that is, P is a wandering point (see Fig. 52).

6.7. Symbolic Dynamics. The structure of the nonwandering set of a vector field v close to v_0 can be described in the following way (Gavrilov and Shil'nikov (1972, 1973)). Let Ω be an invariant subset of a topological Bernoulli shift on three symbols $\{0, 1, 2\}$, defined by the following four conditions:

1) Ω contains the fixed point $\{\ldots, 0, 0, \ldots\}$.

2) The symbol 0 must follow each of the symbols 1, 2.

From these two conditions it follows that to each trajectory in Ω corresponds a sequence $\{\ldots, p_i, p_{i+1}, \ldots\}$ of natural numbers, where p_i is the length of the interval of zeros, contained between two nonzero symbols. To a trajectory which is α-asymptotic (resp. ω-asymptotic) to $\{\ldots, 0, 0, \ldots\}$, there corresponds a sequence $\{p_0, p_1, \ldots\}$ (resp. $\{\ldots, p_{k-1}, p_k\}$), where $p_0 = \infty$ (resp. $p_k = \infty$).

3) There exists a $k \in \mathbb{N}$ such that all $p_i \geqslant k$.

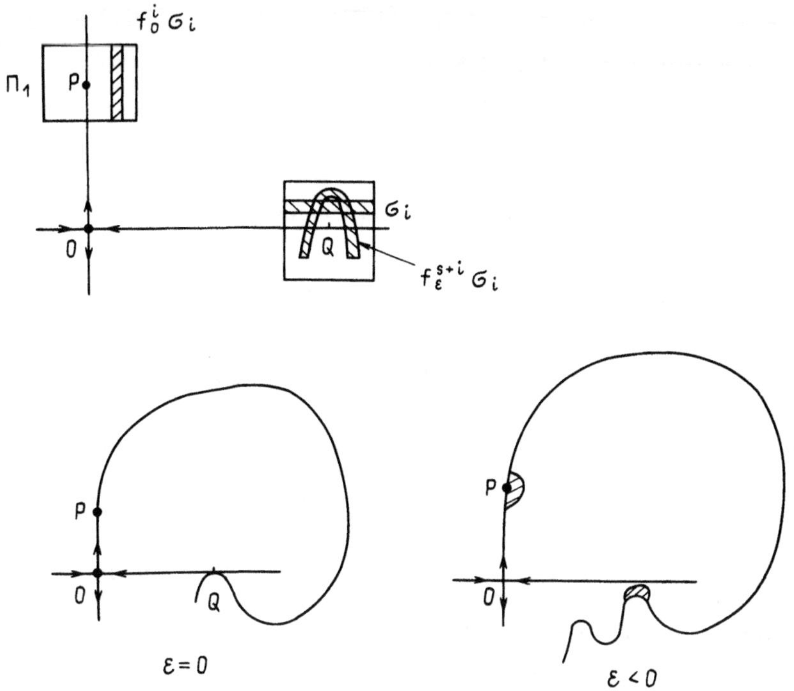

Fig. 52. A Smale horseshoe for a mapping f_ε^{s+i} (the upper figure). A neighborhood of P and its preimage for $\varepsilon = 0$ and $\varepsilon < 0$ (the lower figures)

4) There exist constants $\tilde{\gamma} > 1$, $0 < \tilde{\lambda} < 1$, $d \neq 0$, $v_1 > 0$, $v_2 > 0$, $c \neq 0$ and $e \neq 0$, such that for each trajectory in Ω the inequalities

$$\operatorname{sgn}(d)(v_1 \tilde{\gamma}^{-p_i} - c v_2 \tilde{\lambda}^{p_{i+1}} - e) > 0$$

hold.

Theorem (Gavrilov and Shil'nikov (1972, 1973)). *For any vector field $v \in \mathcal{U}$, there exist constants k, v_1, v_2, $\tilde{\gamma}$, $\tilde{\lambda}$, e, and d, where $\tilde{\gamma}$, $\tilde{\lambda}$ are multipliers of a cycle near to L, $d < 0$, and $e > 0$ for $v \in \mathcal{U}_2$, but $e < 0$ for $v \in \mathcal{U}_1$, such that the following assertion holds: if $\Omega \neq \varnothing$, then there exists a hyperbolic subset of the invariant set in U, the trajectories of which are in one-to-one correspondence with the trajectories in Ω, such that periodic trajectories of the shift $\sigma|\Omega$ correspond to cycles and the asymptotic properties of trajectories are preserved.*

Corollary. For $e > 0$, v has an infinite set of cycles.

Actually, setting $p_i = p_{i+1} = p$, and noticing that the cycle L is dissipative, $d < 0$, and $e > 0$, we obtain the result that the inequality (4) is fulfilled for all sufficiently large p.

We clarify this theorem by returning to the example of Sect. 6.6. Since $\gamma > 0$ ($\lambda < 1$), there exists an $n_1 \in \mathbb{N}$ ($n_2 \in \mathbb{N}$) such that

$$\gamma^{n_1}\varepsilon_0 > y^* + \varepsilon_1, \quad \lambda^{n_1}(x^* + \varepsilon_0) < \varepsilon_1, \quad (\lambda^{-n_2}\varepsilon_1 > x^* + \varepsilon_0, \gamma^{-n_2}(y^* + \varepsilon_1) < \varepsilon_0).$$

For $i \geqslant N = \max(n_1, n_2)$ we define:

$$\sigma_i = \{(x, y) \in \Pi_0 \,|\, |x - x^*| \leqslant \varepsilon_0, |\gamma^i y - y^*| \leqslant \varepsilon_1\}.$$

Obviously, $f_0^i\sigma_i \subset \Pi_1$ and $\sigma_i \cap \sigma_j = \varnothing$, $i \neq j$. The map

$$f_\varepsilon^s \circ f_\varepsilon^i \colon \sigma_i \to \Pi_0((x, y) \mapsto (\bar{x}, \bar{y}))$$

takes the form

$$\bar{x} - x^* = b(y\gamma^i - y^*), \qquad \bar{y} = c\lambda^i x^* + d(y\gamma^i - y^*)^2 + \varepsilon.$$

This map is similar to the well-known map of Hénon (1976). It is easy to check that on each rectangle σ_i, for the corresponding $\varepsilon = \varepsilon_i$, the map acts like a diffeomorphism of a Smale horseshoe.

From the example, it is easy to understand why there exist vector fields in a neighborhood of v_0 that satisfy Smale's Axiom A (see the theorem of Sect. 6.5).

We shall show that there exist values of ε such that all the domains σ_i are mapped like Smale horseshoes. Actually, the distance in the y-direction between σ_i and σ_{i+1} is $O(1/\gamma^{-i})$, and the size in the x-direction of the domain in which all the domains $f_\varepsilon^j\sigma_i$ ($j \geqslant i$) lie is $O(\lambda^i)$. Thus, because the saddle is dissipative, the required values of ε exist (see Fig. 53). From this it follows that all trajectories in a neighborhood of the homoclinic trajectory are hyperbolic, and these are the only newly generated nonwandering trajectories.

Remark. Gavrilov and Shil'nikov (1972, 1973) generalized the above theorem to the case of systems not lying on the boundary of the set of Morse-Smale vector fields, and Gonchenko (1980) also generalized it to the case $n > 3$.

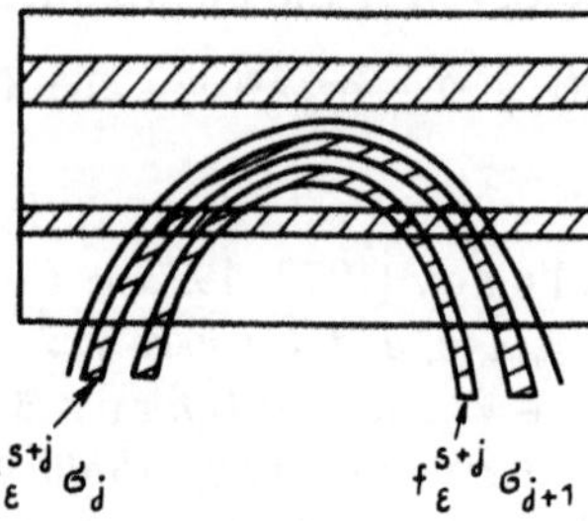

Fig. 53. Images and preimages, under the action of iterates of a diffeomorphism, of "rectangles" lying in a neighborhood of a homoclinic trajectory of a fixed point that is a dissipative saddle

6.8. Bifurcations of Smale Horseshoes. We begin with the example from Sect. 6.6. Here, as ε varies bifurcations connected with the birth of Smale horseshoes appear. It is easily verified that if

$$\varepsilon d > d(y^*\gamma^{-i} - c\lambda^i x^*) + \tfrac{1}{4}(bc\lambda^i - \gamma^{-i})^2,$$

then there are no fixed points of the maps f_ε^{i+s} in σ_i. For

$$\varepsilon_i^1 = y^*\gamma^{-i} - c\lambda^i x^* + \tfrac{1}{4}(bc\lambda^i - \gamma^{-i})^2/d$$

a fixed point of saddle-node type is born. It splits into a saddle and a node, which undergo no further bifurcations as the parameter further decreases down to, and including, the value

$$\varepsilon_i^2 = y^*\gamma^{-i} - c\lambda^i x^* - \tfrac{3}{4}(bc\lambda^i - \gamma^{-i})^2/d.$$

For $\varepsilon = \varepsilon_i^2$ the node (which became a focus and then changes back into a node, but with negative multipliers) undergoes a period-doubling bifurcation (see Fig. 54). The results of this example hold for the general case. Besides this, they are generalized to systems with an n-dimensional phase space (Gonchenko (1984, 1980)). To be more precise, let v_0 be a C^r-smooth vector field ($r \geqslant 4$) on an $(m+2)$-dimensional manifold $\mathbf{M}$ ($m > 1$); moreover, we assume:

1) v_0 has a cycle L of saddle type with multipliers $\lambda_1, \ldots, \lambda_m$, and γ with $|\gamma| > 1 > |\lambda_1| \geqslant |\lambda_j|$ ($j \in \{2, \ldots, m\}$), and λ_1 is not a multiple root of the characteristic equation.

2) the saddle number $|\lambda_1\gamma| < 1$;

3) $W_L^s \cap W_L^u \supset \Gamma$; here Γ is a trajectory of a simple tangency, that does not belong to a nonleading submanifold of the stable manifold W_L^s.[25]

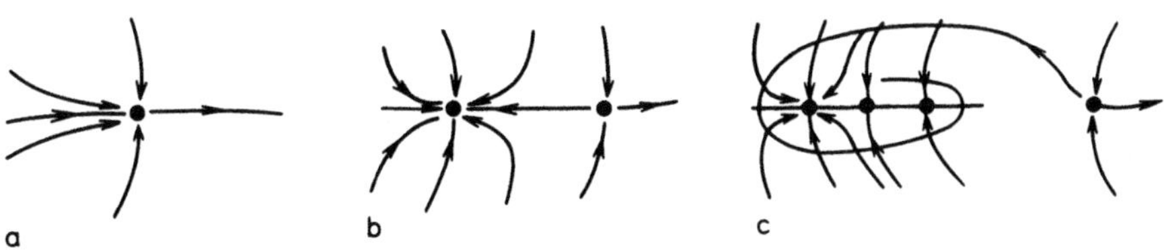

Fig. 54. Bifurcations of periodic points in a neighborhood of a homoclinic trajectory

Remark. In Gonchenko (1984, 1980) there is one more condition on v_0, which does not change its genericity. For the example of Sect. 6.6 this condition is: $c \neq 0$.

Theorem (Gonchenko (1984, 1980)). *For any one-parameter family* $\{v_\varepsilon\}$ *of vector fields in* $\chi^r(\mathbf{M})$ ($r \geqslant 4$) *such that* $\{v_\varepsilon\}$ *is transversal to* $\mathscr{B}_1$ (see Sect. 1.6) *at the point* v_0, *there exists a countable set of intervals* $\{\delta_k^1 \cup \delta_k^2\}$ *on the interval* $[-1, 1]$ *(of* ε*) with*

$$\bar{\delta}_k^1 \cap \bar{\delta}_k^2 = \{\varepsilon_k^2\}; \qquad \partial\delta_k^1 \cup \partial\delta_k^2 = \{\varepsilon_k^1; \varepsilon_k^2; \varepsilon_k^3\},$$

such that for $\varepsilon \in \delta_k^1 \cup \delta_k^2$ *the vector field* v_ε *has a stable limit cycle that is "1-circuiting" for* $\varepsilon \in \delta_k^1$ *and "2-circuiting" for* $\varepsilon \in \delta_k^2$.[26] *For* $\varepsilon \in \bigcup_{i=1}^3 \varepsilon_k^i$, v_ε *has a*

[25] A nonleading submanifold intersects a transversal tangent to the invariant linear subspace corresponding to the multipliers $\lambda_2, \ldots, \lambda_m$ if λ_1 is real, and to $\lambda_3, \ldots, \lambda_m$ otherwise.

[26] Suppose L is a hyperbolic cycle and Γ is a homoclinic trajectory of it. Cycles in the sequence $\{L_n\}$ are called k-circuiting if for any two neighborhoods U of the cycle L, and V of the trajectory Γ, there exist a natural number N and a neighborhood W of L such that for all $n > N$, $L_n \subset U \cup V$, and the set $L_n \setminus W$ consists of k connected components.

nonhyperbolic limit cycle: for $\varepsilon = \varepsilon_k^1$ *it is "1-circuiting" with multiplier* $+1$ *and for* $\varepsilon = \varepsilon_k^2$ (ε_k^3) *it is "1-circuiting" ("2-circuiting") with multiplier* -1.

6.9. Vector Fields on a Bifurcation Surface. In Sect. 6.7, systems in $\mathscr{B}_1$ (see Sect. 1.6) correspond to $e = 0$. In condition (4) the class of admissible pairs of natural numbers p_i and p_{i+1} may change for small changes of λ and γ, even for $e = 0$. One might suppose that bifurcations occur even on the bifurcation surface $\mathscr{B}_1$. This is actually the case, but only for vector fields that are not on the boundary of the set of Morse-Smale vector fields. The following theorem holds:

Theorem (Gonchenko (1984, 1980)). *In a neighborhood of a vector field which satisfies the hypotheses of the theorem in Sect. 6.8, but which does not lie on the boundary of the set of Morse-Smale vector fields, there exists a set of vector fields everywhere dense on the bifurcation surface such that each one has:* 1) *a limit cycle of saddle-node type*; 2) *a limit cycle of nonorientable nodal type* (*with multiplier* -1); 3) *an infinite set of stable limit cycles.*

If the bifurcation surface does lie on the boundary of Morse-Smale vector fields in a neighborhood of the point v_0, then the vector fields $\{v_\varepsilon\}$ are distinguished by a modulus (see Sect. 6.3), but are geometrically "the same": only a homoclinic trajectory of a simple tangency is added to the nonwandering set.

6.10. Diffeomorphisms with an Infinite Set of Stable Periodic Trajectories. In a neighborhood of a diffeomorphism of a two-dimensional surface having a homoclinic trajectory of a simple tangency, there exist diffeomorphisms with an infinite set of stable periodic trajectories. More accurately, we have:

Theorem (Newhouse (1980), Robinson (1983)). *Suppose p is a dissipative fixed point, a hyperbolic saddle of a* C^r*-diffeomorphism*

$$f\colon \mathbf{M}^2 \to \mathbf{M}^2, \qquad r \geqslant 2.$$

Suppose W_p^u *and* W_p^s *have a homoclinic trajectory of a simple tangency. Then there exists a diffeomorphism g arbitrarily* C^r*-close to f for which there exists a neighborhood* $\mathscr{U} \subset \mathrm{Diff}^r(\mathbf{M}^2)$ *and a set of second category* $\mathscr{B} \subset \mathscr{U}$ *such that, for any diffeomorphism* $h \in \mathscr{B}$, *there exist infinitely many stable periodic trajectories.*

The modification of this theorem for one-parameter families of diffeomorphisms is stated in (Newhouse & Palis (1976)):

Theorem. *Suppose* $\{f_\varepsilon\}$ *is a curve of* C^3*-diffeomorphisms on a compact surface* **M** *such that:* 1) *for* $\varepsilon = \varepsilon_0$, f_ε *has a dissipative fixed point p, a saddle, and a homoclinic trajectory of a simple tangency of* W_p^u *and* W_p^s; 2) $\{f_\varepsilon\}$ *intersects* $\mathscr{B}_1$ *transversally at* f_0. *Then there exist values of* $\varepsilon > \varepsilon_0$ *for which* f_ε *has infinitely many stable periodic trajectories.*

Example A The Josephson junction model is

$$\dot{\varphi} = y, \qquad \dot{y} = \rho - \sin\varphi - (1/\sqrt{\beta})(1 + \varepsilon\cos\varphi)y + \alpha\sin\Omega t,$$

where ρ is a dimensionless current and y is a dimensionless voltage. At some (physical) values of the parameters, as is shown by Belykh, Pedersen and Sorensen (1977), this model has a homoclinic trajectory of a simple tangency for a dissipative saddle (fixed point) of the return map of the plane: from $t = 0$ to $t = 2\pi/\Omega$. At approximately these values of the parameters the appearance of nonreproducible voltage-current characteristics was demonstrated experimentally: under the same experimental conditions different voltage-current characteristics were obtained. In the given model one may consider the dependence of $\langle y \rangle$ on ρ to represent the voltage-current characteristics, where

$$\langle y \rangle = \lim_{t\to\infty} \left(\frac{1}{t} \int_0^t y(s)\, ds \right).$$

Since $\dot{\varphi} = y$, we have

$$\langle y \rangle = \lim_{t\to\infty} \frac{\varphi(t) - \varphi(0)}{t}.$$

that is, the "phase rotation number."

It is natural to suppose that the nonreproducibility of voltage-current characteristics is explained by the presence of an infinite limit set (and, in particular, of a countable set of stable limit cycles with different domains of existence with respect to the parameter ρ), containing trajectories with different "phase rotation numbers."

§ 7. Infinite Nonwandering Sets

Here we describe the component of the boundary of the set of Morse-Smale systems that consists of flows with an infinite set of nonwandering trajectories. In all the examples given below, typical points of the boundary are inaccessible. It is unknown if this is true in the general case. In particular, it is unknown if it is true that in a generic one-parameter family of vector fields, the generation of an infinite nonwandering set is preceded by one of the bifurcations described in previous sections (the appearance of a nonhyperbolic singular point or cycle, or of trajectories of a simple tangency, or a nontransversal intersection of the stable and unstable manifolds of the singular point and/or a cycle).

7.1. Vector Fields on the Two-Dimensional Torus. The class of Morse-Smale systems on the two-dimensional torus $\mathbb{T}^2$, as well as on any two-dimensional surface (see Sect. 2), coincides with the class of structurally stable (and rough) systems. Therefore any non-structurally stable system on $\mathbb{T}^2$ lies on the boundary of the set of Morse-Smale systems.

If, for some system on $\mathbb{T}^2$, there exists a global section, a compact transversal to all trajectories of the system, then it is possible to introduce the Poincaré rotation number, irrational values of which correspond to the presence of a nonclosed, Poisson-stable trajectory. By Birkhoff's theorem (see, for example, Nemytskij and Stepanov (1949)) the closure of a nonclosed, Poisson-stable trajectory contains a continuum of nonclosed, Poisson-stable trajectories, each of which is everywhere dense in this closure. In this way, we see that if a system has irrational rotation number, then its nonwandering set contains an infinite set of trajectories.

For any one-parameter family of C^r-smooth ($r \geqslant 1$) vector fields on $\mathbb{T}^2$, continuously depending on the parameter and having a global section for each of its values, the rotation number continuously depends on the parameter. If it changes, then it must take on irrational values. Consequently, systems with infinite nonwandering sets having different rotation numbers must arise in one-parameter families of vector fields which have different rotation numbers, at least for two values of the parameter.

We assume now that for some vector field on $\mathbb{T}^2$ the rotation number is rational. If the vector field is generic, then on $\mathbb{T}^2$ there are an even number of limit cycles, half of which are stable, half unstable. The rotation number may change only after the disappearance of these cycles. Their disappearance is connected with the passage of multipliers through $+1$. Thus, a vector field with an infinite nonwandering set (and with a global section) is a limit of vector fields with cycles having multiplier $+1$. In exactly the same way, the set of vector fields with a given irrational rotation number is a limit for bifurcation surfaces, corresponding to the cycles with multiplier $+1$. As in Arnol'd (1961), it follows that for almost all rotation numbers (in the sense of Lebesgue measure) this set is a smooth submanifold of a Banach space. In the general case the question is open.

For a general two-parameter family of vector fields in which a two-dimensional torus is generated from a cycle with multiplier $e^{\pm i\varphi}$, $\varphi \neq 0, \pi, 2\pi/3, \pi/2$, one can prove that the bifurcation curve corresponding to the vector fields in this family, with some fixed irrational rotation number, is homeomorphic and, for almost all rotation numbers (as follows from Arnol'd (1961)), diffeomorphic to an interval. Whether or not this curve loses its smoothness for some (irrational) rotation numbers is unknown.

7.2. Bifurcations of Systems with Two Homoclinic Curves of a Saddle. For simplicity we describe flows in $\mathbb{R}^3$ (analogous results are true for flows in $\mathbb{R}^n$ having a saddle with a one-dimensional unstable manifold). Let v be a vector field having a saddle at the origin O. Suppose λ_1, λ_2, λ_3 are the roots of the characteristic equation at O, and moreover, $\lambda_3 > 0$ and $\lambda_2 < \lambda_1 < 0$. In this case, the dimension of the unstable (stable) manifold is equal to 1 (2). In the stable manifold there exists a one-dimensional nonleading submanifold W_O^{ss} tangent at O to the eigenvector corresponding to λ_2. W_O^{ss} divides W_O^s into two parts. We assume that $W_O^u \subset W_O^s$ and, moreover, that both components of $W_O^u \backslash O$ lie in the same component of $W_O^s \backslash W_O^{ss}$ (as one may say, this is the case of a "butterfly" and not a "figure eight"). We also assume that any cycle which may be generated from each homoclinic curve has positive multipliers (see Sect. 5). Finally, we assume that the saddle number $\sigma = \lambda_1 + \lambda_3$ is negative (see Fig. 55).

Obviously the vector field v may lie on the boundary of the set of Morse-Smale systems (in a closed sphere of large radius in $\mathbb{R}^3$).

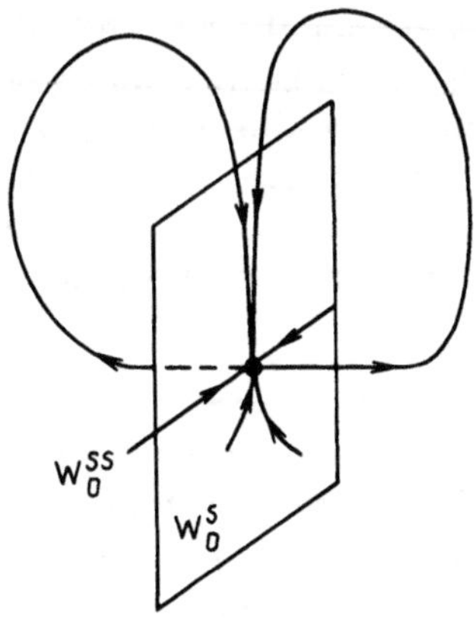

Fig. 55. A vector field with two homoclinic trajectories of a saddle of butterfly type

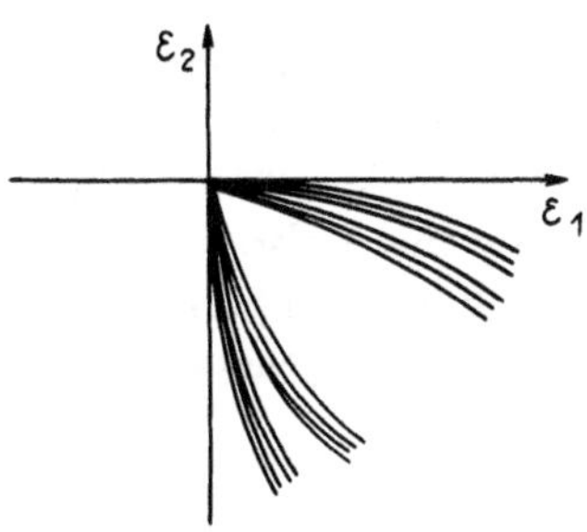

Fig. 56. The bifurcation diagram of a two-parameter family of vector fields having a set of bifurcation curves of cardinality that of the continuum

Theorem. *In an arbitrarily small neighborhood of a vector field v (in the space of C^2-smooth vector fields on $\mathbb{R}^3$) there exist vector fields having nontrivial (that is, different from singular points and limit cycles) Poisson-stable trajectories.*

We fix a neighborhood $\mathcal{N}$ of the point v, and denote the set of vector fields in $\mathcal{N}$ having infinite nonwandering sets (nontrivial Poisson-stable trajectories) by $\mathcal{B}_J$.

Theorem. *If v lies on the boundary of the set of Morse-Smale vector fields, then for any sufficiently small neighborhood of v, the set $\mathcal{B}_J$ also belongs to its boundary and lies in the closure of the set of bifurcation surfaces corresponding to homoclinic curves of a saddle. For a generic two-parameter family passing through v, the bifurcation set is locally homeomorphic (everywhere except on a finite set of curves) to the product of a Cantor set and an interval; moreover, systems with a homoclinic curve of a saddle correspond to boundary points[27] of a Cantor set, and systems with nontrivial Poisson-stable trajectories correspond to interior points* (see Fig. 56).

[27] The boundary points of a Cantor set are the ends of the deleted intervals; the rest are interior points.

It is not known whether or not the bifurcation surfaces corresponding to systems with nontrivial Poisson-stable trajectories, or even those corresponding to bifurcation curves in general two-dimensional systems, are smooth.

We remark that bifurcation "surfaces", corresponding to the presence of an infinite set of nonwandering trajectories, are inaccessible at all points except for v.

The results in this subsection were originally obtained by D.V. Turaev and L.P. Shil'nikov (1985) [9*].

7.3 Systems with Feigenbaum Attractors. It is known that an infinite sequence of period-doubling bifurcations (the so-called Feigenbaum scenario) may lead to the generation of nontrivial Poisson-stable trajectories (see Sect. 6 of Chapt. 2). Moreover, for families of smooth mappings of the interval, this scenario is structurally stable. One can prove that if a family of smooth mappings of "parabolic type" undergoes an infinite sequence of period-doubling bifurcations leading to the generation of a set of nontrivial Poisson-stable trajectories (or a Feigenbaum attractor), then this occurs for any C^2-perturbation of this family. Moreover, up to the moment of generation of a Feigenbaum attractor, the nonwandering set is finite: an endomorphism of the interval with Feigenbaum attractors lies on the boundary of the set of "Morse-Smale endomorphisms".

It is unknown whether or not this is also true for diffeomorphisims of the disk. It is possible that before an infinite set of period-doublings occurs, an infinite set of nonwandering trajectories may be generated because of the tangency of the invariant manifolds of saddle points.

7.4. Birth of Nonwandering Sets. In this and the following three subsections the possible birth of invariant many-dimensional tori "from condensing trajectories" is considered.

Theorem. *In the space of smooth vector fields on a domain in* $\mathbb{R}^n$ *(and on any n-dimensional manifold), there exists, for any* $m < n - 1$, *an open set* $\mathbb{A}$ *(in the* C^1*-topology), the vector fields of which have an invariant m-dimensional torus, and the boundary of* $\mathbb{A}$ *has a nonempty intersection with the boundary of the set of Morse-Smale systems. Moreover, there exists a smooth (highly degenerate) one-parameter family d (for "degenerate"), whose vector fields are Morse-Smale systems for subcritical parameter values, and lie in* $\mathbb{A}$ *for supercritical values.*

Corollary. We consider an arbitrary deformation of the family d, that is, a two-parameter family $\{v\}$ of equations with parameters ε and μ which coincide with d for $\mu = 0$. Then to each small nonzero value of μ, there corresponds a one-parameter family $\{v_\mu\}$ (with parameter ε) and values $\varepsilon^+(\mu)$ and $\varepsilon^-(\mu)$ such that: for $\varepsilon < \varepsilon^-(\mu)$ all equations of the family $\{v_\mu\}$ define Morse-Smale systems, for $\varepsilon > \varepsilon^+(\mu)$ all equations of the family $\{v_\mu\}$ have an invariant torus, and both $\varepsilon^\pm(\mu) \to 0$ as $\mu \to 0$ (see Fig. 57).

Remark. Phenomena taking place on the interval $(\varepsilon^-(\mu), \varepsilon^+(\mu))$ are, for $m > 2$, completely uninvestigated; for $m = 2$ significant results are contained in work of Chenciner (see Sect. 2.3 of Chapter 2). However, in this case, as far as we know,

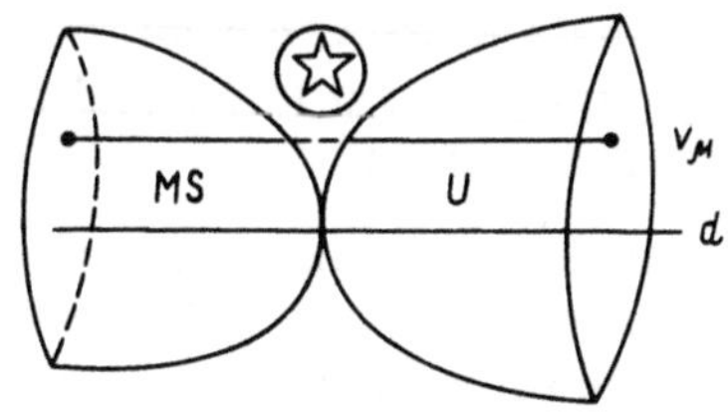

Fig. 57. The left-hand part ($\varepsilon < \varepsilon^-(\mu)$) of the family v_μ consists of Morse-Smale vector fields; the right-hand part ($\varepsilon > \varepsilon^-(\mu)$) consists of vector fields having an invariant torus. The star denotes an uninvestigated interval on which a bifurcation takes place.

no answer to the following question has been found: what happens in a generic family at the first bifurcation that takes it out of the set of Morse-Smale systems?

Apparently, a smooth attracting torus existing in the family $\{v_\mu\}$ for $\varepsilon > \varepsilon^+(\mu)$ loses its smoothness and, before it disappears, becomes a strange attractor.

7.5. Persistence and Smoothness of Invariant Manifolds (following Fenichel (1971)). Roughly speaking, the theorem formulated below asserts that an attracting invariant manifold persists under small perturbations if the speed of approach of trajectories to the manifold from its exterior is greater than the speed of contraction of trajectories on the manifold. The numbers that characterize these speeds are called Lyapunov exponents, and they are defined as follows.

Definition 1. A manifold $\mathbf{M}$ with boundary is called *negatively invariant for a vector field* v if v is tangent to $\mathbf{M}$ at interior points of $\mathbf{M}$ and, at the boundary of $\mathbf{M}$, v is also tangent to $\mathbf{M}$ and is directed to the outside.

Definition 2. A manifold with boundary that is negatively invariant for a field v is said to be *attracting* if there exist a neighborhood of the manifold $\mathbf{M}$, a non-negative function ρ in this neighborhood, and a positive t such that:

$$\rho(x) = 0 \Leftrightarrow x \in g_v^t\mathbf{M}; \qquad L_v\rho < 0 \text{ outside of } g_v^t\mathbf{M}.$$

Suppose that $\mathbf{TM}$ and $\mathbf{N}$ are the tangent and normal bundles to $\mathbf{M}$, respectively, and that $\mathbf{T}$ is the restriction to $\mathbf{M}$ of the tangent bundle to phase space; let $p\colon \mathbf{T} \to \mathbf{N}$ be the projection operator along $\mathbf{TM}$.

Definition 3. The *exponent of attraction* of a manifold $\mathbf{M}$, attracting and negatively invariant for a vector field v, is the number

$$\lambda_N = \sup_{\substack{x \in \mathbf{M} \\ \xi \in \mathbf{T}}} \overline{\lim_{t \to \infty}} \frac{-\ln \|pg_*^t(g^{-t}x)\xi\|}{|t|}.$$

Definition 4. The *exponent of contraction of trajectories on* $\mathbf{M}$ is the number

$$\lambda_T = \sup_{\substack{x \in \mathbf{M} \\ \xi \in \mathbf{TM}\setminus\{0\}}} \overline{\lim_{t \to \infty}} \frac{\ln \|g_*^{-t}(x)\xi\|}{|t|}.$$

The notation emphasizes that the first exponent characterizes the contraction that takes place in the directions normal to **M**, and the that second exponent does the same for the tangent directions.

Example. Consider a hyperbolic singular point of a vector field on $\mathbb{R}^n$ with stable manifold W^s and unstable manifold W^u. The intersection **M** of the unstable manifold W^u with some neighborhood of the singular point of the vector field is negatively invariant for this field. Suppose λ_i are the eigenvalues of the singular point with $\operatorname{Re} \lambda_i < 0$, and μ_j are those with $\operatorname{Re} \mu_j > 0$. Then the exponents of attraction and contraction (of and on **M**, respectively) have the form

$$\lambda_N = -\max_j \operatorname{Re} \lambda_j \qquad \text{and} \qquad \lambda_T = -\min_j \operatorname{Re} \mu_j$$

Theorem. *Suppose v is a smooth vector field,* **M** *is a negatively invariant manifold for v, with boundary, $\lambda_{\mathbf{N}}$ and $\lambda_{\mathbf{T}}$ are the corresponding exponents, and the natural number r satisfies the condition $r\lambda_{\mathbf{T}} < \lambda_{\mathbf{N}}$. Then any vector field C^r-close to v has a C^r-smooth negatively invariant manifold close to* **M**.

Remark. The mechanism of loss of smoothness is the same as that described in Sect. 5.2 of Chapter 2.

7.6. The Degenerate Family and Its Neighborhood in Function Space. Here the theorem stated in Sect. 7.4 is proved. We begin by constructing an auxiliary family of vector fields in the product $\mathbf{I} \times \mathbf{D}$ of the interval $|t| \leqslant 2$ by the $(n-1)$-dimensional ball $\mathbf{D}$: $\|x\| < 1$, $x \in \mathbb{R}^{n-1}$. Consider a smooth vector field v in $\mathbf{D}$, equal to zero in some neighborhood of the boundary of $\mathbf{D}$, having a smooth invariant $(n-2)$-dimensional torus with a positive exponent of attraction, and such that on the torus the field v is diffeomorphic to a constant field that defines a quasi-periodic winding on the torus. From this, it follows that the exponent of contraction of trajectories of the field on the torus is equal to zero and that all the trajectories on the torus are nonwandering.

Let V denote the vector field on the product $\mathbf{I} \times \mathbf{D}$ that is tangent to the vertical fibers: $V = (v, 0)$. Suppose φ_- and φ_+ are two smooth functions on $\mathbf{I}$ such that (a) φ_- (respectively φ_+) is equal to 1 in some neighborhood of the point -1 (respectively, $+1$); (b) the supports of the functions φ_- and φ_+ are disjoint intervals that lie strictly inside of $\mathbf{I}$; and (c) $\varphi_+(-x) = \varphi_-(x)$. Suppose ψ_ε is a smooth family of functions on $\mathbf{I}$ with base $|\varepsilon| \leqslant 1/5$ such that: (α) $\psi_\varepsilon(t) = (t \pm 1)^2 - \varepsilon$ for $|t \pm 1| \leqslant \frac{1}{2}$; and ($\beta$) the functions ψ_ε are even, positive outside of $|t \pm 1| < \frac{1}{2}$, and are equal to 1 in some neighborhood of the end points of $\mathbf{I}$.

We consider the family of fields V_ε in $\mathbf{I} \times \mathbf{D}$ defined by

$$V_\varepsilon = \psi_\varepsilon \partial/\partial t + \varphi_+ V - \varphi_- V.$$

The fields of the family $\{V_\varepsilon\}$ possess the following properties: a) they define a reversible system (the field V_ε changes sign under the symmetry $(t, x) \mapsto (-t, x)$); b) in a neighborhood of the boundary of the product $\mathbf{I} \times \mathbf{D}$ the field V_ε coincides with $\partial/\partial t$; c) for $\varepsilon < 0$ the monodromy transformation of the field V_ε is: $\{-2\} \times \mathbf{D} \to \{2\} \times \mathbf{D}$, and this transformation preserves x because of the symmetry of V_ε. At $\varepsilon = 0$ the field has a semi-stable invariant torus (in fact, two such tori) with a quasi-periodic winding. For $\varepsilon > 0$, the fields V_ε have two repelling tori and two attracting ones with exponents of attraction $\sqrt{\varepsilon}$.

We now construct the degenerate family d on the manifold **M** described in the theorem. For this we take an arbitrary vector field w on **M**, defining a Morse-Smale system, and consider a "flow pencil" **B** of its trajectories (diffeomorphic to the product of a disk and an interval), all phase curves of which pass through it by time 4. We change the field w in this box in the following way. Suppose **H** is a diffeomorphism of $\mathbf{B} \to \mathbf{I} \times \mathbf{D}$ straightening out the field w (transforming it into $\partial/\partial t$), and let $\mathbf{G} = \mathbf{H}^{-1}$. Set

$$v_\varepsilon = \begin{cases} w & \text{outside } \mathbf{B}, \\ \mathbf{G}_* V_\varepsilon & \text{inside } \mathbf{B}. \end{cases}$$

The field V_ε is obviously smooth and depends smoothly upon ε. For $\varepsilon < 0$, the field v_ε gives a Morse-Smale system, since w has this property, and the monodromy transformations of the bottom of the pencil **B** onto its top coincide at the fields V_ε and w. For $\varepsilon \geqslant 0$, the field V_ε has an infinite set of nonwandering trajectories, filling out the four tori. The family d is constructed.

We now investigate a neighborhood of the family d in function space. By virtue of the structural stability of Morse-Smale systems, each of the fields V_ε has a neighborhood consisting of Morse-Smale systems for $\varepsilon < 0$. For $\varepsilon > 0$, each of the fields v_ε has a neighborhood consisting of fields with an invariant $(n - 2)$-dimensional torus. This follows from Fenichel's theorem, since the exponent of attraction to the invariant torus of V_ε is positive for $\varepsilon > 0$, and the exponent of contraction of the trajectories on the torus is equal to zero. The theorem is proved.

7.7. Birth of Tori in a Three-Dimensional Phase Space. We consider a two-parameter family of vector fields on a three-dimensional manifold, in which loss of stability of a limit cycle takes place as a pair of multipliers pass through the imaginary axis, in the case of the degeneracy in the nonlinear terms described in Sect. 2.3 of Chapter 2. If the family is generic, then in a generic one-parameter subfamily (delineated by the solid line in Fig. 22) the following phenomena occur. For all values of the parameter less than some value ε_-, the limit cycle maintains its stability, and for all values greater than some value ε_+, the equations of the family have two invariant tori. The perestroika of the dynamics of the subfamily corresponding to ε varying over the interval $(\varepsilon_-, \varepsilon_+)$ is apparently very complicated (see Sect. 2.3 of Chapter 2), and the first bifurcation occurring on this interval has not yet been described.

§ 8. Attractors and Their Bifurcations

According to a widely disseminated conjecture, the limiting behavior of trajectories of a typical dynamical system on a compact manifold is described as follows. After a finite time each positive semi-trajectory falls into a neighborhood of an attracting set, that is, an attractor. If the attractor is sufficiently massive, that is, if it is different from a finite union of singular points and limit cycles, then the behavior of on phase curves on and close to the attractor is chaotic. There is an analogous conjecture for dissipative systems whose phase spaces are compact manifolds with boundary, on which the vector fields are directed inwards.

These conjectures are unproved. Moreover, a universally adopted definition of an attractor does not exist. The problem of the limiting behavior of trajectories is investigated from two points of view. On one hand, the definition of an attractor is formulated so that every dissipative system (for simplicity, in the remainder of this section we deal only with such systems) has an attractor. Moreover, the attractor should not contain "extra points", but ought to coincide with the "space of stable regimes", which are observed numerically or experimentally. For example, the maximal attractor of a dissipative system, which is the intersection of all shifts of absorbing domains by transformations of the phase flow for positive time (the flow f^t maps **B** into **B** for $t > 0$), may be substantially larger than the "space of stable regimes". In Fig. 58a, a dynamical system with an absorbing annulus, whose maximal attractor is a circle containing two equilibria, one a saddle, the other a node, is shown. The phase curves converging to the saddle come from initial points which form a set of measure zero; almost all (in the sense of Lesbegue measure) phase curves converge to the node. This node should be considered to be the "physical attractor".

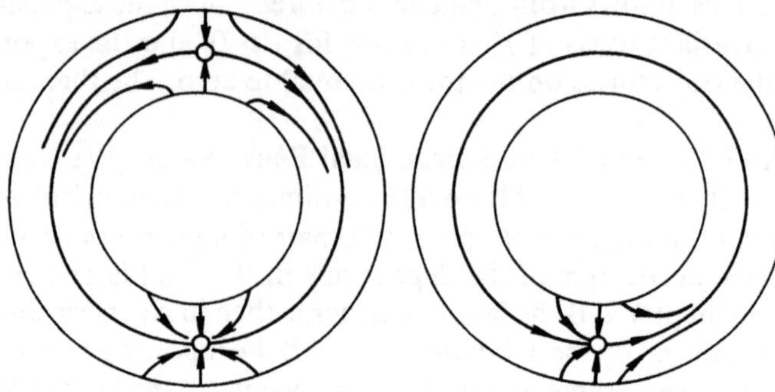

Fig. 58. An absorbing annulus: the maximal attractor in each case is a circle: the likely limit set is a node in case (a), and a saddle-node in case (b).

On the other hand, the definition of an attractor is formulated so as to guarantee the chaotic behavior of trajectories on the attractor (and possibly nearby it). Hyperbolic, stochastic, and other attractors arise from these definitions (Sinai (1979), Sinai and Shil'nikov, eds. (1981), Smale (1967)). However, it is not known whether systems whose trajectories behave chaotically on the attractor are generic in the class of systems whose attractors do not consist of a finite number of points and cycles.

In the remainder of this section we discuss different definitions of attractors, and then describe bifurcations of attractors.

8.1. The Likely Limit Set According to Milnor (1985). Suppose a dissipative system is given on a compact, smooth manifold with boundary. We consider an arbitrary smooth measure on this manifold, that is, a measure having a smooth, positive density, equivalent to Lebesgue measure when restricted to any coordinate neighborhood. The class of measurable sets and sets of measure zero is independent of the choice of a density; this choice is unimportant in what follows.

Definition. The *likely limit set* of a dynamical system is the smallest closed set containing the ω-limit set for almost all points of phase space.

This concept is well defined not only for flows and diffeomorphisms, but for arbitrary smooth mappings as well. The likely limit set is not stable, as is shown in Fig. 58b. In this figure the likely limit set is an equilibrium of saddle-node type.

8.2. Statistical Limit Sets. In numerical experiments limit sets are often photographed. To accomplish this one computes one or several trajectories and plots the values of two functions (for example, two coordinates) at points of these trajectories on the computer screen. On the screen, points flash on and off (more precisely, pixels are turned on and off) as they are computed in time along trajectories. A long-time exposure of the screen is begun a long time after the computation begins. Then those pixels which have been turned on many times during the exposure are seen on the photograph; those pixels that are rarely turned on do not appear in the photograph.

Consider, for example a dynamical system on the sphere with absorbing domain having a maximal attractor in the form of a pair of loops of a hyperbolic saddle (a figure eight; see Fig. 59). In a photograph made in the way just described, one obtains the equilibrium and four intervals of the separatrices (Fig. 59b). The longer the time exposure, the shorter these intervals become, because the relative time spent by the trajectories near the saddle point grows. In this example, the likely limit set is the entire figure eight.

We say that a positive semi-trajectory of a point x under the action of the phase flow $\{g^t\}$ lies in a region $\mathbf{U}$ on average for a positive time if the mean

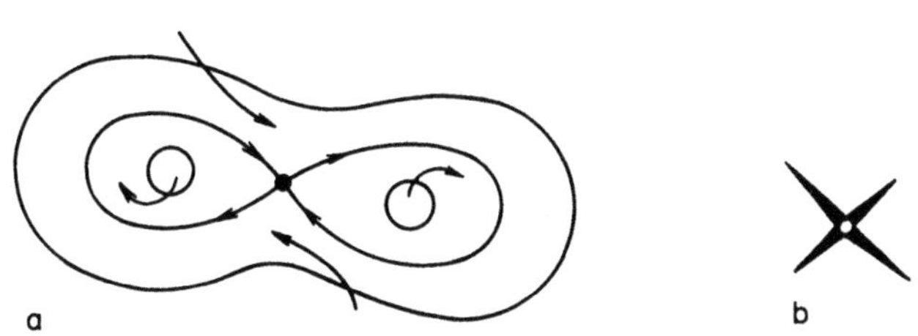

Fig. 59. (a) A vector field with a maximal attractor and a likely limit set of figure-eight type. (b) a statistical limit set that is a saddle

measure of those values of t on the interval $[0, T]$ for which $g^t x \in \mathbf{U}$ has a nonnegative limit superior as $T \to \infty$:

$$\overline{\lim_{T \to \infty}} \frac{1}{T} \int_0^T \chi_{\mathbf{U}} \circ g^t x \, dt > 0.$$

Here $\chi_{\mathbf{U}}$ is the characteristic function of the set $\mathbf{U}$.

Definition (Yu.S. Il'yashenko, 1985 see [2]). Let a dissipative dynamical system be given on a compact manifold with boundary, and let m be a smooth measure with positive density on this manifold. An open set $\mathbf{U}$ is called *essential* if the set of all initial points of positive semi-trajectories that spend on average a positive time in the region $\mathbf{U}$ has positive measure. The *statistical limit set* is the complement of the union of all the nonessential, open subsets of phase space.

In the previous example, the statistical limit set is the saddle point.

Remark. An open set has a nonempty intersection with a statistical limit set if and only if the open set is essential.

Lemma. *The statistical limit set always belongs to the likely limit set.*

◂ Assume the contrary. Suppose some point of the statistical limit set does not belong to the likely limit set. Choose a neighborhood $\mathbf{U}$ of this point, the closure of which does not intersect the likely limit set. This neighborhood is essential; consequently, there exists a set of positive measure in $\mathbf{U}$ such that the positive trajectories issuing from each of its points remain on average for a positive time in $\mathbf{U}$. The ω-limit set of each such point x has a nonempty intersection with the domain $\mathbf{U}$. This contradicts the choice of $\mathbf{U}$. ▸

8.3. Internal Bifurcations and Crises of Attractors. In the remaining part of this section we deal with bifurcations of attractors. Here, by an attractor we mean the maximal attractor in the absorbing domain. We recall a definition. Denote by $\{f^t\}$ the flow generated by a vector field v.

Definition. A domain $\mathbf{B}$ is called *absorbing* if $f^t \overline{\mathbf{B}} \subset \mathbf{B}$ for $t > 0$. The set $\mathbf{A} = \bigcap_{t < \infty} f^t \mathbf{B}$ is called the *maximal attractor* in an absorbing domain $\mathbf{B}$. A set is called an *attractor* if there exists an absorbing domain for which it is the maximal attractor. A set $\mathbf{U}(\mathbf{A})$ is called a *basin* for an attractor $\mathbf{A}$ if it contains all points whose trajectories converge to $\mathbf{A}$ as $t \to \infty$.

The simplest attractors are evidently: a stable equilibrium point, a stable limit cycle, and an attracting two-dimensional torus.

Definition. An attractor is said to be *strange* if it is not a finite union of smooth submanifolds of phase space.

For simplicity, the remainder of this exposition will be carried out in terms of one-parameter families of dynamical systems. Thus, suppose $\{f_\varepsilon^t\}$ is a flow depending on a scalar parameter. We assume that:

1) for $0 \leqslant \varepsilon < \varepsilon^*$ there exists an attractor $\mathbf{A}_\varepsilon$ with absorbing domain $\mathbf{B}_\varepsilon$ such that, for any $\varepsilon' < \varepsilon^*$,

$$\operatorname{int}\left(\operatorname*{lt}_{\varepsilon \to \varepsilon'} \mathbf{B}_\varepsilon\right) = \mathbf{B}_{\varepsilon'};$$

2) ε^* is a bifurcation value of the parameter, and, moreover, $\mathbf{A}_{\varepsilon^*}$ has a nonempty intersection with the bifurcation support (see Sect. 2.8) of the flow $\{f^t_{\varepsilon^*}\}$, where

$$\mathbf{A}_{\varepsilon^*} = \operatorname*{\overline{lt}}_{\varepsilon \to \varepsilon^*} A_\varepsilon.$$

Here $\operatorname{lt} \mathbf{C}_\varepsilon$ and $\overline{\operatorname{lt}}\ \mathbf{C}_\varepsilon$ denote the topological limit and the upper topological limit, respectively, of the family $\{\mathbf{C}_\varepsilon\}$[28].

Definition. A bifurcation value $\varepsilon = \varepsilon^*$ is said to be an *internal bifurcation value* if $\mathbf{A}_{\varepsilon^*}$ is an attractor. Otherwise, it is called a *crisis bifurcation value.* The corresponding bifurcations are called an *internal bifurcation*, and a *crisis* of a family of attractors.

We note that the basin of an attractor may or may not change its topological type when it undergoes an internal bifurcation. For example, for a flow on the disk, upon birth of a stable limit cycle from a focus the basin becomes doubly-connected from simply-connected; but if a point that is a saddle-node arises on a stable limit cycle, the basin is doubly-connected both before and after the bifurcation.

As a result of a crisis, a system transfers to another regime: the point described may move into a "new" domain in phase space.

8.4. Internal Bifurcations and Crises of Equilibria and Cycles. In accordance with the definitions introduced above, in Sect. 8.3, we shall classify bifurcations of cycles as internal and crisis bifurcations. We shall say that the bifurcation is *soft* (*hard*) if the equilibrium or cycle is stable (unstable) at the moment of bifurcation.

Equilibria:

1. A bifurcation of Andronov-Hopf type: in the soft case, this is an internal bifurcation; in the hard case it is a crisis.
2. Formation of a saddle-node and its disappearance: a crisis.

Limit cycles:

1. The merging of a stable cycle with a saddle and their mutual disappearance: a crisis.
2. A period-doubling bifurcation and a bifurcation generating a torus: in the soft case, it is an internal bifurcation; in the hard case it is a crisis.

[28] We recall that a set is called a *topological limit* if it is both the upper and the lower topological limit of a family of sets. A set is called the *upper* (*lower*) *topological limit* of a family of sets $\{\mathbf{C}_\varepsilon\}$ if it is the set of all points p such that for any neighborhood of each p there exists a value of ε arbitrarily close to ε^* such that the intersection of $\mathbf{C}_\varepsilon$ with this neighborhood is nonempty (beginning with some $\varepsilon_0 < \varepsilon^*$, the intersection of $\mathbf{C}_\varepsilon$ with this neighborhood is nonempty for $\varepsilon_0 < \varepsilon < \varepsilon^*$).

3. A limit cycle becomes a homoclinic trajectory of a saddle: a crisis.

4. A limit cycle becomes a homoclinic trajectory of a saddle-node: an internal bifurcation.

5. A blue-sky catastrophe. One can show that in the example of Sect. 2.7 above the upper topological limit of a cycle $L(\varepsilon)$ as $\varepsilon \to \varepsilon^*$ is the whole Klein bottle (ε^* is the bifurcation value corresponding to the disappearance of the cycle); and generally, for any one-parameter family, without an equilibrium on $\mathbb{K}^2$, in which this bifurcation takes place, one may show that the bifurcation is internal. Whether this is so in the general case, or even for systems on a two-dimensional manifold, other than a Klein bottle, is unknown.

Remark. For gradient systems a crisis can be treated as a *catastrophe*.

8.5. Bifurcations of the Two-Dimensional Torus. We assume that a flow $\{f_\varepsilon^t\}$, say, for $0 \leqslant \varepsilon \leqslant \varepsilon^*$, is a Morse-Smale system and has an attracting two-dimensional invariant torus $\mathbb{T}_\varepsilon$. We further assume that, for $0 \leqslant \varepsilon \leqslant \varepsilon^*$, there exists a global section on $\mathbb{T}_\varepsilon$. In this case the rotation number is rational: on $\mathbb{T}_\varepsilon$ there exist an even number of limit cycles, half of which are stable and half of which are unstable (cycles of saddle type in the whole phase space), and $\mathbb{T}_\varepsilon$ is the closure of the union of the two-dimensional unstable manifolds of these cycles. We also suppose that ε^* is a bifurcation value of the parameter and, for $\varepsilon = \varepsilon^*$, one of those bifurcations of codimension 1 considered above takes place. Consequently, this is either a bifurcation of one of the limit cycles lying on $\mathbb{T}_\varepsilon$ for $\varepsilon < \varepsilon^*$, or it is a bifurcation connected with the formation of homoclinic and heteroclinic trajectories on the unstable manifold of one of the saddle cycles.

Internal Bifurcations of a Two-Dimensional Torus (see Afrajmovich and Shil'nikov (1983), Aronson *et al.* (1982) and Sect. 1.6 of Chapter 2).

1. A change of stability of a stable limit cycle on the torus is a period-doubling or a birth of a torus: In this case there exists a value $\varepsilon_1 < \varepsilon^*$ at which the multipliers of the cycle become complex. For $\varepsilon > \varepsilon_1$, $\mathbb{T}_\varepsilon$ is not smooth, the unstable manifold of a saddle cycle winds around a stable cycle, and does not smoothly join with it.

2. The merging of stable and saddle cycles lying on the torus, and the formation of a cycle with multiplier 1, which may be s-critical as well as noncritical: In the first case, if all trajectories on the unstable set are homoclinic, then for $\varepsilon > \varepsilon^*$, a strange attractor may arise (see Sect. 4). If for $0 < \varepsilon < \varepsilon^*$, more than two cycles lie on $\mathbb{T}_\varepsilon$, then for $\varepsilon > \varepsilon^*$, as before, there exists a torus on which there are two fewer cycles.

3. A tangency of the unstable manifold of a saddle cycle with the stable manifold of the same, or of another, saddle cycle: In the first case a homoclinic trajectory, and for $\varepsilon > \varepsilon^*$ a nontrivial hyperbolic set arises, and in the second a heteroclinic trajectory arises, and for $\varepsilon > \varepsilon^*$ the attractor is already not a torus.

Example. The mapping

$$(x, \theta) \to (e^{-r}(x + a \sin \theta), (\theta + r + x + a \sin \theta) \pmod{2\pi})$$

of the annulus $0 < x \leqslant x_0$, where $r \gg 1$, $a \geqslant 0$, $x_0 > a/(1 - e^{-r})$, models the return map in a neighborhood of a "destroyed" torus. It is easy to verify that for any $k \in \mathbb{Z}$ on the plane of parameters (r, a), the curve

$$\mathbf{B}^+ : a = \pm(2\pi k - r)(1 - e^{-r})$$

is a bifurcation curve corresponding to the presence of cycles (fixed points) with multiplier $+1$, and the curve

$$\mathbf{B}^- : a^2 = (2\pi k - r)^2(1 - e^{-r})^2 + 4(1 + e^{-r})^2$$

is a bifurcation curve corresponding to the presence of cycles (fixed points) with multiplier -1. It may also be shown that there exist two bifurcation curves $\mathbf{B}_{1,2}$ corresponding to the tangency of the unstable manifold of a fixed point of saddle type and its stable manifold "from different sides". The corresponding bifurcation diagram is shown in Fig. 60.

Crises:

1. Hard loss of stability of a limit cycle on the torus: for $\varepsilon \to \varepsilon^*$ a saddle cycle of doubled period, or an unstable torus lying on the boundary of the basin of $\mathbb{T}_\varepsilon$

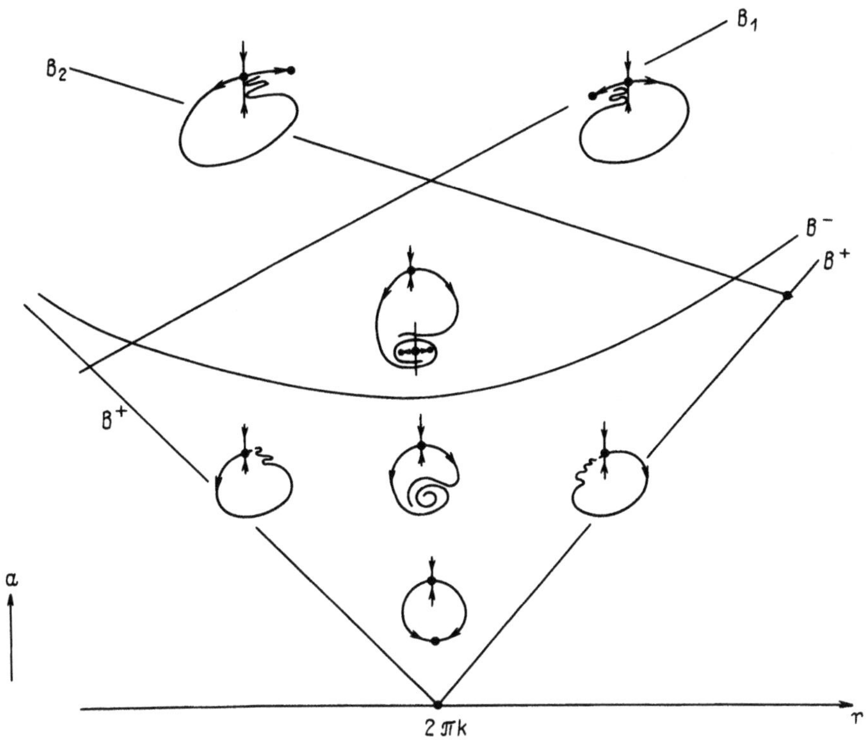

Fig. 60. The bifurcation diagram of a two-parameter family of diffeomorphisms of an annulus. To its different parts there correspond different mechanisms of loss of smoothness, and destruction of a closed invariant curve, as is also shown in this figure.

for $\varepsilon < \varepsilon^*$, is drawn to a stable cycle lying on the torus, and for $\varepsilon = \varepsilon^*$ this unstable torus transfers its instability to the limit cycle.

2. Bifurcation of a stable cycle on the torus during which a saddle cycle, lying on the boundary of the basin of the torus, is attracted to the stable cycle, merges with it and disappears. At this last moment its multiplier becomes $(+1)$.

3. Bifurcation of a saddle cycle on the torus during which, for subcritical (less than critical) values of the parameter, stable cycles do not lose their stability. For $\varepsilon \to \varepsilon^*$ either another cycle or a torus is attracted to the saddle cycle and merges with it for $\varepsilon = \varepsilon^*$. (The multiplier at this last moment may be anywhere on the unit circle: $+1$, or -1, or $e^{i\varphi}$.)

4. A tangency of the unstable manifold of a cycle on the torus with the stable manifold of codimension 1 of an equilibrium or a cycle lying on the boundary of the basin of $\mathbb{T}_\varepsilon$ for $\varepsilon < \varepsilon^*$.

Each of these four cases can obviously be realized in one-parameter families of dynamical systems. We introduce an example illustrating Case 3, see Fig. 61.

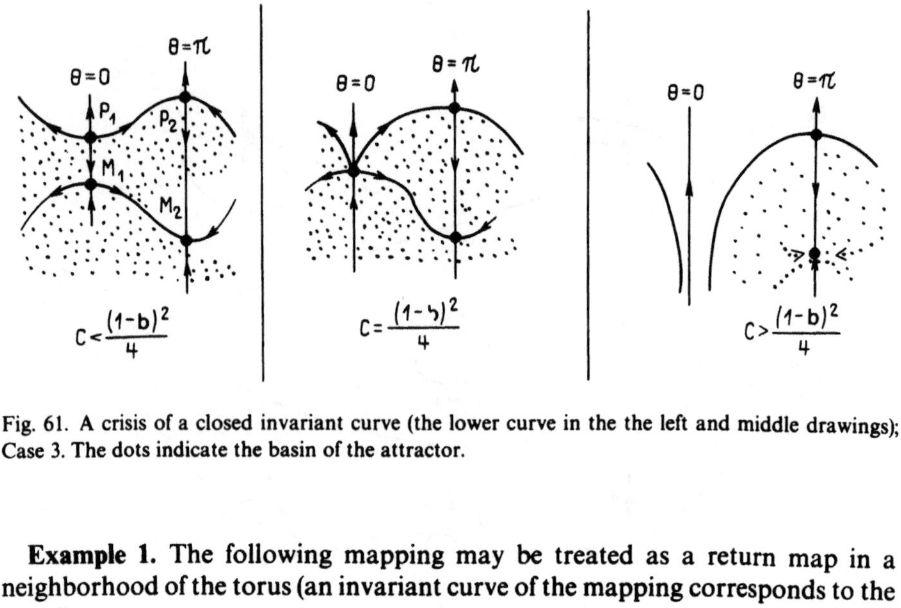

Fig. 61. A crisis of a closed invariant curve (the lower curve in the the left and middle drawings); Case 3. The dots indicate the basin of the attractor.

Example 1. The following mapping may be treated as a return map in a neighborhood of the torus (an invariant curve of the mapping corresponds to the torus):

$$(x, \theta) \to (bx + x^2 + c\cos\theta, (\theta + a\sin\theta) \pmod{2\pi}), \qquad (a > 0, b > 0, c > 0).$$

It is not difficult to verify that for $c < (b-1)^2/4$ this mapping has an invariant closed curve on which there are two fixed points

$$\mathbf{M}_1 = (\tfrac{1}{2}(1 - b - \sqrt{\mathbf{D}^-}), 0), \qquad \mathbf{M}_2 = (\tfrac{1}{2}(1 - b - \sqrt{\mathbf{D}^+}), \pi),$$

where $\mathbf{D}^\pm = (1-b)^2 \pm 4c$. Moreover, $\mathbf{M}_1$ is a saddle, and $\mathbf{M}_2$ is a stable node. In the boundary of the basin of this curve there are two fixed points

$$\mathbf{P}_1 = (\tfrac{1}{2}(1 - b + \sqrt{\mathbf{D}^-}), 0), \qquad \mathbf{P}_2 = (\tfrac{1}{2}(1 - b + \sqrt{\mathbf{D}^+}), \pi);$$

moreover, $\mathbf{P}_1$ is an unstable node, and $\mathbf{P}_2$ is a saddle. For $c = (1 - b)^2/4$ the points $\mathbf{P}_1$ and $\mathbf{M}_1$ merge, and for $c > (1 - b)^2/4$, $\mathbf{P}_1$ and $\mathbf{M}_1$ disappear. The invariant curve also disappears; see Fig. 61.

One can show that a C^2-perturbation of the given family of mappings (in which the parameters a and b are fixed, and c changes) undergoes a similar bifurcation, that is, the family is structurally stable in a neighborhood of the point $c = (1 - b)^2/4$ of parameter space and in some ring in phase space.

Example 2. An analogous example for Case 2 is given by the mapping

$$(x, \theta) \to (bx + x^2 + c \sin \theta, (\theta + a \cos \theta)(\text{mod } 2\pi)), \qquad (a > 0, b > 0, c > 0).$$

It is easy to show that for $c < (1 - b)^2/4$ there exists an invariant curve (it is the closure of the unstable manifold of the fixed point

$$(\tfrac{1}{2}(1 - b - \sqrt{\mathbf{D}^+}), 3\pi/2)),$$

on which the stable fixed point $\mathbf{P} = (\tfrac{1}{2}(1 - b - \sqrt{\mathbf{D}^-}), \pi/2)$ lies). For $c \to (1 - b)^2/4$ the saddle point $\mathbf{Q} = (\tfrac{1}{2}(1 - b + \sqrt{\mathbf{D}^-}), \pi/2)$ is attracted towards $\mathbf{P}$; for $c = (1 - b)^2/4$, $\mathbf{P}$ and $\mathbf{Q}$ merge; and for $c > (1 - b)^2/4$, $\mathbf{P}$ and $\mathbf{Q}$ disappear; see Fig. 62.

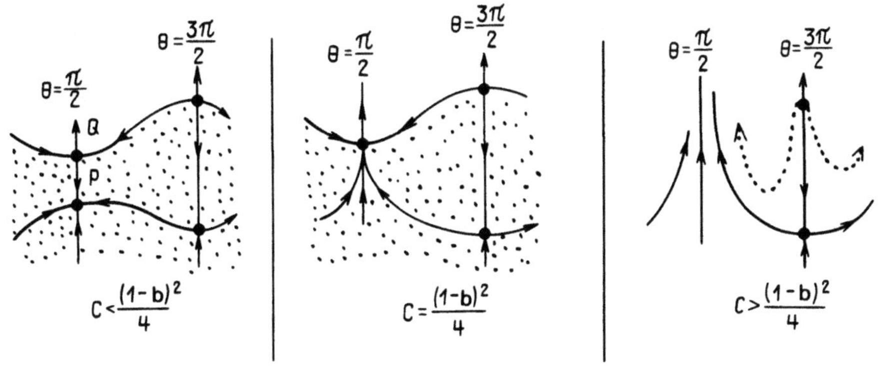

Fig. 62. A crisis of a closed invariant curve; Case 2

Remark. Bifurcations of strange attractors also can be classified into internal bifurcations and crises (see, for example, Afrajmovich (1984) and Grebogi, Ott and Yorke (1983)). However, these bifurcations occur in the class of systems which have an infinite set of cycles, and their description is outside the scope of this survey.

Chapter 4
Relaxation Oscillations

In bifurcation theory one usually considers systems with parameters whose values are independent of time. However, in applications, situations are often encountered where the parameters slowly evolve over time. In this situation new phenomena may arise. For example, a stable equilibrium may, as a parameter changes, disappear or become unstable; and then the state of the system must change rapidly (compared with the rate of change of the parameter) to a new state of motion (an attractor).

For systems with slowly varying parameters there are characteristically two scales of time and two speeds: fast motions defined approximately by a "frozen" system, in which the values of parameters (slow variables) are fixed; the evolution of the parameters with time is termed a "slow motion" (the character of which may, however, depend on the state of the fast motion).

Among systems with fast and slow motions, those in which the fast motion leads to a stable state of equilibrium are distinguished. A system with one fast variable (that is, with a one-dimensional phase space of fast motion) may serve as an example. For fixed values of the slow parameters, such a generic system quickly passes to a stable equilibrium state. The process that quickly establishes this equilibrium is called relaxation. As the slow variables change, the stable equilibrium may (after a long interval of time in the time scale of fast motion) disappear or lose stability. Then again relaxation occurs (a jump to a new equilibrium state), etc. The process that arises, consisting of periods during which the fast system is in a state of quasi-equilibrium (relaxed) and of almost instantaneous (in comparison with the length of these periods) jumps from one state of equilibrium of the fast system to another, is called a process of relaxation oscillations. This term was introduced by van der Pol (1926).

Relaxation oscillations may be periodic. However, more complex behavior is possible (for example, stochastic relaxation oscillations, when the sequence of intervals between consecutive jumps appears to be a random sequence).

In the study of relaxation oscillations, the greatest interest is usually in the asymptotics of the slow motion as a small parameter ε converges to zero. This small parameter ε defines a relationship between the speeds of the fast and slow motions. Moreover, it is important to investigate the behavior of the system over a time during which the slow variables change substantially (for periodic relaxation oscillations this time is several periods in length). If one chooses the time scale so that the characteristic time of fast motion is of order 1 ("fast time"), then one must study the slow motion for a time at least of order $1/\varepsilon$ in order to observe changes in the slow variables.

If we go to "slow time", $\tau = \varepsilon t$, then in the limit as $\varepsilon \to 0$, the relaxation takes place instantaneously and the motion of the system is described as being piecewise continuous. Usually in applications, it is most important to describe the limiting finite discontinuities. In the intervals between jumps the system is described by differential equations having a phase space of lower dimension than the original one (this lower dimension is the number of slow variables), however the jumps at the moments of discontinuity are defined by the fast motion.

The description of relaxation oscillations is widely applied in natural science and in engineering (models of structures with braking in mechanics, amplifiers in electronics, the Belousov-Zhabotinskij reaction in chemistry, the functioning of nerve cells in biology, etc.).

The first two sections of this chapter were written by V.I. Arnol'd, Sect. 3 (except for Sect. 3.7) by Yu.S. Il'yashenko, Sect. 3.7 by N.Kh. Rozov, Sect. 4 by A.I. Nejshtadt, and Sect. 5 by A.K. Zvonkin.

§ 1. Fundamental Concepts

We begin with an example.

1.1. An Example: van der Pol's Equation. The simplest example of a system performing relaxation oscillations is the van der Pol oscillator

$$\dot{x} = y - x^3 + x, \qquad \dot{y} = -\varepsilon x.$$

For $\varepsilon = 0$, the slow variable y becomes a parameter. The fast variable x then relaxes to a stable state of equilibrium, see Fig. 63.

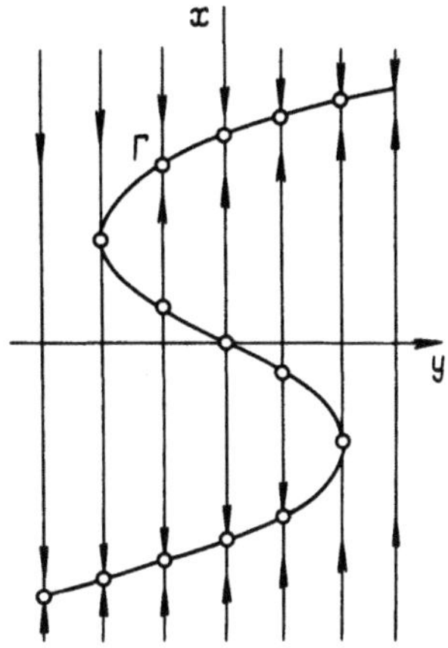

Fig. 63. The van der Pol system for $\varepsilon = 0$

If $\varepsilon > 0$, but is small, then the relaxed system slowly moves along the curve of equilibrium states of the fast motion (denoted by Γ in Fig. 64), to the left on the upper part of Γ and to the right along its lower part. At points where the tangent to Γ is vertical, the system makes jumps. As a result, a limit cycle is formed; see Fig. 64.

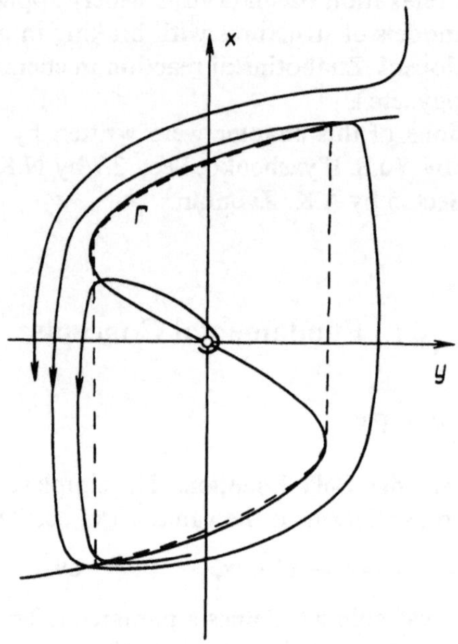

Fig. 64. The van der Pol system for small $\varepsilon > 0$

1.2. Fast and Slow Motions. Consider a system of ordinary differential equations depending upon parameters. In other words, suppose a smooth bundle $E \to B$ and a vertical (tangent to the fibers) vector field on E are given.

Definition. The equations given by the vertical field are called the *unperturbed system* or the *fast equations*.

For the van der Pol system, E is the phase plane, and B is the y-axis. The fast equations are

$$\dot{x} = y - x^3 + x, \qquad \dot{y} = 0.$$

In the general case, (local) coordinates on E can be chosen so that the unperturbed system of fast motion takes the form

$$\dot{x} = f(x, y), \qquad \dot{y} = 0 \tag{1}$$

($y \in \mathbb{R}^m$ are coordinates on the base, and $x \in \mathbb{R}^l$ are coordinates along the fibers). The x variables are the fast variables; the y variables are the slow variables.

Definition. A one-parameter deformation of the fast system is called the *perturbed system* or the *equations of fast and slow motions*.

Such a deformation is given by a one-parameter family of fields containing the original vertical field at the zero value of the parameter. In coordinate form the perturbed system is[29]

$$\dot{x} = F(x, y, \varepsilon), \qquad \dot{y} = \varepsilon G(x, y, \varepsilon), \tag{2}$$
$$F(x, y, 0) = f(x, y).$$

1.3. The Slow Surface and Slow Equations

Definition. The set of singular points of the fast equations is called the *slow surface*.

For the van der Pol system the slow surface is the cubic parabola Γ. For a generic vertical field the slow surface is a smooth manifold. The dimension of this manifold is equal to the dimension of the base of the bundle (the number of slow variables). At generic points the slow surface is locally a section of the bundle; that is, it may be projected diffeomorphically onto the base.

However, this projection is not globally diffeomorphic in general. For example, the cubic parabola in the van der Pol system has two points with a vertical tangent.

Consider the points in neighborhoods of which the slow surface may be projected diffeomorphically. By the Implicit Function Theorem, at these points all the eigenvalues of the linearization of the fast system, on a fixed fiber (that is, for fixed values of the slow variables), are nonzero. Such points are called *regular* points.

At regular points on the slow surface a vector field is naturally defined, the field of slow speeds. It is defined by projection of the perturbations of the original vertical field onto the tangent plane of the slow surface along the fibers of the bundle.

Definition. At a regular point of the slow surface the derivative with respect to ε at $\varepsilon = 0$ of the projection of a vector of the perturbed field onto the tangent plane of the slow surface along a fiber of the bundle is called the *slow speed vector*.

In this way, the slow surface is furnished with the slow speed vector field at regular points. This field provides a *slow system* on the slow surface. In the coordinates introduced above, this slow system takes the form

[29] Actually, relaxation oscillations take place in all systems that are close to the original unperturbed system. Consequently, one should simply study a neighborhood of the unperturbed field in a suitable function space. However, here and in other problems of perturbation theory, for the sake of mathematical convenience, in the statements of the results of an investigation such as an asymptotic result, we introduce (more or less artificially) a small parameter ε and, instead of neighborhoods, we consider one-parameter deformations of the unperturbed systems. The situation here is as with variational concepts: the directional derivative (Gateaux differential) historically preceeded the derivative of a mapping (the Fréchet differential).

$$f(x, y) = 0, \qquad dy/d\tau = g(x, y),$$
$$g(x, y) = G(x, y, 0). \tag{3}$$

Example. The slow system for the van der Pol system is:

$$y = x^3 - x, \qquad dy/d\tau = -x;$$

here $\tau = \varepsilon t$ is slow time.

The slow system is the system of evolution of the slow variables subject to the condition that the fast variables remain in equilibrium. The fundamental idea of the theory of relaxation oscillations is to construct the asymptotics of the true perturbed motion as a sequence of interchanging intervals of fast and slow motion.

As one approaches a generic nonregular (or singular) point (a fold of the projection) along the slow surface, the speed of slow motions (with respect to slow time) tends to infinity at a rate inversely proportional to the distance, along the slow surface, from the fold.

1.4. The Slow Motion as an Approximation to the Perturbed Motion. Consider the slow system on the slow surface:

$$f(x, y) = 0 \qquad dy/d\tau = g(x, y).$$

Assume that (1) solutions of this system are defined on an interval $0 \leqslant t \leqslant T$ and (2) all points on a phase curve of the slow system on the slow surface are attracting hyperbolic points for the fast system (the eigenvalues of the singular points of the fast system lie in the left half-plane).

Example. For the van der Pol system these conditions are fulfilled if the interval of the phase curve of the slow system that corresponds to $0 \leqslant t \leqslant T$ does not intersect the arc of the slow curve Γ (joining nonregular points), where the tangent to Γ is vertical. In other words, the slow motion under study takes place entirely on either the highest or the lowest of the three branches of the slow curve.

Theorem (Vasil'eva (1969), Gradshtejn (1953), Tikhonov (1952). *Suppose the conditions* (1) *and* (2) *hold. Then, there exists a neighborhood of the phase curve (independent of* ε*) of the slow system such that, for sufficiently small* $\varepsilon > 0$, *the solution of the perturbed system, with any initial condition from this neighborhood that lies in the fiber over the initial point of the phase curve of the slow system under consideration, is defined for* $0 \leqslant t \leqslant T$. *This solution differs from the phase curve being considered by an amount not greater than* $C_1\varepsilon$ *on the whole interval of time except for a short (in slow time) initial interval* $0 \leqslant \tau \leqslant C_2\varepsilon|\ln \varepsilon|$, *in which the solution is close to the fast motion along the initial fiber. Further, the constants* C_1 *and* C_2 *are independent of* $\varepsilon > 0$ *and the initial point; moreover, under natural conditions of the uniformity of attraction, they are independent of the original phase curve of the slow system. The excluded initial interval arises because the fast system requires a time* $O(\ln(1/\varepsilon))$ *for relaxation to a nondegenerate equilibrium.* ▲

1.5. The Phenomenon of Jumping. Besides stable equilibria of the fast motion, the slow surface generally contains unstable equilibria. Therefore a phase curve of the slow system may (after a finite slow time) arrive at the boundary of stability of the fast motion, and then the previous theorem becomes inapplicable.

Example. For the van der Pol oscillator, the motion along the upper branch of the slow curve is to the left and arrives at a nonregular point, after which the slow motion cannot be prolonged. In this example the perturbed motion, after its moment of arrival to a nonregular point, loses its connection with the slow curve: a "jump" of the trajectory of the slow curve occurs (accompanied by a relaxation to another equilibrium state, that is, a jump to the lower branch).

A phenomenon analogous to jumping takes place in other generic systems. According to the general theory, a loss of stability by an equilibrium point of a generic system depending upon parameters (in the present case, the fast system), takes place on two hypersurfaces of parameter space (in the present case, in the space of the slow variables). One of these hypersurfaces corresponds to the collision of a stable equilibrium point with an unstable one, after which both equilibria disappear (they become complex). On the slow surface this phenomenon is observed at nonregular points (critical points of the projection of the slow surface onto the base); at these points, the linearization of the fast system in a fiber has a zero eigenvalue. For example, in the van der Pol system, the jump takes place at points of vertical tangency of the slow curve.

The second hypersurface associated with loss of stability corresponds to the passage from the left to the right half-plane of the two complex conjugate eigenvalues of the linearization of the fast system at the equilibrium. This behavior is observed, in general, at certain regular points of the slow surface that form a submanifold of codimension 1 (details about this are given in Sect. 4 below). Still more complicated phenomena occur on submanifolds of greater codimension than 1, for example, the combination of a zero eigenvalue and a purely imaginary pair, etc.

In both cases, the perturbed motions jump from the slow surface after loss of stability by the equilibrium state of the fast system; but their fates, in general, differ. In the first case (disappearance of equilibria) the loss of stability is always "hard": the fast system leads the phase point to some other attractor (and sometimes it moves off to "infinity", which physically denotes the explosive character of the process. This attractor may prove to be, for example, a limit cycle or a torus; and then to study the further motions one may use the averaging method (N.N. Bogolubov and Yu.I. Mitropolskij; see Ref. 17 in Arnol'd and Il'yashenko (1985)).

The new attractor may also simply turn out to be a stable equilibrium lying at the side of the fast system. Indeed, this is the case for the van der Pol system and, generally, for systems with one fast variable (because typical motions of a generic system with a one-dimensional phase space approach a nondegenerate stable equilibrium state).

In the case of many fast variables, the motion relaxes to equilibrium if the fast system is a gradient system:

$$\dot{x} = \partial U(x, y, \varepsilon)/\partial x, \qquad \dot{y} = \varepsilon G(x, y, \varepsilon).$$

Takens (1976) called this system a "constrained system".

A second case, namely, loss of stability by a singular point of the fast system as a pair of eigenvalues pass through the imaginary axis, is investigated in §4.

§2. Singularities of the Fast and Slow Motions

Here we present normal forms of various objects close to nonregular points on the slow surface, where jumps may occur. We consider generic systems, and we show what results the general theory of singularities gives if it is applied to the study of relaxation phenomena.

2.1. Singularities of Fast Motions at Jump Points of Systems with One Fast Variable. The singularities of the projection of a slow surface onto the base are described by the general theory of singularities of differentiable mappings (see Arnol'd, Varchenko and Gusejn-Zade (1982)). If there is one fast variable, there is a finite list of such singularities that occur in generic systems. In suitable (smooth, analytic) local coordinates the slow surface may be written in the Whitney normal form $P_\mu(x, y) = 0$, where

$$P_\mu = x^{\mu+1} + y_1 x^{\mu-1} + \cdots + y_\mu.$$

Here $(y_1, \ldots, y_\mu)$ are the slow variables (perhaps only a subset of them); the value of μ must not exceed the dimension of the base, but it may be smaller.

The transformation of the slow surface to the Whitney normal form is achieved by a local fiber-preserving diffeomorphism, that is, by a diffeomorphism of the bundle space, transforming fibers into fibers: $x = h(X, Y)$, $y = k(Y)$.

Example. In generic systems with one fast and one slow variable, only a fold is realized ($x^2 + y = 0$), as at the points with a vertical tangent on the slow curve of the van der Pol system.

In generic systems with one fast and two slow variables both a fold ($x^2 + y_1 = 0$), and a pleat ($x^3 + xy_1 + y_2 = 0$), are realized. In this case the nonregular points form a smooth curve – *the fold curve* – on the slow surface. At discrete *pleat points* this curve is vertical (tangent to a fiber of the bundle; see Fig. 65). The set of critical values of the projection (onto the plane of the slow variables y) has cusp points at projections of pleats. In a neighborhood of a cusp, the curve of critical values of the projection is diffeomorphic to a semi-cubical parabola.

In a neighborhood of the slow surface, a generic analytic fast system may be transformed by an analytic fiber-preserving diffeomorphism to the normal form $\dot{x} = P/E$, where P is a function in Whitney normal form, and $E = \pm 1 + C(y)x^\mu$.

For example, in the case of one fast and one slow variable:

$$\dot{x} = C(y)x \qquad \text{(a regular point)},$$
$$\dot{x} = (x^2 + y)(1 + C(y)x)^{-1} \qquad \text{(a jump point)}.$$

The functional parameter (modulus) C in this case is necessary, since the sum of the residues of the differential form dt at points close to zero is an invariant of diffeomorphisms of the x-axis.

If there are two slow and one fast variables, then in a neighborhood of a generic point of the fold, the normal form of the fast system is

$$\dot{x} = (x^2 + y_1) + (C + y_2)x^3,$$

where C now is not a function but a number.

Smooth analogs of the above assertions hold.

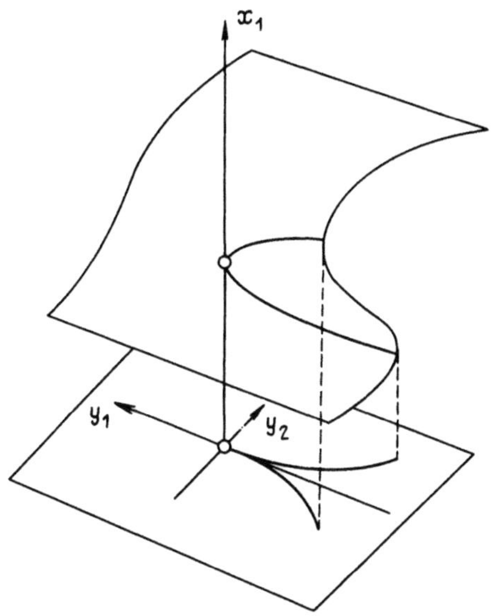

Fig. 65. The pleat on a slow surface

2.2. Singularities of Projections of the Slow Surface. If the number of fast variables k is greater than 1, the normal form of the slow surface of a generic system is the same as the one above (one just adds the equations $x_2 = 0, \ldots, x_k = 0$) if the dimension of the kernel of the projection of the slow surface onto the space of slow variables at the point considered is equal to 1, that is, if the zero eigenvalue of the linearized fast system at the equilibrium point for fixed values of the slow variables is simple.

The last condition (a one-dimensional kernel) is automatically fulfilled for generic systems for any number of fast variables, provided the number of slow variables does not exceed 3. Thus, in typical systems with one, two or three slow variables, the equations of the slow surface are, after a local fiber-preserving diffeomorphism, equivalent to one of the forms:

$$x_1 = \cdots = x_k = 0 \qquad \text{(a regular point)},$$

$$x_1^2 - y_1 = x_2 = \cdots = x_k = 0 \qquad \text{(a fold)},$$

$$x_1^3 + y_1 x_1 + y_2 = x_2 = x_k = 0 \qquad \text{(a pleat)},$$

$$x_1^4 + y_1 x_1^2 + y_2 x_1 + y_3 = x_2 = \cdots = x_k = 0.$$

In the last case, the set of critical values of the projection is a surface in $\mathbb{R}^3$ called a swallowtail (see Fig. 5b). If there are four slow variables, then more complicated singularities arise (with two-dimensional kernels); however, the list of singularities remains finite for up to and including 9 slow variables.

2.3. The Slow Motion for Systems with One Slow Variable.

Theorem. *In a neighborhood of a jump point, the equation of slow motions of a generic system with one fast and one slow variable is reduced by a fiber-preserving diffeomorphism to the form*

$$y = x^2, \qquad \dot{y} = \pm 1 + xA(y);$$

moreover, A is invariant under the group of such diffeomorphisms. The transforming diffeomorphism is smooth (*analytic*) *for a smooth* (*analytic*) *system.*

Corollary. The slow motion may be reduced to the form $y = \pm t + t^{3/2}a(t)$; and if one admits a C^∞-change of time, it may be reduced to the form:

$$y = \pm t + t^{3/2}.$$

Proof of the Theorem and its Corollary:

1°. We may write the slow equation as

$$y = x^2, \qquad \dot{y} = b(y) + xa(y).$$

Here we have separated $g(x, x^2)$ into its even $b(x^2)$ and odd $xa(x^2)$ parts. For generic systems, $b(0) \neq 0$. With a change of variables

$$y = z(c_0 + c_1 z + \ldots), \qquad x = u(d_0 + d_1 u^2 + \ldots)$$

the slow equation may be put in the form

$$z = u^2, \qquad \dot{z} = \pm 1 + uA(z).$$

Two slow equations with different A's cannot be transformed into one another by a fiber-preserving diffeomorphism of the slow curve, since the ratio $(1 + uA(z))/(1 - uA(z))$ is invariant relative to such diffeomorphisms. This proves the theorem.

2°. The equation of slow motion can be written in the form (we now write (x, y) instead of (z, u)):

$$y = x^2, \qquad dt/dx = 2x/(\pm 1 + xA(y)).$$

Hence

$$t(x) = \pm x^2 + x^3 B(x), \qquad y = x^2.$$

By a theorem of Dufour (1979), using the C^∞-change of variables $z = z(y)$, $\tau = \tau(t)$ one can reduce the pair of functions t, y (for a generic system) to the form

$$\tau(x) = x^2 + x^3, \qquad z(x) = x^2.$$

The corollary is proved.

Remark. An analogous normalization with the aid of an analytic change of variables, as a rule, is impossible in the general case. (Voronin (1982)).

2.4. The Slow Motion for Systems with Two Slow Variables. In this case one can study the family of phase curves of the slow system in full detail: the problem in this case may be reduced to one of the theory of implicit differential equations. For simplicity we shall assume that there is just one fast variable. The slow motion for generic systems with an arbitrary number of fast variables, but just two slow variables, is the same as in the case of one fast variable. Indeed, for generic systems with fewer than four slow variables the kernel of the projection of the slow surface is one-dimensional.

Therefore, all the fast variables except one may be chosen so that they are all equal to zero on the whole slow surface (see Sect. 2.2 above). The behavior of the system for nonvanishing values of these variables does not affect the slow field. Thus, for investigating the slow motion, we may forget about these variables.

Thus, suppose the space E of the bundle $E \to B$ is three-dimensional, with a two-dimensional base and one-dimensional fibers. At each point of this three-dimensional space there is a vertical direction (tangent to the fiber along which both the slow variables are constant). At nonsingular points the *perturbing* field (i.e., the value of the derivative of the perturbed field with respect to the small parameter ε at $\varepsilon = 0$) has its own direction, distinct from that of the vertical field. For a generic system the singular points of the perturbing vector field do not lie on the slow surface; therefore, we do not consider them. At the points under consideration, two direction fields are defined; they correspond to the vertical and the perturbing vector fields.

For generic systems these fields are collinear only at points of some smooth curve, and this curve transversally intersects the slow surface at regular points. These points of intersection are the equilibria of the slow system. Since they are regular points, they are usual singular points of the smooth (slow) vertical field on the slow surface (nodes, saddles, foci). For their investigation, the usual local theory (Arnol'd and Il'yashenko (1985)) is applicable.

We are interested in the singular points of projections of the slow surface. At these points the directions of our fields are not collinear. Consequently, they generate a smooth plane field.

A generic smooth plane field gives rise to a contact structure (if the field is defined as the field of zeros of a 1-form α, then the 3-form $\alpha \wedge d\alpha$ is nondegenerate).

The points where the contact structure defined by a generic plane field (defined on three-dimensional space) is degenerate form a surface. For a generic system this *surface of degeneration of the contact structure* transversally intersects the slow surface along a curve. Moreover, it may transversally intersect the smooth curve of nonregular points of the projection of the slow surface (the curve of folds) at discrete points. For a generic system the points of intersection will be fold points and not pleats.

We now examine the traces on the slow surface of the plane field in the three-dimensional space constructed above. A plane of this field intersects the tangent plane to the slow surface at a regular point in the direction of the slow field. Therefore, the traces of the constructed plane field form, in the regular part of the slow surface, exactly the direction field of the slow motion.

This direction field of the slow motion may be extended, even on the curve of critical points of the projection, to a smooth direction field. It has singularities only at those places where a plane of the field is tangent to the slow surface. This can happen for a generic system only at discrete points. Such points must lie on the fold curve, since the plane of the field contains the vertical direction. For a generic system these discrete points will be neither pleats nor points of degeneration of the contact structure.

Thus, the phase curves of the slow motion are pieces of integral curves of the field of traces on the slow surface of the planes constructed above. This direction field on the slow surface is vertical on the curve of critical points of the projection (since both the plane field and a tangent to the slow surface at these points contain the vertical direction), and may have, in addition, discrete singular points on this curve (neither pleats nor points of degeneracy of contact structure).

In the next subsection, normal forms are given, to which integral curves of the direction field constructed on the slow surface (and consequently phase curves of the slow system as well) are transformed by fiber-preserving diffeomorphisms.

2.5. Normal Forms of Phase Curves of the Slow Motion. In a neighborhood of a fold point the slow surface is mapped by a fiber-preserving diffeomorphism to the form $y = x^2$, where x is a fast variable, y is a slow variable; we denote the second slow variable by z.

Thus the integral curves we study lie on the surface $y = x^2$ in three-dimensional space (x, y, z) and are projected along the x-axis onto the (y, z)-plane.

Theorem (V.I. Arnol'd, 1984). *In a neighborhood of a fold point of the projection of the slow surface of a generic system with two slow and one fast variable, the family of integral curves of the slow system is mapped by a fiber-preserving diffeomorphism of the slow surface into one of the following normal forms:*

$$z = x^3 + c \quad \textit{(a generic point of a fold)}, \tag{4}$$

$$z = x^3 z + x^5 + c \quad \textit{(a point of degeneration of the contact structure)}, \tag{5}$$

$$x\,dx = (2x + az)\,dz \quad \textit{(singular points of the direction field)}, \tag{6}$$

where the slow surface has the equation $y = x^2$.

On the slow surface, singular points of the last type may be foci $(a < -1)$, nodes $(-1 < a < 0)$ or saddles $(0 < a)$.

In the first case, the projections onto the (y, z)-plane of slow variables of the integral curves on the slow surface have semi-cubical singularities $y^3 = (z - c)^2$; see Fig. 66.

In the case of a point of degeneration of the contact structure, the projections of the integral curves are illustrated in Fig. 67.

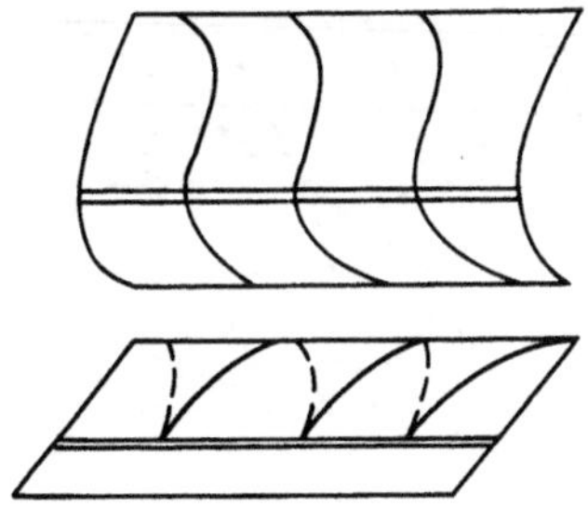

Fig. 66. Phase curves of the slow equation in a neighborhood of a generic fold point on the slow surface: normal form

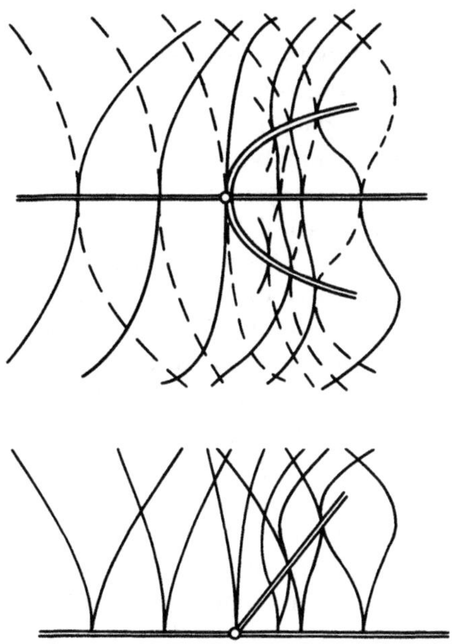

Fig. 67. Phase curves of the slow equation in a neighborhood of a point of degeneration of the contact structure. The set of points of tangency of the integral curves with their reflections is illustrated by the doubled line arcs.

These projections may be described as follows. Consider the surface (the folded umbrella of Whitney) $u^2 = v^3w^2$ in three-dimensional space. Project the level curves of the function $u + v + w$ restricted to this surface onto the v, w-plane. One obtains the desired family of curves on the plane.

The projections of integral curves in a neighborhood of a focus, node or a saddle onto the plane of the slow variables are illustrated in Fig. 68. These are folded singularities; their normal forms were found by A.A. Davidov (1984 [1*]; see Arnol'd (1984), and Arnol'd and Il'yashenko (1985)). They are given by an implicit differential equation, of the form

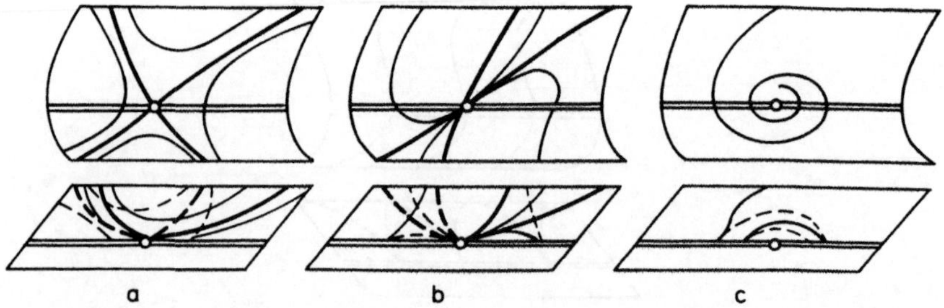

Fig. 68. Typical singular points of the slow equation at a fold of the slow surface

$$y = (dy/dz + kz)^2$$

In neighborhoods of pleat points, the projections are described as follows. Consider the surface of the swallowtail: $\{\lambda: x^4 + \lambda_1 x^2 + \lambda_2 x + \lambda_3$ has multiple roots$\}$. The planes $\lambda_1 = \text{const.}$ partition the swallowtail into curves. The projections of the integral curves in a neighborhood of a pleat (cusp) point of the projection of the slow surface of a generic system are obtained from this standard family of plane sections of the swallowtail by a generic smooth mapping of three-dimensional space onto a plane. This mapping has rank 2 at the vertex of the swallowtail. Consequently, a three-dimensional neighborhood of the vertex is smoothly fibered into one-dimensional fibers (preimages of points of the plane). The direction of the fiber at the vertex is transversal to both the plane $\lambda_1 = 0$ and the tangent plane to the tail ($\lambda_3 = 0$) for generic mappings. Depending on how this direction intersects these two planes, the form of the projection clearly changes; see Fig. 69. Moreover, the projections of the integral curves have (less noticeable) topological functional moduli, if one takes into account the projections of the integral curves from all three branches of the slow surface. If one only considers the branches corresponding to the stability of an equilibrium in the fast system, then there are three topologically different generic pictures; see Fig. 70 (the first two correspond to the stability of an equilibrium on the outside branches, the third one to the stability of an equilibrium on the middle branch).

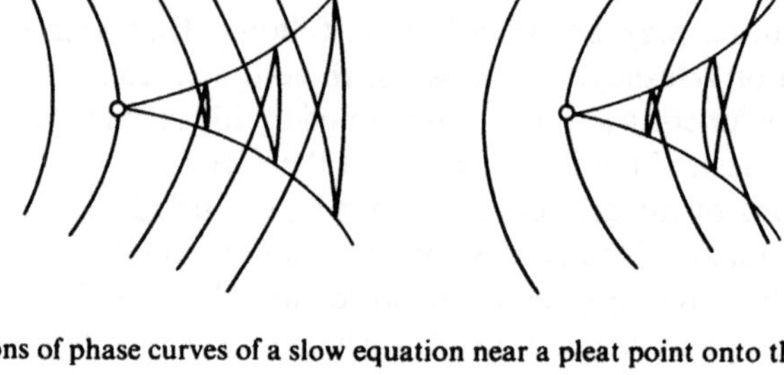

Fig. 69. Projections of phase curves of a slow equation near a pleat point onto the plane of the slow variables

Fig. 70. Projections of phase curves from the stable branch of the slow surface near a pleat point onto the plane of the slow variables

A topological classification of the singularities of slow motions in generic systems with two slow and one fast variables, taking into account only the stable equilibria of the fast motion, was given by Takens (1976).

2.6. Connection with the Theory of Implicit Differential Equations. Consider a point where our plane field is nondegenerate (defines a contact structure[30]). The fibers of our bundle are tangent to the planes of our field. This means that it is a Legendre bundle (consisting of integral manifolds of maximal dimension). All Legendre bundles in a contact space of fixed dimension are locally *contactomorphic* (they are transformed into each other by a contact-structure-preserving diffeomorphism in a neighborhood of each point of the space of the bundle). Consequently, our three-dimensional space of fast and slow variables with the introduced contact structure is transformed by a local fibered diffeomorphism (over the plane of slow variables) into the three-dimensional space of 1-jets of functions of a single variable over the space of 0-jets, with its natural contact structure.

In coordinates, the 1-jet is defined by the value of an independent variable (denoted by z), a dependent variable (denoted by y), and a derivative dy/dz (denoted by p). The contact structure of the space of 1-jets in these coordinates is written in the form $dy = pdz$, the projection onto the 0-jet (eliminating p) is: $(z, y, p) \mapsto (z, y)$.

The surface of slow motions, and its projection onto the plane of slow variables along the axis of fast variables under the indicated map into the space of 1-jets, map into a surface defined by an implicit differential equation

$$F(z, y, p) = 0, \qquad p = dy/dz,$$

and its projection along the p-axis into the space of 0-jets, respectively. The integral curves of our field on the slow surface are transformed into integral curves of the implicit equation.

Therefore, the theorem stated above follows from results of the theory of implicit differential equations in all cases except for the case of a degeneration

[30] For more about contact structures and Legendre bundles, see the detailed article by V.I. Arnol'd and A.B. Givental in EMS 4: Dynamical Systems IV; the remainder of Sect. 2 may be read independently.

of the contact structure. Moreover, the normalizing diffeomorphism reduces not only integral curves of traces of the contact plane field on the surface to a normal form, but the whole contact plane field as well. In terms of the theory of relaxation oscillations, this means that the projections of a direction of the perturbing field onto the plane of slow variables may be transformed by fiber-preserving diffeomorphism to a standard form, not only on the slow surface, but in a whole neighborhood of the point under consideration in three-dimensional space.

2.7. Degeneration of the Contact Structure. In this case one may reason as follows. Folding defines an involution of the slow surface, exchanging both points of one fiber. In a neighborhood of a fold point the slow surface is reduced to the normal form $y = x^2$ by a fiber-preserving diffeomorphism (y is the slow, and x the fast variable). We shall use local coordinates (x, z) on the slow surface, where z is the second slow variable. Then the involution indicated above becomes $(x, z) \mapsto (-x, z)$.

The integral curves on the slow surface form a smooth family in a neighborhood of the point under consideration, and they can be defined as the family of level curves of some function: $\Phi(x, z) = \text{const}$. At fold points, these curves are tangent to the kernel of the folding map, that is the direction of the x-axis. This means that the function Φ can be reduced to the form $z + x^2\mathbf{A}(x, z)$, changing neither the slow surface nor the family of level curves.

Changing the slow variable z to a function of z and $y = x^2$, we obtain, without changing the slow surface, a fiber-preserving diffeomorphism (y is the slow variable). Choosing z in this way, we can annihilate the whole even part (in x) of $\mathbf{A}$. We thus have reduced Φ to the form $z + x^3\mathbf{B}(x^2, z)$.

Consider now, together with the family of curves $\Phi(x, z) = \text{const.}$, its image under the involution that changes the sign of x. At fold points ($x = 0$) the curves of both families are tangent to each other; moreover, the order of the tangency is even (like the tangency of a straight line and a parabola of odd degree). If $\mathbf{B} \neq 0$ at the point under study, then the tangency is of second order.

It is easy to verify that the conditions above define a nondegeneracy (contactness) condition on the plane field, whose traces define our integral curves on the slow surface.

At a point of degeneration $\mathbf{B}(0, z)$ takes the value 0. For a generic system, the zero is of first order, and the order of the tangency of an integral curve with its reflection jumps from second up to fourth order. This allows us to reduce the equation of the integral curves to the form

$$z + x^3z\mathbf{C}(x^2, z) + x^5\mathbf{D}(x^2) = \text{const.}, \qquad \mathbf{C}(0, 0) \neq 0, \qquad \mathbf{D}(0) \neq 0.$$

By the odd change of variable $x' = x[\mathbf{C}(x^2, z)]^{1/2}$ and a rescaling we may reduce the equation to the form

$$z + x^3z + x^5 + x^7\mathbf{E}(x^2, z) = c.$$

Finally, the term with $\mathbf{E}$ may be eliminated altogether, by combining a C^∞-diffeomorphism of the (x, z) plane, which commutes with the involution that changes the sign of x, with a C^∞-change of numbering of the curves (the parameter c).

To accomplish this, it is sufficient to consider the set where the integral curves are tangent to their images under the involution. This set is symmetric with respect to the x-axis; and, in addition to the x-axis, it contains a curve resembling a parabola: $z = -5/3x^2 + \cdots$ (see Fig. 67). There exist two involutions on this curve: one interchanges x and $-x$, the other interchanges two points on one integral curve. The difference between the two involutions is of order x^4.

Such a pair of involutions may be reduced by a single local C^∞-diffeomorphism of the curve to the following normal form (for example, $x \mapsto -x$ for the first, and $x \mapsto x'$ for the other, where $x^2 + x^5 = x'^2 + x'^5$; Dufour (1979)). In the analytic case such a pair of involutions, in spite of the simple formal normal form, has functional moduli (S.M. Voronin (1982)).

Choose the coordinate x on the curve of tangency so as to normalize both involutions. We shall also use this coordinate to number the integral curve tangent at this point of its image under the involution. This numeration gives a correspondence between integral curves of the families with $\mathbf{E} \neq 0$ and those with $\mathbf{E} = 0$: the corresponding curves are tangent to the reflected curves at the points with the same number.

A point of intersection of an integral curve of the family with $\mathbf{E} \neq 0$ and number x_1 with a reflected curve with number x_2 is associated to the (topologically analogous) point of intersection of the curves of the standard family ($\mathbf{E} = 0$) with the same numbers. The resulting correspondence may be extended to a diffeomorphism, commuting with the involution and sending the family of curves with $\mathbf{E} \neq 0$ to the standard family.

We project the standard family of curves

$$z + x^3 z + x^5 = c$$

onto the plane of slow variables (y, z). The family of projections satisfies the equation

$$(z - c)^2 = y^3(z + y)^2 \qquad \text{since } y = x^2.$$

We lift each projection to its level c. We obtain a surface in the three-dimensional space with coordinates (y, z, c).

In this space we choose new coordinates

$$z - c = u, \qquad y = v, \qquad z + y = -w.$$

The equation of the surface is now $u^2 = v^3 w^2$.

If a value $c = c_0$ is fixed, then on the resulting plane curve we have $u + v + w = -c_0$. Hence,

$$y = v, \qquad z = u + c_0 = -u - w.$$

Therefore, the family of projections of integral curves onto the plane of slow variables (y, z) is locally diffeomorphic to the family of projections of plane sections $u + v + w = \text{const.}$ of the folded umbrella $u^2 = v^3 w^2$ onto the (v, w)-plane along the w-axis.

The decomposition of the folded umbrella $u^2 = v^3w^2$ into the level curves of a generic smooth (analytic) function is (C^∞, analytically) diffeomorphic to the decomposition on the plane section $u + v + w = \text{const}$.

The folded umbrella first appeared in the theory of singularities by another route, as a singularity of a bicaustic swept out by the cuspidal edges of moving caustics; see Arnol'd (1982). The analysis produced above, in these terms, is the investigation of the decomposition of a bicaustic into instantaneous cuspidal edges of caustics. This surface is called a folded umbrella because it is obtained from a cylinder over a semi-cubical parabola lying in three-dimensional space by the generic fold map of $\mathbb{R}^3$ into $\mathbb{R}^3$. The folded umbrella also appears as one of the components of the boundary of the variety of fundamental systems of solutions of scalar linear equations (M.Eh. Kazaryan (1985)).

§ 3. The Asymptotics of Relaxation Oscillations

In this section we study the asymptotic behavior, with respect to a parameter, of solutions of the equations of fast and slow motions, as the parameter tends to zero. Here, only those systems are considered in which the singular points of the equation of fast motions loses stability as the slow variable changes, as a result of the vanishing of exactly one of the eigenvalues of the linearized fast equations. In other words, the center manifolds of the system of fast motions are one-dimensional for each value of the slow variable. In this case, the slow surface splits into stable and unstable parts, separated by "jump points", critical points of the projection of the slow surface along the space of fast variables onto the space of slow variables. We shall call such equations *equations of type* 1 in accordance with the dimensions of their center manifolds.

3.1. Degenerate Systems. Consider the type 1 system (2):

$$\dot{x} = F(x, y, \varepsilon), \qquad \dot{y} = \varepsilon G(x, y, \varepsilon). \tag{2}$$

Corresponding to this system is the degenerate system of type 1

$$f(x, y) = 0, \qquad \dot{y} = g(x, y),$$

where

$$f(x, y) = F(x, y, 0), \qquad g(x, y) = G(x, y, 0).$$

A phase curve of the degenerate system is an orientable curve that consists of alternating parts of phase curves of fast and slow motions; moreover, the orientation of time on the fast and slow parts coincides with that on the whole phase curve.

The phase curves of degenerate systems subdivide into *regular phase curves* and degenerate duck-like curves or simply "*degenerate ducks*". A regular phase curve contains only those arcs of the phase curves of the slow motion which lie on the stable part of the slow surface; degenerate ducks contain arcs of slow phase curves which lie on the unstable part of the slow surface.

Until recently the equations studied in the theory of relaxation oscillations were fast-slow ones of type 1 whose phase curves in the neighborhood of a jump point converge, as $\varepsilon \to 0$, to regular phase curves of the corresponding degenerate systems. However, not long ago it was discovered that for some fast-slow systems the phase curves, close to jump points, may, as $\varepsilon \to 0$, approach degenerate ducks. Details on this are presented in Sect. 5 below.

Theorem (Mishchenko and Rozov (1975), Pontryagin (1957)). *Suppose $(x,y) = p$ is a fold point of the slow surface of a fast-slow system of type 1 of the form (2) (that is, a system with center manifolds of dimension at most 1 at equilibria of the fast motions). Suppose that the vector $G(x, y, 0)$ is transversal to the projection of the fold onto the base along the fibers (that is, the projection of the fold onto the space of slow variables along the space of fast variables). Moreover, suppose that this vector is directed to the exterior, relative to the projection of the slow surface on the plane of the slow variables. Then, there exists a neighborhood U of the point p in phase space such that for any point $q \in U$, the connected component of the intersection of the neighborhood U with a positive semi-trajectory of the system (2) with initial point q converges, as $\varepsilon \to 0$, to a regular phase curve of the degenerate system.*

Under the conditions of the theorem the asymptotics of solutions close to jump points have been computed up to terms that are $O(\varepsilon)$; see Mishchenko and Rozov (1975) and Pontryagin (1957).

3.2. Systems of First Approximation. We introduce a change of scale in a neighborhood of a jump point. The coefficients of the stretching and the size of the neighborhood depend upon the parameter ε of the system (2) in such a way that, as $\varepsilon \to 0$, the stretched image of the neighborhood contains any compact set, beginning with some sufficiently small (depending on the compact set) value of the parameter. The goal of this construction is to obtain, after a change of variables (time and parameter) a system (the so-called *system of first approximation*) in which all the motions proceed in the same time scale.

Example 1. The case of one fast and one slow variable.

Proposition 1. *A typical system (2) with one fast and one slow variable may, by a fiber-preserving diffeomorphism, be transformed into a form such that the change of variables and parameters*

$$x = \mu x_1, \qquad y = \mu^2 y_1, \qquad \tau = \mu t, \qquad \varepsilon = \mu^3$$

leads to the system

$$\dot{x}_1 = x_1^2 - y_1 + O(\mu), \qquad \dot{y}_1 = -1 + O(\mu),$$

defined in the region $|x| \leqslant 1$ and $|y| \leqslant 1$. The corresponding system of first approximation is

$$\dot{x} = x^2 - y, \qquad \dot{y} = -1.$$

Proposition 1 is proved in (Pontryagin (1957) and Mishchenko and Rozov (1975, § 8)).

Example 2. The case of one fast and two slow variables.

Proposition 2 (Yu.S. Il'yashenko (1985)). *A typical system (2), with one fast and two slow variables, may be reduced by a fiber-preserving diffeomorphism, in a neighborhood of a jump point on the fold of the slow surface, to a form that, after a linear change of coordinates f_μ, a change of time $\tau = \mu t$ and a change of parameter $\varepsilon = \varepsilon(\mu)$, leads to a system coinciding with the system of first approximation up to terms $O(\mu)$ in the cube $|x_1| < 1, |y| < 1, |z_1| < 1$. The changes f_μ and $\varepsilon(\mu)$ and also the system of first approximation are written down below (see Table 1).* ▲

Table 1

Type of point on the fold	The equation of the slow surface	Phase curves of the slow system:	The change of variables f_μ: $(x, y, z) \mapsto (\xi, \eta, \zeta)$	The change of parameter $\varepsilon(\mu)$	The system of first approximation
1. A nonsingular point at which the contact structure is nondegenerate	$y = x^2$	$z - x^3 = c$	$x = \mu\xi$, $y = \mu^2\eta$, $z = \mu^3\zeta$	$\varepsilon = \mu^3$	$\xi' = \xi^2 - \eta$, $\eta' = -1$, $\zeta' = -\xi$
2. A point of degeneration of the contact structure	$y = x^2$	$(1 - x^3)z - x^5 = c$	$x = \mu\xi$, $y = \mu^2\eta$, $z = \mu^4\xi + \mu^5\zeta$	$\varepsilon = \mu^3$	$\xi' = \xi^2 - \eta$, $\eta' = -1$, $\zeta' = (a - \delta)\cdot$ $\cdot\xi^3 - a\xi\eta$, $\delta = \pm 1$
3. A singular point	$y = x^2$	$xdx =$ $(2x + az)\,dz$	$x = \mu\xi$, $y = \mu^2\eta$, $z = \mu\zeta$	$\varepsilon = \mu^2$	$\xi' = \xi^2 - \eta$, $\eta' = 2\xi$ $+ a\zeta$, $\zeta' = \pm 1$

In Table 1, x is the fast variable, y and z are the slow variables, the z-axis is directed along the fold of the slow surface, and the y-axis is perpendicular to it. In the second and third columns the normal forms presented in Sect. 2.5 are shown; the phase curves of the slow system are given either by a first integral, or by the corresponding direction field. Proposition 2 is proved in Sect. 3.3 and 3.4, which follow.

Remark. The equation $d\xi/d\eta = \xi^2 - \eta$ is one of the simplest equations that is not integrable by quadratures.

3.3. Normalizations of Fast-Slow Systems with Two Slow Variables for $\varepsilon > 0$

Theorem (Yu.S. Il'yashenko (1985)). *A typical fast-slow system with two slow variables can, in a neighborhood of any point on the fold of the slow surface for all sufficiently small values of ε, be transformed by a fiber-preserving diffeo-*

morphism, that smoothly depends upon ε, into the system

$$\dot{x} = F(x, y, z, \varepsilon), \qquad \dot{y} = G_1(x, y, z, \varepsilon), \qquad \dot{z} = G_2(x, y, z, \varepsilon)$$

for which the surface $F = 0$ *has the form* $y = x^2$, *and the direction field obtained on this surface as the trace of the plane field generated by the frame vector field*

$$(\partial/\partial x, (G_1\partial/\partial y + G_2\partial/\partial z)_{|y=x^2})$$

coincides with one of the normal forms corrresponding to the families (4), (5) *or to equation* (6), *as set forth in Arnol'd's theorem in Sect.* 2.5 *above.*

◀ One may assume that for $\varepsilon = 0$ the surface $F = 0$ and the direction field on it are already normalized. Close to the origin, the family of surfaces obtained by a deformation of the surface $y = x^2$ in (x, y, z)-space, is transformed by a fiber-preserving diffeomorphism, depending smoothly upon the deformation parameter, into the constant family $y = x^2$. This allows one to normalize the surface $F = 0$ for small ε. The direction fields described in the theorem are obtained by a small perturbation of one of the standard ones. The genericity requirement imposed on the the vector field in the proof of Arnol'd's theorem in Sect. 2.5, defines an open set in the corresponding function space. Therefore, all fields that are close to one of the normalized fields given by equations (4)–(6) of Sect. 2.5 can be reduced to the same normal forms. The forms (4), (5) have no parameters; the form (6) has one. The vector field

$$(G_1\partial/\partial y + G_2\partial/\partial z)|_{y=x^2}$$

depends on ε. Therefore, the normal form of the direction field near the singular point contains an ε dependent parameter $a(\varepsilon)$. The diffeomorphisms that normalize the fields may be chosen to depend smoothly on the parameter of the deformations. This is easy to conclude from the reasoning in Sect. 2.5–2.7 above. ▶

Corollary. A typical fast-slow system with two slow variables and one fast variable can be transformed by a fiber-preserving diffeomorphism, smoothly depending upon ε, into one of the following systems, in a sufficiently small neighborhood of any point of the fold of the slow surface:

1. Near a typical point of the fold (where the contact structure is nondegenerate, and where there is no singular point of the slow system):

$$\dot{x} = (x^2 - y)\mathbf{A}, \qquad \dot{y} = \varepsilon\mathbf{G}_1, \qquad \dot{z} = \varepsilon[x\mathbf{G}_1 + (x^2 - y)\mathbf{B}], \quad \text{with } \mathbf{A}(0) = -\mathbf{G}_1(0) = 1; \tag{7}$$

here and below **A** and **B** are smooth functions of x, y, z, ε.

2. Near a point of degeneration of the contact structure:

$$\dot{x} = (x^2 - y)\mathbf{A}, \qquad \dot{y} = \varepsilon\mathbf{G}_1, \qquad \dot{z} = \varepsilon\left[\frac{(\delta x^3 + 3kxz/2)}{(1 - kx^3)}\mathbf{G}_1 + (x^2 - y)\mathbf{B}\right], \tag{8}$$

with $\mathbf{A}(0) = -\mathbf{G}_1(0) = 1$ $\delta = \pm 1$.

3. Near a singular point on the fold of the slow surface:

$$\dot{x} = (x^2 - y)\mathbf{A}, \qquad \dot{y} = \varepsilon(2x + a(\varepsilon)z\mathbf{G}_2 + (x^2 - y)\mathbf{B}), \qquad \dot{z} = \varepsilon\mathbf{G}_2, \quad \text{with } \mathbf{A}(0) = |\mathbf{G}_2(0)| = 1. \tag{9}$$

Moreover, preserving the parameter and the slow variables by a smooth changes of variables in (7)–(9) one may achieve the equality $\mathbf{A} = 1 + \mathbf{C}(y, z, \varepsilon)x$. ▲

◀ The surface $F = 0$ has the form $y = x^2$. Consequently, close to the origin $F = (x^2 - y)\mathbf{A}$, where $\mathbf{A}$ is a smooth function of x, y, z, ε. The direction field on the slow surface, as described in the theorem, coincides with the field of zeros of the form $(\mathbf{G}_1\, dz - \mathbf{G}_2\, dy)|_{y=x^2}$. On the other hand, one may assume that this vector field has one of the normal forms (4)–(6). Hence, it coincides in cases 1, 2 and 3 with the fields of zeros of the forms ω_1, ω_2, and ω_3 respectively, which are defined as follows on the surface $y = x^2$ (on which $dy = 2xdx$):

$$\omega_1 = (dz - 3x^2\, dx)|_{y=x^2} = (dz - (3/2)x\, dy)|_{y=x^2},$$

$$\omega_2 = dz - \frac{5x^4 + 3x^2 z}{1 - x^3}\, dx|_{y=x^2} = dz - \frac{5x^3 + 3xz}{2 - 2x^3}\, dy|_{y=x^2}$$

$$\omega_3 = [2(2x + az)\, dz - 2x\, dx]|_{y=x^2} = [2(2x + az)\, dz - dy]|_{y=x^2}.$$

Consequently, in case 1

$$\mathbf{G}_2 = (3/2)x\mathbf{G}_1 + (x^2 - y)\mathbf{B};$$

here and below $\mathbf{B}$ (different in each of the cases 1, 2, 3) is a smooth function of x, y, z, ε. Therefore, the fast-slow system has the form

$$\dot{x} = (x^2 - y)\mathbf{A}, \qquad \dot{y} = \varepsilon\mathbf{G}_1, \qquad \dot{z} = \varepsilon[(3/2)x\mathbf{G}_1 + (x^2 - y)\mathbf{B}].$$

For a generic system, $\mathbf{A}(0)\mathbf{G}_1(0) \neq 0$. With a stretching of the axes and a change of time one may obtain the equalities $|\mathbf{A}(0)| = |\mathbf{G}_1(0)| = 1$. Reversing the orientation of x, if necessary, we obtain the equality $\mathbf{A}(0) = 1$. Since 0 is a jump point, the phase curves on the stable part of the slow surface ($x < 0$) approach the fold curve. Thus, $\mathbf{G}_1(0) < 0$; and, this means $\mathbf{G}_1(0) = -1$. By stretching the z-axis we change the coefficient in the last equation from 3/2 to 1. This proves the Corollary of the theorem in case 1.

In case 2 an analogous argument gives

$$\mathbf{G}_2 = \frac{5x^3 + 3xz}{2 - 2x^3}\mathbf{G}_1 + (x^2 - y)\mathbf{B}.$$

Stretching the coordinate, time and parameter axes (the stretching coefficients of the time and parameter axes are positive), one achieves $\mathbf{A}(0) = \mathbf{B}(0) = -\mathbf{G}_1(0) = 1$, and the coefficient of εx^3 in the equation for z has modulus 1. This proves the Corollary in case 2.

Finally, in case 3,

$$\mathbf{G}_1 = 2[2x + a(\varepsilon)z]\mathbf{G}_2 + (x^2 - y)\mathbf{B}.$$

The rest of the proof is analogous to that in case 2.

The last assertion in the Corollary follows from the previous reasoning and the theorem in Sect. 5.7 of Chap. 2. ▶

3.4. Derivation of the Systems of First Approximation. Proposition 2 is proved here. The fiber-preserving diffeomorphism in Proposition 2 is that diffeomorphism that takes the system into one of the normal forms (7)–(9). The changes of variables shown in Table 1 transform these systems into the following

$$\xi' = \xi^2 - \eta + O(\mu), \quad \eta' = -1 + O(\mu), \quad \zeta' = -\xi + O(\mu); \tag{7'}$$

$$\xi' = \xi^2 - \eta + O(\mu), \quad \eta' = -1 + O(\mu), \quad \zeta' = (a - \delta)\xi^3 - a\xi\eta + O(\mu), \tag{8'}$$

where $a = \dfrac{\partial B}{\partial x}(0)$;

$$\xi' = \xi^2 - \eta + O(\mu), \quad \eta' = (2\xi + a(\varepsilon)\zeta) + O(\mu), \quad \zeta' = \pm 1 + O(\mu). \tag{9'}$$

Taking limits as $\mu \to 0$, we obtain the system of first approximation from Proposition 2.

3.5. Investigation of the Systems of First Approximation. A phase curve γ of the system of first approximation is called *approximating* if it possesses the following property. Suppose $\{f_\mu^{-1}\}$ is the family of contractions inverse to the stretchings with whose help the system of first approximation was obtained from the fast-slow system. Then, there exists a neighborhood of the origin whose intersection with $f_\mu^{-1}\gamma$ converges, as $\mu \to 0$, to an arc of a regular phase curve of the corresponding degenerate system. The solution corresponding to the approximating phase curve is called *approximating.*

Proposition 3. *The system*

$$\xi' = \xi^2 - \eta, \qquad \eta' = -1$$

has a unique approximating phase curve.

◂ Pontryagin (1957) (see also Mishchenko and Rozov (1975, §§9, 10)) it is proved that this system has a solution $(\xi(\tau), \eta(\tau))$ of the form

$$\eta(\tau) = -\tau, \ \xi(\tau) = -\sqrt{(-\tau)} + O(1) \qquad \text{for } \tau \leqslant 0, \qquad \text{and } \xi(\tau) \to \infty \text{ as } \tau \to \tau_0,$$

where τ_0 is some positive constant. We denote the corresponding phase curve by γ; we denote by γ^+ and γ^- the positive and negative semi-trajectories of γ, respectively. The curve γ is called separating, since all solutions with initial conditions above γ tend to infinity as $\tau \to -\infty$, while all solutions with initial conditions below γ diverge to infinity in a finite time as τ decreases; see Fig. 71.

The semi-trajectory γ^+ is situated in a vertical half-strip belonging to the upper half-plane. Under the action of the diffeomorphism f_μ^{-1} this half-strip contracts and in the limit becomes the positive ξ half-axis, which is a phase curve of the fast system.

Under the action of the diffeomorphism f_μ^{-1} the semi-trajectory γ^- becomes the curve

$$\{(-\sqrt{y} + O(\mu), y \mid y \geqslant 0\}$$

and as $\mu \to 0$ it approaches the stable part of a slow curve. ▸

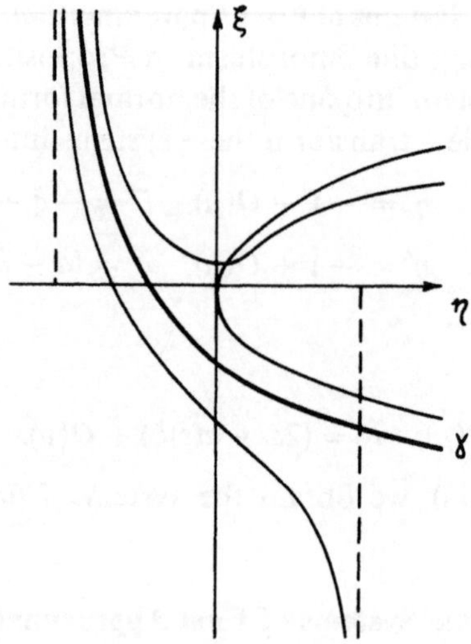

Fig. 71. The phase curves of the system of first approximation in the case of one fast and one slow variable. An approximating phase curve is singled out.

Proposition 4. *The system*

$$\xi' = \xi^2 - \eta, \qquad \eta' = -1, \qquad \zeta' = -\xi$$

has an approximating phase curve that is defined uniquely up to a translation along the ζ-axis.

◂ The required solution of the above system has the form:

$$(\xi(\tau), \eta(\tau), \zeta(\tau)),$$

where $(\xi(\tau), \eta(\tau))$ is the approximating solution from Proposition 3, and

$$\zeta(\tau) = -\tfrac{2}{3}(-\tau)^{3/2} + O(\tau) \qquad \text{as} \qquad \tau \to -\infty.$$

This equality follows from the formula for $\xi(\tau)$, and the equation $\zeta' = -\xi$. The phase curve of the slow system corresponding to the system (7) which approaches the origin on the stable part of the slow surface, has the form

$$\{(x, x^2, \tfrac{2}{3}x^3) | x \leqslant 0\}.$$

The rest of the proof is like the previous one. ▸

Remarks 1. In case 2, the system of first approximation does not have an approximating phase curve, but it does have a family of negative semi-trajectories; one is obtained from another by a translation along the ζ-axis, which, under the action of the contraction f_μ^{-1}, converge as $\mu \to 0$ to the curve

$$\{(x, x^2, \tfrac{2}{5}\delta x^5) | x \leqslant 0\}.$$

This curve has a tangency of high order with the phase curve of the slow system corresponding to the system (8):

$$\{(x, x^2, (2/5)\delta x^5)(1 - kx^3)^{-1} | x \leqslant 0\},$$

ending at the origin; their 7-jets at the origin coincide.

2. Close to a singular point on the fold of the slow surface, as $\varepsilon \to 0$, the phase curves of the fast-slow system may converge to a degenerate duck; see Sect. 5 below.

3.6. Funnels

Definition (Takens, (1976)). A point p on the boundary of the stable part of the slow surface is called a *funnel* if in any neighborhood of p there exists a region through whose points pass phase curves of the degenerate system that jump from the surface of slow motions at the point p.

Example. A singular point of the slow system of folded nodal type (see Fig. 68b) is a funnel. The results of Sect. 2.5 show that such funnels cannot be removed by small perturbations of the fast-slow system.

Theorem (Takens (1976)). *Generic constrained systems in the case of three or more slow variables have no funnels.*

3.7. Periodic Relaxation Oscillations in the Plane

Theorem (Zheleztsov (1958)). *Suppose the right-hand sides of the two-dimensional system*

$$\dot{x} = F(x, y, \varepsilon), \qquad \dot{y} = \varepsilon G(x, y, \varepsilon) \qquad (x, y \in \mathbb{R}) \qquad \text{(2 bis)}$$

are continuously differentiable, the function F twice and the function G once. Assume that the corresponding degenerate system

$$f(x, y) = 0, \qquad \dot{y} = g(x, y) \qquad \text{(3 bis)}$$

has a closed phase curve L_0. Then there exists a number $\varepsilon_0 > 0$ such that, for each value of the parameter $\varepsilon \in (0, \varepsilon_0)$, there is a neighborhood of the trajectory L_0, in which a unique stable limit cycle L_ε of the system (2 bis) lies. Moreover, L_ε converges to L_0 as $\varepsilon \to 0$.

For example, the van der Pol equation has one unique stable limit cycle, near the dashed curve illustrated in Fig. 64.

For the limit cycle L_ε of the last theorem there exist asymptotic approximations to an arbitrary prescribed order of smallness in ε.

Theorem (Mishchenko and Rozov (1975)). *Suppose that the right-hand sides of the generic two-dimensional system (2 bis) are infinitely differentiable. Assume that the corresponding degenerate system has a closed trajectory L_0, and assume that at each jump point p the condition $f_{xx}(p) \neq 0$ is satisfied. Then the period T_ε of the relaxation oscillation corresponding to the limit cycle L_ε has the following asymptotic expansion (valid as $\varepsilon \to 0$)*

$$T_\varepsilon = T_{00} + \sum_{k=2}^{\infty} \varepsilon^{k/3} \sum_{\nu=0}^{\pi(k-2)} T_{k,\nu} \ln^\nu(1/\varepsilon),$$

where π is an integer-valued function on the natural numbers given by

$$\pi = [k/3] + \begin{cases} 0 & \text{if } k \not\equiv 1 \pmod 3, \\ 1 & \text{if } k \equiv 1 \pmod 3, \end{cases}$$

and the $T_{k,\nu}$ are numerical coefficients that can be effectively computed (without integrating the system (2 bis) or (3 bis)) as values of well defined functionals of the functions $f(x, y)$ and $g(x, y)$ and the curve L_0. The asymptotic expansions of the amplitude X_ε of the x-component, and the amplitude Y_ε of the y-component of the relaxation oscillation, have the same form and structure as the asymptotic expansion of the period T_ε.

The first four terms of the expansion for T_ε give the asymptotic formulae of Dorodnitsyn (1947) and Zharov, Mishchenko & Rozov (1981) for the period T_λ and amplitude X_λ of the periodic solution of the van der Pol equation

$$\ddot{x} - \lambda(1 - x^2)\dot{x} + x = 0:$$

$$T_\lambda = 1.613706\lambda + 7.014432\lambda^{-1/3} - (2/3)((\ln \lambda)/\lambda)) - 1.3233\lambda^{-1} + O(\lambda^{-5/3}),$$

$$X_\lambda = 2 + 0.77937\lambda^{-4/3} - (16/27)((\ln \lambda)/\lambda^2)) - 0.8762\lambda^{-2} + O(\lambda^{-8/3}).$$

In the case of a system (2) of order $n > 2$, the asymptotic representations as $\varepsilon \to 0$ of a closed trajectory L_ε of a relaxation oscillation and its period T_ε are calculated up to terms of order $O(\varepsilon)$ in Pontrygin (1957) and Mishchenko and Rozov (1975); moreover, it is assumed there that the jump points are generic (see Sect. 3.2 above).

Recently the asymptotics of the solution near a generic jump point on a fold structure up to any power of ε were announced in the general case by A.Yu Kolesov and E.F. Mishchenko [5*] and in the case of two slow variables by I.V. Teperin [8*].

Translator's Note. H. Bavinck and J. Grasman (Int. J. of Nonlinear Mechanics 9 (1974), 421–434) obtained the formulae:

$$T_\lambda = (3 - 2\ln 2)\lambda + 3 3\alpha_1\lambda^{-1/3} - (2/3)((\ln \lambda)/\lambda))$$
$$+ (\ln 2 - 3 + 3b_1 - 1 - \ln \pi - 2\ln[Ai'(-\alpha_1)]\lambda^{-1} + o(1/\lambda),$$

$$X_\lambda = 2 + (\alpha_1/3)\lambda^{-4/3} - (16/27)(\ln \lambda)/\lambda))$$
$$+ [(1/3)b_1 - (1 - 2\ln 2 + 8\ln 3)/9]\lambda^{-2} + O(\lambda^{-8/3}),$$

wherein $\alpha_1 \approx 2.33810741$ is the absolute value of the first negative zero of the Airy function $Ai(x)$, b_1 is a definite integral whose value is numerically calculated to be $b_1 \approx 0.17235$, and $\ln[Ai'(-\alpha_1)] \approx -0.35494670$. Using these numerical values one finds that Bavinck and Grasman's formulae agree with those of A.A. Dorodnitsyn except that they yield -1.32321 not -1.3233 for the coefficient of

λ^{-1} in T_λ. However, the extensive, highly precise, numerical studies reported by C.M. Anderson and J.F. Geer in the S.I.A.M. J. Appl. Math. 42 (1982), 678–693, did not agree with the coefficient of $-2/3$ of the $(\ln \lambda)/\lambda$ term in T_λ: the values obtained by Anderson and Geer were approximately $-9(-2/3)$, and they were unable to explain this discrepancy. A later paper by M. Dadfar, J. Geer and C. Anderson in the S.I.A.M. J. Appl. Math. 44 (1984), 881–895, tends to confirm Bavinck and Grasman's formula for X_λ.

§ 4. Delayed Loss of Stability as a Pair of Eigenvalues Cross the Imaginary Axis

In this section we consider systems with slowly varying parameters in which, as a parameter varies, an equilibrium state loses stability as a pair of eigenvalues crosses the imaginary axis. A delay occurs because the actual departure of a phase point from the equilibrium state that has lost stability takes place not immediately afterwards, but rather after a time during which (in an analytic system) the parameter changes by a finite amount. Analyticity is critical in this situation since in typical systems with finite (and even infinite) smoothness, jumping occurs over a time interval in which the parameter is not noticeably changed.

More generally, we consider fast-slow systems for which a singular point of the fast system loses stability as the slow variables change, as a pair of eigenvalues crosses the imaginary axis. For generic analytic systems, the positive semi-trajectories originating in some region of phase space converge, as $\varepsilon \to 0$, to phase curves of the degenerate system having arcs of comparable length, one of which lies on the stable part of the slow surface, the other lying on its unstable part. Motions of this description are like the "ducks" considered in Sect. 5 below.

4.1. Generic Systems. Consider the class of fast-slow systems (2), whose slow surfaces consist of regular points (this surface is projected diffeomorphically onto the space of slow variables along the space of fast variables). We also require that the set of nonhyperbolic equilibria of the fast system consist of points with two-dimensional center manifolds, and a pair of nonzero eigenvalues on the imaginary axis. We shall call such systems *systems of type* 2. These systems form an open set in a suitable function space.

The slow surface of a type 2 system splits into two domains, stable and unstable. The first consists of the stable equilibria of the fast system; the second consists of the unstable equilibria; their common boundary is called the boundary of stability. On the stable part of the slow surface (for a generic system of type 2), the following set of points forms an open set: those points from which phase curves of the slow system exit, that transversally intersect the boundary of stability, and are also such that as the parameter moves along these slow curves, a pair of eigenvalues of a singular point of the fast system crosses the imaginary axis transversally with positive speed. We shall call such points *proper* points.

In the sequel, we consider only proper points that lie on the stable part of the slow surface.

4.2. Delayed Loss of Stability. A phase point of a system of type 2, lying on the stable part of the slow surface and originating not too far from a proper point, quickly (for a time of order $|\ln \varepsilon|$) enters an $O(\varepsilon)$-neighborhood (a neighborhood of diameter $O(\varepsilon)$) of the slow surface; see Fig. 72. Then this phase point remains close to a slow trajectory, at least up to the moment when this trajectory reaches the boundary of stability. If the fast-slow system (2) is analytic, then an interesting, and somewhat unusual phenomenon, must occur during the future motion of this phase point: delayed loss of stability of the fast motion. Namely, the phase point still moves along the unstable part of the slow surface in an $O(\varepsilon)$-neighborhood of this slow trajectory for a time of order $1/\varepsilon$ after the slow trajectory intersects the boundary of stability. Moreover, during this $O(1/\varepsilon)$-time interval, the slow trajectory goes beyond the boundary of stability for a distance that is $O(1)$. Only then may the jump occur; that is, the fast motion, after a time of order $|\ln \varepsilon|$ (during which the slow variable changes by a small amount of order $\varepsilon|\ln \varepsilon|$), moves an $O(1)$-distance from the slow surface; see Fig. 72. This phenomenon was discovered and investigated in an example by Shishkova (1973); the general case is considered by Nejshtadt (1985), see [6*], [7*]).

If the original system has only finite smoothness (or even infinite smoothness, but is not analytic), then no long delay of the loss of stability generally occurs. In

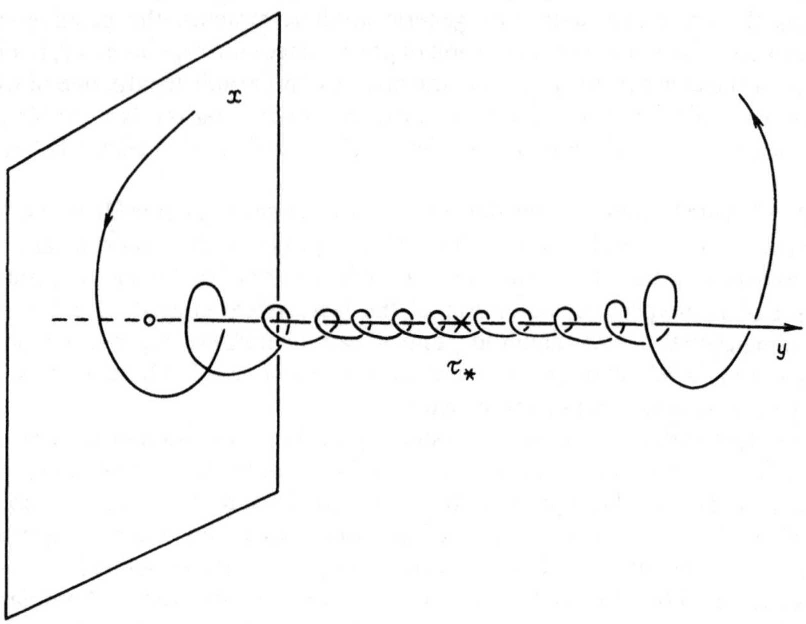

Fig. 72. Delayed loss of stability in a system of type 2 for one slow and two fast variables. The slow curve coincides with the y-axis. The cross marks the boundary of stability.

the class $C^k (k \geqslant 1)$, there is an open set of systems for which points leave the slow surface for an $O(1)$-distance, going beyond the boundary of stability by an amount $O(\sqrt{[\varepsilon|\ln \varepsilon|]})$. If $k = \infty$, then they travel away from the boundary of stability by an amount less than $M(\varepsilon)\sqrt{[\varepsilon|\ln \varepsilon|]}$, where $M(\varepsilon) \to \infty$ as $\varepsilon \to 0$, but $M(\varepsilon)$ may grow at an arbitrarily slow rate. In these systems, the jump from the slow surface takes place near the boundary of stability, at a distance of order ε^p $(p < \frac{1}{2})$. On the other hand, in a system of class C^k, one can guarantee that a point remains in an $O(\varepsilon)$-neighborhood of the slow surface, while going no less than an $O(\sqrt{[\varepsilon|\ln \varepsilon|]})$ distance from the boundary of stability. For $k = \infty$, $\sqrt{[\varepsilon|\ln \varepsilon|]}$ changes to $M(\varepsilon)\sqrt{[\varepsilon|\ln \varepsilon|]}$.

4.3. Hard Loss of Stability in Analytic Systems of Type 2. The loss of stability in analytic systems of type 2, described in the preceding subsection, is always hard, independently of whether or not the loss of stability in the family of fast equations is hard or soft as the parameter of the family (the slow variable) changes along a phase curve of the slow motion. Consider the following example, corresponding to any open set of fast-slow systems of type 2. Without loss of generality, one may assume that the slow surface has the form $x = 0$, since it is diffeomorphically projected onto the space of slow variables along the space of the last variables; suppose that in coordinates this projection takes the form $(x, y) \mapsto y$. Let $y(\tau)$ be a solution of the slow system, a phase curve of which at the moment $\tau = 0$ passes through the boundary of stability from the stable to the unstable part of the slow surface. Suppose that in the corresponding fast system $\dot{x} = f(x, y(\tau))$, with parameter τ, there is a soft loss of stability at $\tau = 0$ together with the birth of a stable limit cycle. The limiting behavior of a trajectory of the fast system in the fiber $y = y(\tau)$, with initial condition near to zero, changes continuously with τ: for $\tau \leqslant 0$ the phase curves converge to the equilibrium state $y(\tau)$, and for $\tau > 0$ they wind onto the limit cycle depending continuously upon τ. For small ε, a phase point of such a trajectory jumps off the slow surface, not at $\tau = 0$, but later, when the cycle has diameter of order 1. Therefore, a self-oscillatory regime arises suddenly with an amplitude of order 1; that is, the jump from equilibrium is hard.

4.4. Hysteresis. The following example illustrates the phenomenon of hysteresis in fast-slow systems of type 2. Consider the "triangular" system

$$\dot{x} = F(x, y, \varepsilon), \qquad \dot{y} = \varepsilon G(y, \varepsilon).$$

This system factorizes: its phase curves lie over phase curves of the slow system for any $\varepsilon > 0$. Together with the previous fast-slow system, we consider the system

$$\dot{x} = F(x, y, \varepsilon), \qquad \dot{y} = -\varepsilon G(y, \varepsilon).$$

We shall assume that there is only one slow variable in each of these systems. Let $y(\tau)$ be a solution of the first slow equation, passing at $\tau = 0$ from the stable part of the slow curve into the unstable part. Then there exist positive numbers τ_0 and τ^* such that the solution of the first system with the initial condition

$(x_0, y(-\tau_0))$, near to $y(-\tau_0)$, jumps away from the limit cycle near to the point $y(\tau^*)$. This cycle already has radius of order 1. A solution of the second system, with initial condition close to this cycle, drifts along cycles of the fast system corresponding to the parameter values $y(-\tau)$, and exits onto the slow surface near the point $y(0)$. Under the same assumptions, the evolution of a phase point of the second system does not reduce to the evolution of a phase point of the first system with the aid of reversal of time, which is in contrast to the evolution of attractors of the corresponding fast systems (hysteresis is observed).

4.5. The Mechanism of Delay. Here we shall show how the delay in loss of stability occurs. Consider a fast-slow system of type 2 with slow surface $x = 0$. The equations for the fast variables have the form

$$\dot{x} = A(y, \varepsilon)x + \varepsilon h(y, \varepsilon) + O(|x|^2). \tag{10}$$

Next, make a change of variables of the form

$$\tilde{x} = x + \varepsilon A^{-1}h.$$

The system for $\tilde{x}$ has the same form as (10), but the new function h has order ε. The next iteration of this transformation yields $h = O(\varepsilon^2)$, etc. This sequence of consecutive changes of variables diverges in general. Estimates show that after completing $O(1/\varepsilon)$ such changes of variables, one may reduce an analytic system to one the form (10) in which h is exponentially small: $h = O(\exp(-c/\varepsilon))$, $c =$ const. > 0; moreover, the composition of the changes of variables differs from the identity by $O(\varepsilon)$ (for analogous estimates, see Nejshtadt (1984)).

Consider now the motion of a phase point of the system (10) with h exponentially small and with initial point (x_0, y_0). Suppose that the slow curve with initial point y_0 exits to the boundary of stability in a slow time of order 1. Then, if x_0 is sufficiently small, the phase curve beginning at (x_0, y_0) exits to the boundary of stability (more exactly, crosses over it) in a slow time of order $1/\varepsilon$. Moreover, in the beginning $|x(t)|$ quickly decreases, becomes exponentially small, and remains that small up to its passage through the boundary of stability. Afterwards, $|x(t)|$ may quickly grow initially, but in order to grow from an exponentially small value to one of order ε, a fast time of order at least $1/\varepsilon$ is required. Conseqently, the loss of stability is delayed. If at some moment of time $|x(t)| \sim \varepsilon$, then in a fast time of order $|\ln \varepsilon|$, $|x(t)| \sim 1$ becomes true; that is, a jump occurs.

If the system has a finite degree of smoothness, then the above sequence of changes of variables stops after a finite number of steps, so that one may only achieve $h = O(\varepsilon^l)$. Then, upon passing through the boundary of stability, x will be $O(\varepsilon^{l+1})$. In order to grow from ε^{l+1} to ε (or from ε to 1) requires a fast time of order $\sqrt{[|\ln \varepsilon|/\varepsilon]}$.

4.6. Computation of the Moment of Jumping in Analytic Systems. Consider a fast-slow system of type 2, and fix its slow trajectory. We shall consider only those solutions of the fast-slow system whose initial conditions lie above an $O(\varepsilon)$-neighborhood of the fixed trajectory. A moment of slow time τ is called the asymptotic moment of jumping of a solution if, in an $O(\varepsilon|\ln \varepsilon|)$-neighborhood of τ, there lies an interval at whose left-hand end point the phase point $(x(\tau), y(\tau))$ is at a distance $O(\varepsilon)$ from the slow surface, and at whose right-hand end point $(x(\tau), y(\tau)) = O(1)$. A moment of slow time is called an *asymptotic moment of falling* if under a reversal of time it becomes an asymptotic moment of jumping.

In analytic systems one may express the asymptotic moment of jumping in terms of the asymptotic moment of falling, or at least estimate it from below (Shishkova (1973) and Nejshtadt (1985)). To achieve this requires a construction connected with analysis of solutions for complex values of time. So as to simplify the exposition, we give it under the hypotheses that the fast system is 2-dimensional. The formulation in the general case is given in a paper by Nejshtadt (1985).

Suppose $\tau \mapsto y(\tau)$, $\tau = \varepsilon t$, is a fixed solution of the slow system with $y(\tau_0) = y_0$. Suppose that for some τ_* its phase curve intersects the boundary of stability, and, moreover, at that moment a pair of complex conjugate eigenvalues $\lambda_1(\tau)$, $\lambda_2(\tau)$, crosses the imaginary axis with nonzero speed. We introduce the complex phase

$$\Psi(\tau) = \int_{\tau_*}^{\tau} \lambda_1(y(s))\, ds.$$

The restriction of Re $\Psi(\tau)$ to the real axis obviously has a nondegenerate minimum at τ_*. Therefore, on a sufficiently small interval of the real time axis adjoining τ_* from the left, a function Π can be defined assigning to the moments $\tau < \tau_*$ the moments $\Pi(\tau) > \tau_*$, such that

$$\mathrm{Re}\ \Psi(\tau) = \mathrm{Re}\ \Psi(\Pi(\tau)).$$

In the complex plane of slow time, the points τ and $\Pi(\tau)$ are joined by an arc L of the level curve Re $\Psi(\tau) =$ const.; see Fig. 73. If $\tau < \tau_*$ is sufficiently close to τ_*, then in a domain K, symmetric with respect to the real axis, bounded by L and by the complex conjugate of L, the following conditions are satisfied: 1) the slow trajectory is analytic, f and g are analytic at points of the slow trajectory, 2) $\lambda_{1,2} \neq 0$, 3) $\lambda_1 \neq \lambda_2$, 4) no tangent to the curve Re $\Psi(\tau) =$ const. is vertical. We define τ^- to be a lower bound of values $\tau < \tau_*$ for which the conditions (1)–(4) are satisfied. We define $\tau^+ = \Pi(\tau^-)$.

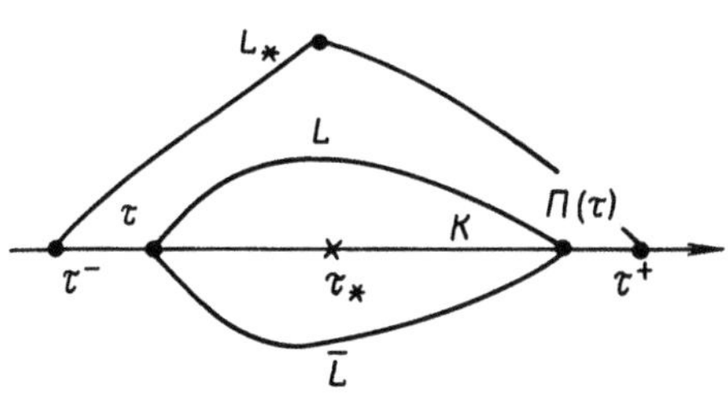

Fig. 73. Construction of the moment of jumping via the moment of falling ("the entrance-exit function")

Theorem (Nejshtadt (1985)). *Consider a fast-slow system of type 2 with a fixed slow solution. Suppose τ^- and τ^+ are defined as above, and suppose $\tau_0 \in (\tau^-, \tau^+)$ is an asymptotic moment of falling of a solution of the fast-slow system. Then $\Pi(\tau_0)$ is an asymptotic moment of jumping of this solution. On the interval $(\tau_0 + c\varepsilon|\ln\varepsilon|, \Pi(\tau_0) - c\varepsilon|\ln\varepsilon|)$ of slow time the fast-slow solution lies in an $O(\varepsilon)$-neighborhood of the fixed slow solution.*

If the moment of falling lies to the left of τ^-, then it is possible to show that the phase curve remains close to the slow solution, at least while $\tau < \tau^+ - \delta(\varepsilon)$, where $\delta(\varepsilon) \to 0^+$ as $\varepsilon \to 0$. For the computation of the asymptotic moment of jumping in this case, it is sufficient to investigate the behavior of the slow solution and eigenvalues on the curve L_*, an arc of the level curve $\operatorname{Re}\Psi(\tau) = \text{const.}$ having endpoints τ^- and τ^+.

In a number of examples τ^+ turns out to be an asymptotic moment of jumping for all motions whose asymptotic moment of fall lies to the left of τ^-.

Example 1. Consider the linear inhomogeneous system

$$\dot{z} = (y - i)z + \varepsilon a(y), \qquad z = x_1 + ix_2, \qquad \dot{y} = \varepsilon.$$

Suppose the function $a(y)$ is analytic for $|\operatorname{Im} y| < 2$. The slow solution $z = 0$, $y = \varepsilon t = \tau$ intersects the boundary of stability at $\tau = 0$. The eigenvalue $\lambda_1 = \tau - i$ becomes zero for $\tau = i$, and the arc L_* consists of two segments which join the points $\tau^- = -1$ and $\tau^+ = +1$ with the point $\tau^c = \mathrm{i}$ (the superscript c stands for critical); see Fig. 73. Suppose that the asymptotic moment of falling, τ_0, for the fast-slow solution $z(t)$ lies to the left of -1. Then $z(-1/\varepsilon) = O(\varepsilon)$. In order to compute $z(1/\varepsilon)$, it is convenient to move into the complex τ-plane from the point -1 to $+1$ along the arc L_*. A linear equation with a purely imaginary eigenvalue (reducing to zero for $\tau = i$) is obtained for z on L_*. Far from the point i, the value of $|z|$ only undergoes oscillations of order ε. The substantial change in $|z|$ takes place in a neighborhood of the point i, and is easily calculated by the method of stationary phase. One obtains

$$z(1/\varepsilon) = \sqrt{(2\pi\varepsilon)}|a(i)| + O(\varepsilon).$$

If $a(i) \neq 0$, then $\tau^+ = 1$ is an asymptotic moment of jumping.

Example 2. We add nonlinear terms to the previous system, and we set $a = 1$:

$$\dot{z} = (y - i)z + \varepsilon + \gamma z|z|^2, \gamma = \text{const.}, \qquad \dot{y} = \varepsilon.$$

It is proved (Shishkova (1973)) that nonlinearity turns out not to have a great influence on the jump from the slow solution $z = 0$, $y = \tau$. Again $\tau^+ = 1$ is an asymptotic moment of jump for motions for which the asymptotic moment of falling lies to the left of $\tau^- = -1$. In the slow system the loss of stability is soft for $\gamma < 0$, and is hard for $\gamma > 0$. In the full system, the difference caused by a change in the sign of γ is not displayed.

The arc L_* is characterized by one of the conditions (1)–(4) introduced above being violated on it. In the examples just considered the condition $\lambda_1 \neq 0$ is violated. The corresponding systems are not generic, since, for them, the slow surface remains regular as λ_1 goes to zero. For a generic system at the point where $\lambda_1(y(\tau_0)) = 0$ the slow solution comes to a fold of the slow surface and has branching of type $\sqrt{(\tau - \tau_0)}$. It is likely in this case that, under some genericity conditions, the phase points, attracted to the slow surface for $\tau < \tau^-$, jump from it almost simultaneously at $\tau = \tau^+$; this conjecture was expressed by Nejshtadt (1985).

4.7. Delay Upon Loss of Stability by a Cycle. Analogous delay phenomena accompany loss of stability by a limit cycle in analytic systems. Suppose that for each y, the fast system has a nondegenerate cycle L_y. The evolution of the system for y is obtained by averaging the phase of the motion of the subsystem for y in (2) along the cycle L_y (Pontryagin and Rodygin (1960)). We denote by $y = Y(\tau)$, $\tau = \varepsilon t$, the solution of the evolving system with $Y(\tau_0) = y_0$.

Suppose the cycle L_{y_0} is stable, that is its multipliers lie inside the unit circle. A phase point of the fast-slow system that begins its motion sufficiently close to this cycle for $\tau = \tau_0$, moves quickly, along the slowly evolving cycle $L_{Y(\tau)}$, remaining in its $O(\varepsilon)$ neighborhood all the time that the cycle is preserving its stability (Pontryagin and Rodygin (1960)). Let us assume that for some $\tau = \tau_*$ the cycle $L_{Y(\tau)}$ loses its stability either because a pair of multipliers intersect the unit circle in complex conjugate points, or because a single multiplier passes through the point -1, and the remaining multipliers remain in the unit circle. It turns out that the loss of stability for $\tau > \tau_*$ is delayed: for $\tau - \tau_* \sim 1$ the phase point still remains in the $O(\varepsilon)$-neighborhood of the cycle $L_{Y(\tau)}$, and only afterwards does the jump take place. In a nonanalytic system such a long delay, generally speaking, does not occur.

4.8. Delayed Loss of Stability and "Ducks". The phenomena investigated above fit the theory of "ducks"; see Sect. 5. A fast-slow trajectory of duck type also moves for a long time along the unstable part of the slow surface after passing through a curve of degenerate equilibria. But "ducks" are relatively rarely found in systems depending upon an additional parameter: they exist for an exponentially small interval of values of this parameter.

For a delay upon loss of stability by a nondegenerate equilibrium (cycle) it is not necessary to select parameters. On the other hand "ducks" exist in systems with finite smoothness, and the delays studied in Sect. 4.1–4.6, generally speaking, occur only in analytic systems.

§ 5. Duck Solutions

In some special cases the phase curves of fast-slow equations may converge, as $\varepsilon \to 0$, to curves that consist not only of portions of the fast motion and stable arcs of a slow curve, but of unstable arcs as well. These limiting curves are called *ducks* because of their shapes; see Fig.75. The codimension of the corresponding set in the function space of relaxation systems with one fast and one slow variable is equal to 1 (this set includes those with a self-intersecting slow curve, and with a singular point of the slow system on a fold of the slow curve). Thus, ducks occur in a generic fashion by small perturbations of the parameter (at discrete values of the parameter) in generic one-parameter families of relaxation systems with one fast and one slow variable (besides the slow parameter ε, it is necessary to have one more parameter for ducks to arise).

In cases of a large number of slow variables, ducks are found even in generic systems.

5.1. An Example: A Singular Point on the Fold of the Slow Surface. Consider the family

$$\dot{x} = y - f(x), \qquad \dot{y} = \varepsilon(a - x) \tag{$11_{\varepsilon,a}$}$$

with parameters ε and a, which, for each value of the fixed parameter a, represents a fast-slow system of the form (2) of Sect. 1.

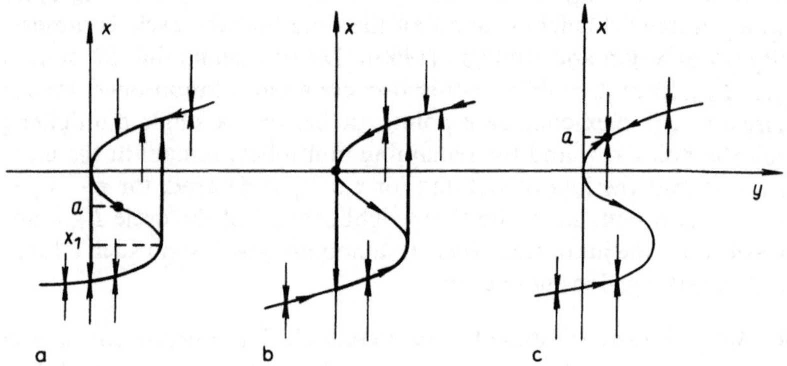

Fig. 74. Phase curve of a degenerate system: (a) $x_1 < a < 0$ (a large cycle). (b) $a = 0$ (?). (c) $a > 0$ (a stable singular point)

Suppose the graph of the function f resembles the graph of a cubic polynomial; see Fig. 74. Then the phase curves of the degenerate system are like those in Fig. 74a, b, and c for $x_1 < a < 0$, $a = 0$, and $a > 0$ respectively. For $a = 0$ many different phase curves of the degenerate system may pass through a single point: a phase curve originating at a singular point on the fold of the slow curve may coincide with this point, with the cycle shown by the heavy curve in Fig. 74b, going through it an infinite number of times, and may stop at the singular point on the fold after going around this cycle a finite number of times.

Consider the one-parameter subfamily (11_a) of the given family ($11_{\varepsilon,a}$), corresponding to the set $\varepsilon =$ const. Then, if $a > 0$ is large in comparison with ε, all solutions of the corresponding system converge to a stable equilibrium; but, for values of a lying on the interval between the x-extremes of the slow curve and sufficiently far from the end points of this interval, all solutions converge to a stable cycle, situated at a distance $O(\varepsilon^{2/3})$ from the cycle drawn in Fig. 75a. It turns out that as a changes in the family $(11_{\varepsilon,a})$ for $\varepsilon = \text{const.}$, the attracting solution changes continuously; and for a changing over a small $O(\exp(-1/\varepsilon))$-interval, the attracting solutions are close to duck solutions of the degenerate systems. Such solutions of the system $(11_{\varepsilon,a})$ also are called duck solutions. In Fig. 75 perestroikas of duck solutions in the family $(11_{\varepsilon,a})$ are shown for $f(x) = x^3 + x^2$, and in Fig. 76 for $f(x) = x^5 - x^3 + x^2$. In Fig. 77 the region of parameter space (ε, a) covered by slashes corresponds to the duck solutions. As

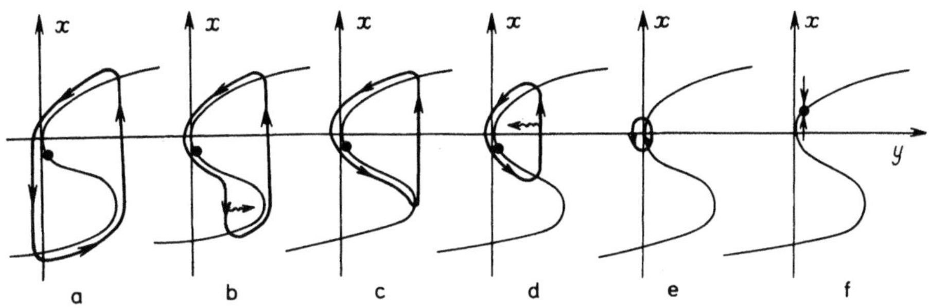

Fig. 75. Evolution of a cycle for $f(x) = x^2 + x^3$: (a) a large cycle; (b) "a duck with a head"; (c) & (d) "ducks with no heads"; (e) & (f) a bifurcation with the disappearance of a cycle

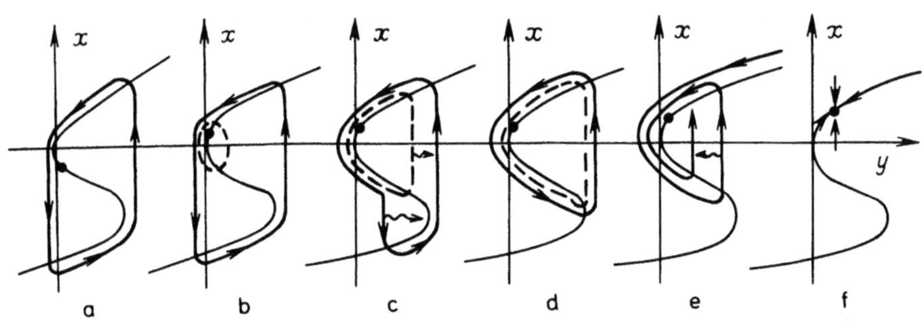

Fig. 76. Evolution of a cycle for $f(x) = x^2 - x^3 + x^5$: (a) a large cycle; (b) and (c) birth of a (small unstable) cycle; (c) an asymptotically unstable cycle inside an asymptotically stable cycle; (d) the merger of a stable and an unstable cycle and their disappearance; (e) a "staircase" with fixed steps; (f) a stable singular point.

a passes through 0 in the negative direction, the singular point (0, 0) loses stability. Whether the loss of stability is hard or soft depends upon the sign of $f'''(0)$: for $f'''(0) > 0$ the loss of stability is soft and a small, stable cycle is born, for $f'''(0) < 0$ it is hard, and a small, unstable cycle disappears. This bifurcation takes place outside the domain of parameters that correspond to the duck solutions shown in Fig. 75d and Fig. 75e and Fig. 76b and Fig. 76a.

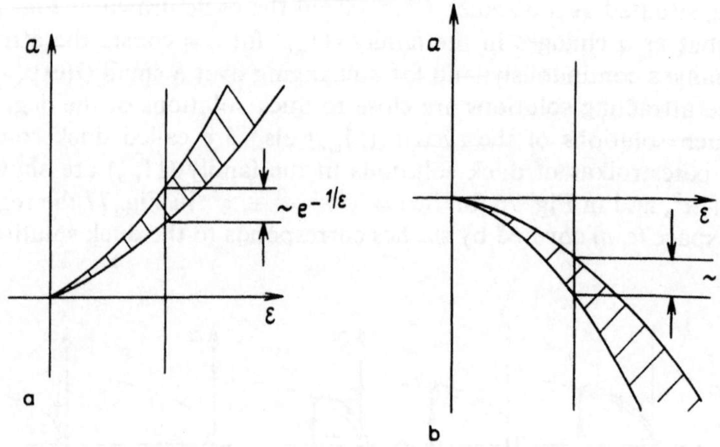

Fig. 77. The region of "duck" values of the parameters. Below this region a large cycle exists; above it a stable singular point. The line $a = 0$ is the bifurcation line of birth and death of a cycle. $f'(0) > 0$: (a) $f'''(0) < 0$. (b) $f'''(0) > 0$

5.2. Existence of Duck Solutions

Definition. *A simple, degenerate duck* is an oriented, connected curve, consisting of three arcs: the first and the last are intervals of the phase curve of the fast system, and the second is an arc of the slow curve, consisting of connected stable and connected unstable parts; see Fig 78. The stable part of this second arc is traversed first and the unstable part is traversed second.

Theorem 1 (Existence of ducks in a neighborhood of a fold). *Consider the system*

$$\dot{x} = f(x, y, a), \qquad \dot{y} = \varepsilon g(x, y, a). \tag{$12_{\varepsilon,a}$}$$

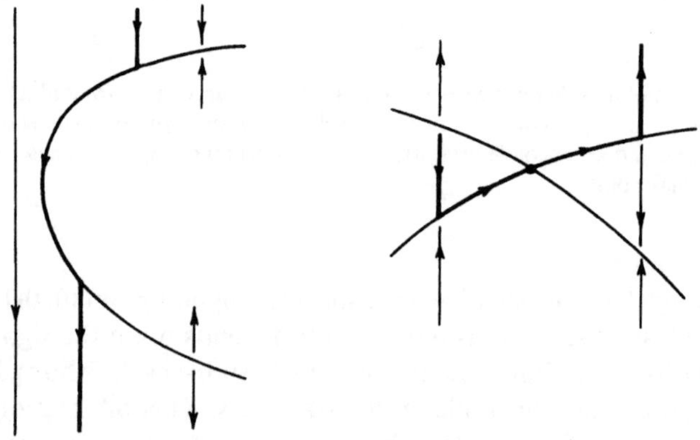

Fig. 78. Simple degenerate ducks

Suppose that for any fixed a *the slow curve of the corresponding fast-slow system has a fold point through which a singular point of the system passes with nonzero speed as the parameter* a *passes through* 0. *Then, for each simple degenerate duck passing through the fold point, there exists a function* $A\colon \varepsilon \mapsto A(\varepsilon)$, $A(0) = 0$, *such that the system* $(12_{\varepsilon, A(\varepsilon)})$ *has a solution, the phase curve of which converges to the degenerate duck as* $\varepsilon \to 0$.

Theorem 2 (Existence of ducks in a neighborhood of a self-intersection). *Suppose for* $a = 0$ *the slow curve of the system* $(12_{\varepsilon,a})$ *has a simple point of self-intersection, and at this point the condition* $\partial f/\partial a \neq 0$ *is satisfied. Suppose the function* g *is nowhere zero. Then, for each simple degenerate duck passing through the point of self-intersection, the conclusion of Theorem* 1 *holds.*

Theorem 3 (The life of ducks is short). *Suppose that* A_1 *and* A_2 *are two functions of either Theorem* 1 *or Theorem* 2, *corresponding to two degenerate ducks. Then there exists a* $c > 0$ *such that*

$$|A_1(\varepsilon) - A_2(\varepsilon)| < e^{-1/(c\varepsilon)}$$

for all sufficiently small ε.

All the functions $A(\varepsilon)$ corresponding to ducks have the same asymptotic expansion in powers of ε. There exists an algorithm for calculating the coefficients of this expansion, involving the derivatives of f and g at the critical point. The analogous assertion is correct for the duck solutions themselves: near the slow curve they are exponentially close.

Moreover, suppose there are two simple, degenerate ducks, two (possibly coinciding) functions $A_1(\varepsilon)$ and $A_2(\varepsilon)$, and two families of solutions of the system $(12_{\varepsilon, A_i(\varepsilon)})$, $i = 1, 2$, the phase curves of which converge to the corresponding degenerate ducks. Choose pieces of these phase curves, converging to an arc of the slow curve, which is formed by the intersection of slow arcs of two degenerate ducks, with fixed neighborhoods of the endpoints eliminated from this intersection. Then there exists a $c > 0$ such that one piece of the phase curve lies in an $e^{-1/(c\varepsilon)}$-neighborhood of the other piece for all sufficiently small ε. All the slow pieces of all duck solutions have the same asymptotic expansion in powers of ε. There exists an algorithm for calculating the coefficients in the expansions, involving the functions f and g and their derivatives.

5.3. The Evolution of Simple Degenerate Ducks. Fix an initial point (x_0, y_0) not lying on the slow curve, and such that a piece of the phase curve of the fast motion, starting from this point, falls onto a stable branch of the slow curve. Suppose ε is fixed, and a changes, passing through an interval on which the ducks are generated. Then taking the limit as $\varepsilon \to 0$, we obtain the evolution of simple degenerate ducks, shown in Figs. 79 and 80.

The limit of all positive trajectories may consist either of a finite or of an infinite number of simple degenerate ducks glued to one another (this is correct, in particular, for duck-cycles); the question: how are simple ducks glued together?

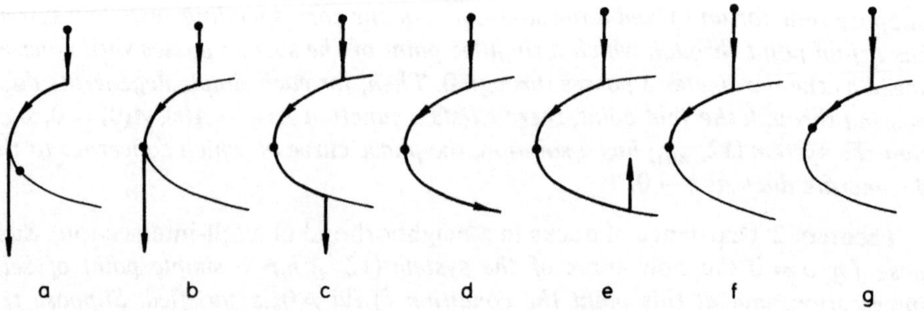

Fig. 79. The evolution of simple degenerate ducks at a fold point (ducks are present in c), d), e))

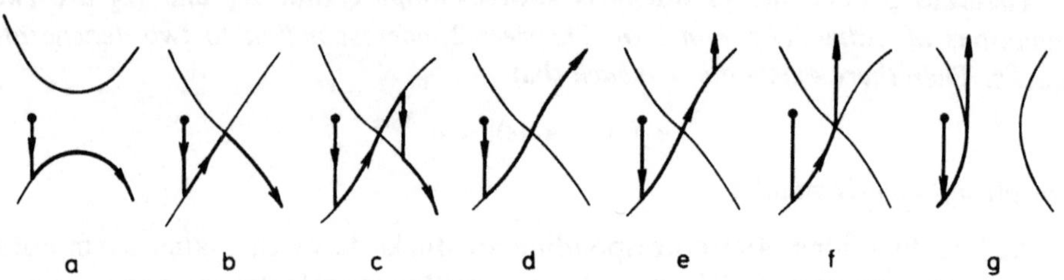

Fig. 80. The evolution of simple degenerate ducks close to points of self-intersection (ducks are present in c), d), e))

is answered with the aid of a so-called "entrance-exit" function; see Zvonkin and Shubin (1984) and their Ref. 2, and Benoit, Callot, Diener and Diener (1980).

5.4. A Semi-local Phenomenon: Ducks with Relaxation. Suppose that for $a = 0$ the slow curve has two fold points P and Q with the same coordinate y_0, and, moreover, the segment PQ contains no other points of the slow curve. Suppose the fast motion goes from P to Q. Next, we change the definition of a simple, degenerate duck by including an additional segment of the phase curve of the fast system between the stable and unstable parts of the slow curve; the new duck is called *a duck with relaxation.* We assume that as a passes through 0 the

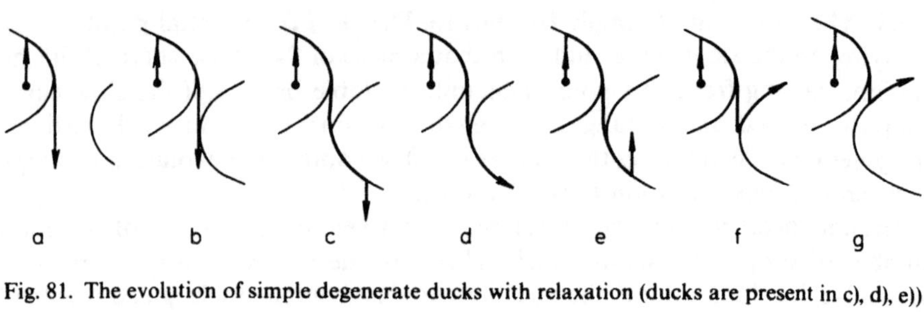

Fig. 81. The evolution of simple degenerate ducks with relaxation (ducks are present in c), d), e))

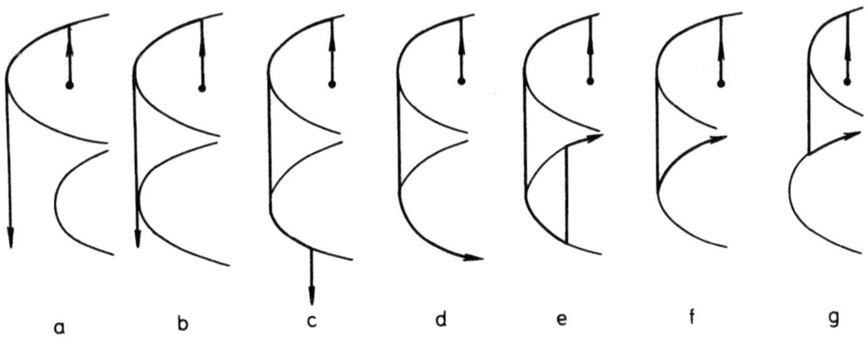

Fig. 82. The evolution of simple degenerate ducks with relaxation, a second variant (ducks are present in c), d), e))

y-coordinates of the points P and Q pass through each other with nonzero speed. Suppose that the function g is nonzero in neighborhoods of P and Q; and, moreover, assume that the sign of g is such that the slow motion in a neighborhood of P is directed towards P and, in a neighborhood of Q, is directed away from Q. Under these assumptions the analogues of Theorems 1 and, probably, Theorem 3 hold for ducks with relaxation. The evolution of ducks with relaxation is shown in Figs. 81 and 82.

5.5. Ducks in $\mathbb{R}^3$ and $\mathbb{R}^n$. In dimension 3 and higher, ducks exist in fast-slow systems with one fast variable even for generic systems (and not only for one-parameter families of equations, as in the two dimensional case). Consider the system

$$\dot{x} = f(x, y), \qquad \dot{y} = \varepsilon g(x, y), \tag{13_ε}$$

where $x \in \mathbb{R}$ and $y \in \mathbb{R}^{n-1}$, $n \geqslant 3$.

We change the definition of a duck yet again: *A simple degenerate duck* is an oriented, connected curve, consisting of three arcs: the first and the last are intervals of the phase curve of the fast system, and the second is a curve $\gamma = \gamma_1 \cup \{p\} \cup \gamma_2$, where p is a critical point, and γ_1 and γ_2 are intervals of phase curves of the slow system located on the stable and unstable branches of the slow surface (initially γ_1 is traversed and then γ_2). If the curve γ is smooth (at the point p), then we shall call the duck smooth.

A generic slow system on a two-dimensional surface in $\mathbb{R}^3$ may have singularities of three types: folded nodes, saddles and foci (see Sect. 2). Degenerate ducks exist only for folded saddles and for some folded nodes; see Fig. 83. In the case of a folded saddle (under additional nondegeneracy conditions, which we do not explicitly formulate here) the analogue of Theorem 1 holds (Benoit (1983)): for any simple degenerate duck passing through a folded saddle, the system (13_ε) has a solution whose phase curve converges to the degenerate duck as $\varepsilon \to 0$.

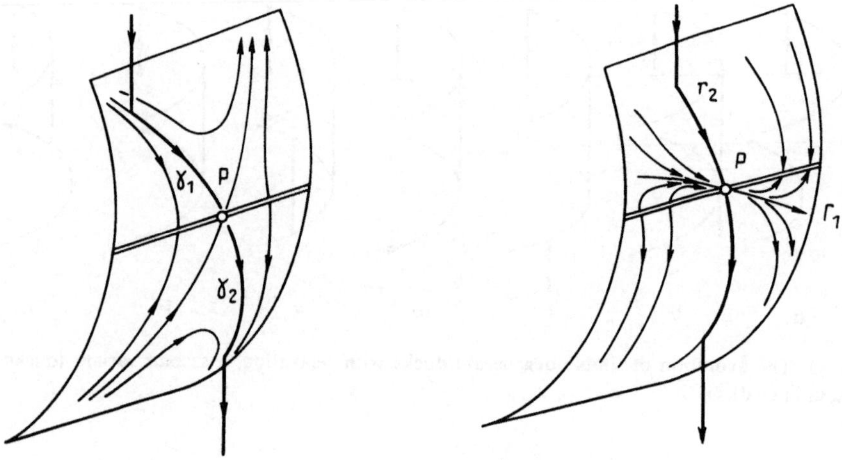

Fig. 83. Ducks in $\mathbf{R}^3$, passing through a folded saddle (a) and a folded node (b)

This principle is first violated in the case of a folded node: not every degenerate duck is a limit of a solution of the system (13_ε).

Example. Consider the system

$$\dot{x} = -(x^2 + y_1), \qquad \dot{y}_1 = \varepsilon(ay_2 + bx), \qquad \dot{y}_2 = \varepsilon. \tag{14_ε}$$

For $b^2/a > 8$, the slow system has a folded node at the origin: for $b > 0$ a degenerate duck passes through this node. It turns out (Benoit (1985)) that if the ratio between the eigenvalues of the linearization of the slow system is not an integer, then the smooth degenerate duck is a limit of phase trajectories of the system (14_ε) if and only if the arc of the duck lying on the slow surface is an arc of either the curve Γ_1 or the curve Γ_2, where Γ_1 and Γ_2 are analytic phase trajectories of the slow system; see Fig. 83b. At the same time, in this example, for some values of the parameters a and b there exist nonsmooth degenerate ducks that are limits of phase curves of the system (14_ε).

Ducks with relaxation have also been studied in $\mathbb{R}^3$ (Benoit (1983)). For simple smooth degenerate ducks in $\mathbb{R}^n$, S.N. Samborskij (1985) obtained necessary and sufficient conditions for the existence of a small deformation (of order ε) of the functions f and g in the system (13_ε) such that the deformed system has a solution that converges to a given degenerate duck as $\varepsilon \to 0$. These conditions are conditions on the joining at the critical point p and consist of the following: if the tangent to γ at the point p is not vertical, then $g(p) \neq 0$, but if it is vertical then $\partial g/\partial x(p) \neq 0$.

The credit for the discovery and investigation of ducks (1977) is due to a group of French mathematicians, namely to E. Benoit, J.-L. Callot, F. Diener and M. Diener. A survey and bibliography may be found elsewhere (Zvonkin and Shubin (1984) and Ref. 1 of that paper).

Recommended Literature

The literature on the theory of bifurcations is enormous: for example, the bibliography in S.-N. Chow and J.K. Hale (1982) contains over 700 entries. K. Shiraiwa has collected a bibliography of 4405 entries on dynamical systems (*Bibliography of Dynamical Systems*, compiled by K. Shiraiwa, Dept. of Math., College of General Education, Nagoya University, preprint series of 1985, No. 1, Nagoya, 390 pp.). Unfortunately, this work rather incompletely represents the theory of bifurcations, especially its early period (before 1970).

In his dissertation and in works on the theory of equilibria of a rotating fluid and on celestial mechanics, H. Poincaré established an informal basis for the theory of bifurcations, including, for example, the theory of versal deformations and the technique of normal forms. The formal foundations of bifurcation theory were created by Andronov and his students (Andronov (1933), (1956), Andronov, Vitt and Khajkin (1959), Andronov and Leontovich (1938), (1939) (1965), (1970), Andronov, Leontovich Gordon and Majer (1966), (1967) and Andronov and Lyubina (1935)). Their theory was motivated by their investigations of applied problems. In particular, they studied in detail the birth of cycles at bifurcations when an equilibrium loses its stability, the case of "Hopf bifurcation", so named by a misunderstanding. Unfortunately, these works of A.A. Andronov (1933), Andronov and Leontovich (1938, 1939, 1965) are not very widely known in the West.

N.N. Semenov (1934) and Ya.B. Zel'dovich (1941) investigated important practical applications of the theory of generic bifurcations, also including families of bifurcations; that is, what are now called "imperfect" bifurcations (Golubitsky and Schaeffer (1979)).

Papers by Yu.I. Nejmark (1959), N.N. Brushlinskaya (1961), V.K. Mel'nikov (1962) and R. Sacker (1965) belong to the early period of investigations of the birth of cycles or tori. Mel'nikov and Sacker in their papers corrected a mistake made by Nejmark, who discovered the birth of a torus by bifurcation upon the loss of stability by a self-oscillation, but omitted the case of a strong resonance. Mel'nikov (1962) and Sacker (1965) predicted "principal systems", as well as the basic characteristics of their versal deformations, in the cases of weak and strong resonances of orders different from 4. Modern formulations of these results were published by F. Takens in 1974 (up to the present his proof has not appeared), and for 1 : 1 resonances by V.I. Arnol'd (1972) (the proof was published by R.I. Bogdanov (1976)). Cases of strong resonance were investigated by V.I. Arnol'd (1977). The proofs for resonances of orders different from 4 were published only by E.I. Khorozov (1979). On bifurcations of self-oscillations near to 1: 4 resonances see Arnol'd (1978, 1977), Berezovskaya and Khibnik (1979, 1980) and Nejshtadt (1978).

N.N. Brushlinskaya (1968, 1965) applied the theory of bifurcations of tori to the hydrodynamic equations of Navier-Stokes – an area which became fashionable only after D. Ruelle and F. Takens announced its connections with turbulence (1981) (see, in addition, the addresses of A.N. Kolmogorov "Experiment and mathematical theory in the study of turbulence" and of N.N. Brushlinskaya (1965), both given at the meeting of the Moscow Mathematical Society on May 18, 1965). A survey of the contemporary status of the theory of bifurcations of tori was written by L.J. Broer; see Bruter *et al.* (1983). The "Hopf" bifurcation in hydrodynamics was also investigated by V.I. Yudovich (1965) and is discussed in detail by J.E. Marsden and M. McCracken in their book (1976). This book is also valuable for its extensive list of references.

A numerically oriented exposition of the theory and application of the "Hopf" bifurcation is contained in the book by Hassard, Kazarinoff and Wan (1981). Bifurcations in distributed systems and their applications to the theory of combustion are dealt with in the surveys by Vol'pert (1983, 1982). On the bifurcations of tori born at the loss of stability by self-oscillations see Afrajmovich and Shil'nikov (1983) and Aronson, Chory, McGehee and Hall (1982).

The analysis of bifurcations of phase portraits close to equilibria in generic one-parameter families of higher-dimensional systems was accomplished after the general reduction theorem of A.N. Shoshitajshvili (1975) appeared, which transferred the investigation of arbitrary local families to the study of their reductions to center manifolds. It is important to note that genericity of a reduced family is equivalent to the genericity of the original family; this is also proved in (Shoshitajshvili (1975)). The existence of the center manifold itself was established earlier by V.A. Pliss (Arnol'd (1972):

Ref. 70) (for the case of no unstable manifold) and the general case by A. Kelley (Marsden and McCracken (1976): Ref. 1] and by M.W. Hirsch, C.C. Pugh and M. Shub (1971) (a detailed exposition is in Hirsch *et al.* (1977))

Bifurcations of phase portraits close to equilibria in generic two-parameter families were completely investigated for the case of two zero eigenvalues by R.I. Bogdanov (1976). The study of bifurcations in the case of two purely imaginary pairs of eigenvalues or the case of one zero and one pure imaginary pair of eigenvalues, after a transformation to amplitudes, leads to the investigation of bifurcations in families of vector fields on the plane with an invariant pair of lines or an invariant line, respectively. The difficulties of this investigation turned out to be severe. After a series of attempts (Arnol'd (1972), Gavrilov (1978) and (1980), Khorozov (1981), Guckenheimer (1984) and Guckenheimer and Holmes (1983)), these difficulties were overcome by H. Żołądek; see Sects. 4.5 and 4.6 of Chapter 1 and (Żołądek (1983) and (1987)). Investigations of bifurcations of phase portraits in local three-parameter families of vector fields, containing germs with two zero eigenvalues at a singular point and with a supplementary degeneracy in their nonlinear terms, were mainly performed by Dumortier, Roussarie and Sotomayor (1985), Yakovenko (1984), by F.S. Berezovskaya and A.I. Khibnik in a paper in the collection [*Methods of the qualitative theory of differential equations*, Gorkij, 1985, pp. 128–138], and was finished by R. Roussarie (1986).

The greatest complications in the investigation of bifurations of equilibria in the plane are presented by the problem of the birth of limit cycles. As a rule, the basic part of the solution of this problem leads to investigations of Abelian, or similar to Abelian, integrals over the phase curves of special Hamiltonian systems. These investigations are produced by purely real methods (Bogdanov (1976), Markley (1979), and Nejshtadt (1978)), or with the aid of analytical continuation into the complex domain, and with the help of an application of the Picard-Lefschetz theorem and of the theory of elliptic integrals and Picard-Fuchs equations (Il'yashenko (1977), (1978), Petrov (1984), Khorozov (1979), Yakovenko (1984), Dumortier *et al.* (1985), and Sanders and Cushman (1985)).

The "dangerous" and "safe" parts of a stability boundary were investigated by N.N Bautin (1949); see also the paper by Bautin and Shil'nikov in the Russian translation of Marsden and McCracken (1976). The sizes of "dangerous" and "safe" deviations from a boundary of stability near all its strata up to and including those of codimension 3 were evaluated by L.G. Khazin and Eh. Eh. Shnol' (1985).

Normal forms of local vector fields and of diffeomorphisms (in relation to the analyticity, local smoothness and, above all, to finitely differentiable changes of variables) have been investigated by Bryuno (see Arnol'd and Il'yashenko (1985): Ref 18), Belitskij (1979), Gomozov (1976), Kostov (1984), Samovol (1982) and Takens (1971). On Feigenbaum universality, see the survey by Vul, Sinai and Khanin (1984) and the book by Collet and Eckmann (1980), which contains an extensive bibliography.

The early period of investigation of nonlocal bifurcations of vector fields on the plane and the sphere is summed up in the books by Andronov, Leontovich, Gordon and Majer (1966) and Bautin and Leontovich (1976). Structural stability and bifurcations of vector fields on two-dimensional surfaces other than the plane and the sphere were investigated comparatively recently (Peixoto (1962), Sotomayor (1973a,b), (1974)). The paper by Malta and Palis (1981) is closely related to the hypothesis on global bifurcations of vector fields in one-parameter families of vector fields on the sphere (Sect. 2.2 of Chapter 3).

Nonlocal bifurcations of higher-dimensional system have been investigated mainly by mathematicians of A.A. Andronov's school. On bifurcations of homoclinic trajectories of a nonhyperbolic saddle, see the works of L.P. Shil'nikov (1963, 1966, 1969). On bifurcations of homoclinic trajectories of a nonhyperbolic cycle, see Afrajmovich (1974), Afrajmovich and Shil'nikov (1974), (1982) and Newhouse, Palis and Takens (1983), of a hyperbolic saddle see Shil'nikov (1967, 1968, 1970), and Gaspard (1984). On bifurcations of contours (called cycles in the West), see Afrajmovich and Shil'nikov (1972), Gavrilov (1973), Gavrilov and Shil'nikov (1973), Gonchenko (1983), de Melo, Palis and van Strien (1981), Newhouse (1974, 1980), Newhouse and Palis (1976), Newhouse, Palis and Takens (1983) and Palis (1978). Nonlocal bifurcations in generic two-parameter families have been studied by Bykov (1977), (1980), Gonchenko (1983, 1984, 1980), Luk'yanov (1982) and Luk'yanov and Shil'nikov (1978). On cascades (chains) of bifurcations beginning with a point attractor and ending with a Lorenz attractor, see Afrajmovich, Bykov and Shil'nikov (1982), Ya.G. Sinai and L.P. Shil'nikov, eds. (1981) and Marsden and McCracken (1976). On different definitions of attractors, see

Sinai (1979), Sinai and Shil'nikov, eds. (1981), Guckenheimer and Holmes (1983), Marsden and McCracken (1976), Milnor (1985), Nitecki (1971) and Smale (1967).

Differentiable dynamics (see Nitecki (1971) and Smale (1967)) and symbolic dynamics (we especially mention V.M. Alekseev's book (see Anosov *et al.* (1985): Ref. 1) as containing a set of general references) are very useful tools in the nonlocal theory of bifurcations.

The term "relaxation oscillation" was introduced by van der Pol (1926). The early period of the development of the theory of relaxation oscillations is summarized in the book by Andronov, Vitt and Khajkin (1939) and (1957), where numerous applications are discussed. The connection between slow motions in systems of relaxation type and true motions was treated in the works of A.N. Tikhonov (1952), A.B. Vasil'eva (1952), and Gradshteijn (1953). On the asymptotics of solutions close to the moment of a jump, see the works of L.S. Pontryagin (1957), E.F. Mishchenko and N.Kh. Rozov (1975) and others. The phenomenon of delayed loss of stability in analytic fast-slow systems as a pair of eigenvalues of a singular point of the fast system crosses the imaginary axis is described for a particular example in a paper by L.S. Pontryagin's student, M.A. Shishkova (1973). For generic equations this phenomenon was investigated by A.I. Nejshtadt (1985).

Duck solutions of fast-slow systems were discovered and investigated by Benoit (1983), (1985) and Benoit, Callot, Diener and Diener (1980), and Samborskij (1985); see also the survey by Zvonkin and Shubin (1984).

As an example of a study of stochastic relaxation oscillations we point out an article by N.N. Chentsova (1982).

On applications of the method of averaging in the theory of relaxation oscillations see the book by N.N. Bogolyubov and Yu.A. Mitropolskij (1974).

In this survey we have ignored the widening and rapidly developing theory of bifurcations of systems with symmetries. The abundance of different groups of symmetries and of their applications, and also the wide range of problems with symmetry in applications, make this field very attractive; here already in the case of a small number of parameters complicated bifurcation diagrams are again typical. One may become acquainted with the contemporary state of this theory through the articles and the book of M. Golubitsky and D. Schaeffer (1979a,b), (1982) and (1985); see also Danglmayr and Armbruster (1983), Field (1984a,b), Golubitsky, Keyfitz and Schaeffer (1979) and Sattinger (1978), (1979) and Schaeffer and Golubitsky (1981).

References*

Afrajmovich, V.S. [1974] Certain global bifurcations, connected with the appearance of a countable number of periodic motions. Gor'kov. Gos. Univ., Gor'kij

Afrajmovich, V.S. [1984] Strange attractors and quasi-attractors. In: *Nonlinear and turbulent processes in physics.* R.Z. Sagdeev (ed.), *3* (Kiev, 1983), 1–34, Zbl.532.58018

Afrajmovich, V.S., Bykov, V.V., Shil'nikov, L.P. [1982] On structurally unstable attracting limit sets of Lorenz attractor type. *Tr. Mosk. Mat. Obsher. 44*, 50–212; English transl.: Trans. Mosc. Math. Soc. 1983, No. 2, 153–216 (1983), Zbl.506.58023

Afrajmovich, V.S., Shil'nikov, L.P. [1972] Singular trajectories of dynamical systems. *Usp. Mat. Nauk 27*, No. 3, 189–190, Zbl.274.54032

Afrajmovich, V.S., Shil'nikov, L.P. [1974a] Certain global bifurcations connected with the disappearance of fixed points of saddle-node type. *Dokl. Akad. Nauk SSSR. 219*, 1281–1284; English transl. Sov. Math., Dokl. 15, 1761–1765 (1975), Zbl.312.34035

Afrajmovich, V.S., Shil'nikov, L.P. [1974b] The accessible transitions from Morse-Smale systems to systems with several periodic motions. *Izv. Akad. Nauk SSSR*, Ser. Math. *38*, No. 6, 1248–1288;

*For the convenience of the reader, references to reviews in Zentralblatt für Mathematik (Zbl.), compiled using the MATH database, have, as far as possible, been included in this bibliography.

English transl.: Math. USSR, Izv. 8, 1235–1270 (1976), Zbl.322.58007

Afrajmovich, V.S., Shil'nikov, L.P. [1982] On bifurcations of codimension 1, leading to the appearance of fixed points of a countable set of tori. *Dokl. Akad. Nauk SSSR 262*, No. 4, 777–780; English transl. Sov. Math., Dokl. 25, 101–105 (1982), Zbl.509.58033

Afrajmovich, V.S., Shil'nikov, L.P. [1983] On invariant two-dimensional tori, their disintegration and stochasticity. In: *Methods of the qualitative theory of differential equations*. Gor'kov. Gos. Univ., Gor'kij, 3–26, Zbl.568.34001

Andronov, A.A. [1933] Mathematical problems of the theory of self-oscillations, *The First All-Union Conference on Auto-oscillations. Moscow-Leningrad*, GTTI, 1933, 32–71. (Also in: *Collected works*, Izd. Akad. Nauk SSSR, Moscow, 1956, 538 pp.)

Andronov, A.A. [1956] *Collected works*. Izdat. Akad. Nauk SSSR, Moscow, 538 pp.

Andronov, A.A., Vitt, A.A., Khajkin, S.Eh. [1937] *Theory of Oscillations*. 2nd ed., Fizmatgiz, Moscow, 1959, 916 pp., 1st ed. Moscow-Leningrad, 520 pp. English transl.: 1st ed., Princeton Univ. Press, 1949, ix + 358pp., 2nd ed., Pergamon Press 1966, 815 pp., Zbl.85,178

Andronov, A.A., Leontovich, E.A. [1938] Sur la théorie de la variation de la structure qualitative de la division du plan en trajectoires. *Dokl. Akad. Nauk SSSR* (2) *21*, 423–426, Zbl.22,22

Andronov, A.A., Leontovich, E.A. [1939] Some cases of dependence of limit cycles on parameters. *Uchen. Zap. Gor'kov. Univ.*, No. 6, 3–24

Andronov, A.A., Leontovich, E.A. [1965] Dynamical systems of the first degree of nonroughness on the plane. *Mat. Sb.*, Nov. Sev. *68* (110), No. 12, 328–372; English transl.: Am. Math. Soc., Transl., II. Ser. 75, 149–199 (1968), Zbl.143,179

Andronov, A.A., Leontovich, E.A. [1970] Sufficient conditions for nonroughness of first degree for dynamical systems on the plane. *Differ. Uravn. 6*, No. 12, 2121–2134; English transl.: Differ. Equations 6, 1610–1618 (1973), Zbl.218,216

Andronov, A.A., Leontovich, E.A., Gordon, I.I., Majer, A. G. [1966] *Qualitative theory of dynamical systems of second order*. Nauka, Moscow, 1966, 568 pp.; English transl.: J. Wiley and Sons, N.Y., 1973, 524 pp, Zbl.168,68

Andronov, A.A., Leontovich, E.A., Gordon, I.I., Majer, A.G. [1967] *The theory of bifurcations of dynamical systems on the plane*. Nauka, Moscow, 487 pp, Zbl.257.34001

Andronov, A.A., Lubina, A.D. [1935] Application of Poincaré's theory on bifurcation points and exchanges of stability to simple auto-oscillating systems. *Zh. Ehksper. Teoret. Fiziki 5*, No. 5, 3–4, 296–309, Zbl.12,130

Andronov, A.A., Pontryagin, L.S. [1937] Systèmes grossières. *Dokl. Akad. Nauk SSSR 14*, 247–250, Zbl.16,113

Anosov, D.V. [1960] On limit cycles of systems of differential equations with small parameters in the highest derivatives. *Mat. Sb., Nov. Ser. 50* (92), No. 3, 299–334; English transl.: Am. Math. Soc., Transl., II. Ser. 33, 233–275 (1963), Zbl.128,86

Anosov, D.V. [1977] Introductory article, in: *Smooth Dynamical Systems*. Mir, Moscow, 7–31

Anosov, D.V., Aranson, S.Kh., Bronshtejn, I.U. and Grines, V.Z. [1985] *Smooth dynamical systems*. Itogi Nauki Tekh., Ser. Sovrem. Probl. Math. Fundam. Napr. 1, 151–242, English transl.: Dynamical Systems I, Encycl. Math. Sci. 1, 149–233 (1988), Zbl.605.58001

Aranson, S.Kh. [1968] On the absence of nonclosed, Poisson-stable semi-trajectories and trajectories which are doubly asymptotic to a double limit cycle in dynamical systems of the 1st degree of nonroughness on orientable two-dimensional manifolds. *Mat. Sb.*, Nov. Ser. *76* (118), No. 2, 214–230; English transl.: Math. USSR, 56.5, 205–219 (1968), Zbl.159,119

Aranson, S.Kh. [1970] Dynamical systems on two-dimensional manifolds. *Proc. Int. Conf. Nonlinear Oscillations*. Kiev, 1969, vol. 2, 46–52, Zbl.286.34074

Aranson, S.Kh., Zhuzhoma, E.V., Malkin, M.I. [1984] On connections among smoothness and topological properties of transformations of the circle (theorems of the type of Denjoy). Gor'kov. Gos. Univ., Gor'kij, 152 pp.

Aranson, S.Kh. [1986] Funkts. Anal. Prilozhen 20, no. 1, 62–63; English translation: Functional Anal. Appl. 20

Armbruster, D., Dangelmayr, G., Gütinger, W. [1985] Imperfection sensitivity of interacting Hopf and steady-state bifurcations and their classification. *Physica D16*, 99–123, Zbl.579.58017

Arnol'd, V.I. [1961] Small divisors. I. On mappings of the circle onto itself. *Izv. Akad. Nauk SSSR, Ser. Mat. 25*, 21–86; English transl.: Am. Math. Soc. Transl., II. Ser. 46, 213–284 (1965), Zbl.135,426

Arnol'd, V.I. [1972] Lectures on bifurcations and versal families. *Usp. Math. Nauk 27*, No. 5, 119–184; Enghish transl.: Recrs. Math. Surv. 27, No. 5, 54–123 (1973), Zbl.248.58001

Arnol'd, V.I. [1977] Loss of stability of auto-oscillations close to resonance and versal deformations of equivariant vector fields. *Funkts. Anal. Prilozh. 11*, No. 2, 1–10; English transl.: Funct. Anal. Appl. 11, 85–92 (1977), Zbl.356.57029

Arnol'd, V.I. [1978] *Supplementary chapters of the theory of ordinary differential equations.* Nauka, Moscow, 304 pp; French transl.: Mir, Moscow, 323 pp., Zbl.455.34001; English transl.: Geometric methods in the theory of ordinary differential equations, Springer, 1983

Arnol'd, V.I. [1982] Evolution of singularities of potential flows in a collision-free medium and caustic metamorphoses in three-dimensional space. *Tr. Semin. Im. I.G. Petrovskogo 8*, No. 8, 21–57; English transl.: J. Sov. Math. 32, 229–258 (1986) Zbl.504.58032

Arnol'd, V.I. [1983] Remarks on perturbation theory for problems of Mathieu type. *Usp. Mat. Nauk 38*, No. 4, 189–203; English transl.: Russ. Math. Surv. 38, No. 4, 215–233 (1983), Zbl.541.34035

Arnol'd, V.I. [1984] *Ordinary differential equations.* 3rd ed., Nauka, Moscow, 272 pp; English transl.: MIT Press, London, (1980), Zbl.243.34001, Zbl.577.34001

Arnol'd, V.I., Varchenko, A.N. & Gusejn-Zade, S.M. [1982] *Singularities of differentiable mappings I. Classification of critical points, caustics and wave fronts.* Nauka, Moscow, 304 pp; English transl.: Birkhäuser, Boston etc. (1985), Zbl.513.58001

Arnol'd, V.I., Il'yashenko, Yu.S. [1985] *Ordinary differential equations.* Itogi Nauki Tekh., Ser. Sovrem. Probl. Mat., Fundam. Napr. *1*, 7–149; English transl.: Dynamical System I. Encycl. Math. Sci. 1, 1–148 (1988) Zbl.602.58020

Arnol'd, V.I., Kozlov, V.V., Nejshtadt, A.I., [1985] *Mathematical aspects of classical and celestial mechanics.* Itogi Nauki Tekh., Ser. Sovrem. Probl. Mat., Fundam. Napr. *3*, 5–304; English transl.: Dynamical Systems III, Encycl. Math. Sci. 3, 1–129 (1987), Zbl.612.70002

Aronson, D.G., Chory, M.A., McGehee, R.P., Hall, G.R. [1982] Bifurcation from an invariant circle for two parameter families of maps of the plane. *Commun. Math. Phys. 83*, 303–354, Zbl.499.70034

Bautin, N.N. [1949] *The behavior of dynamical systems near to the boundaries of stability.* Gostekhizdat., Moscow-Leningrad, 164 pp, 2nd. ed. Nauka, Moscow (1984), Zbl.537.34001

Bautin, N.N., Leontovich, E.A. [1976] *Methods and procedures of qualitative investigations of dynamical systems on the plane.* Nauka, Moscow, 496 pp.

Bautin, N.N., Shil'nikov, L.P. [1980] The behavior of of dynamical systems near to the boundaries of stability of equilibria ("dangerous" and "safe" boundaries), pp. 294–316 of the Russian Trans., *Bifurcations giving birth to a cycle and its applications.* Mir, Moscow, 368 pp., of *The Hopf bifurcation and its applications* by J.E. Marsden & M.McCracken, Appl. Math. Sci., Vol. 19, Springer-Verlag: New York, Heidelberg, Berlin, 1976, xiii, 408 pp, Zbl.346.58007

Belair, J., Glass, L. [1985] Universality and self-similarity in the bifurcations of circle maps. *Physica D16*, 143–154, Zbl.593.58030

Belitskij, G.R., [1978] Equivalence and normal forms of germs of smooth mappings. *Usp. Mat. Nauk 33*, No. 1, 95–155; English transl.: Russ. Math. Surv. 33, No. 1, 107–177 (1978), Zbl.385.58007

Belitskij, G.R. [1979] *Normal forms, invariants and local mappings.* Naukova Dumka, Kiev, 176 pp. Zbl.479.58001

Belyakov, L.A. [1980] On the bifurcation set in a system with a homoclinic saddle curve. *Mat. Zametki 28*, 911–922; English transl.: Math. Notes 28, 910–916 (1981), Zbl.454.34022

Belykh, V.N., Pedersen, N.F., Soerensen, O.H. [1977] Shunted Josephson junction model, II., The nonautonomous case. *Phys. Rev. B 14*, 4860–4871

Benoit, E. [1983] *Canards de* $\mathbb{R}^3$. Thèse, Paris, Zbl.489.34061

Benoit, E. [1985] Enlacements de canards. *C.R. Acad. Sci., Paris*, Sér. I. *300*, No. 8, 225–230, Zbl.577.34027

Benoit, E., Callot J.-L, Diener, F., Diener, M. [1981] Chasse au canards. *Collect. Math. 31*, 37–119, Zbl.529.34016

Berezovskaya, F.S., Khibnik, A.I. [1979] On the problem of bifurcations of auto-oscillations, near to a 1 : 4 resonance. Preprint, N.I.V.Ts., Akad. Nauk SSSR, Pushchino, 24 pp.

Berezovskaya, F.S., Khibnik, A.I. [1980] On bifurcations of separatrices in the problem of loss of stability by auto-oscillations near to a 1 : 4 resonance. *Prikl. Mat. Mekh. 44*, 938–942, English transl.: J. Appl. Math. Mech. 44, 663–667 (1981), Zbl.485.58015

Bogdanov, R.I. [1976] Bifurcations of a family of vector fields on the plane. *Tr. Semin. Im. I.G. Petrovskogo, 2*, 23–36; Versal deformation of a singular point of a vector on the plane in the case of zero eigenvalues, ibid, 37–65; English transl.: Sel. Math. Sov. 1, 373–387 and ibid, 389–421 (1981), Zbl.518.58029 and Zbl.518.58030

Bogoljubov, N.N., Mitropol'skij, Yu. A. [1974] *Asymptotic methods in the theory of relaxation oscillations.* Nauka, Moscow, 503 pp, Zbl.303.34043

Braaksma, B.L., Broer, H.W., Takens, F. (eds.) [1985] *Dynamical systems and bifurcations. Proceedings,* Gronigen, 1974, Lect. Notes Math. 1125, Springer-Verlag: New York, Heidelberg, Berlin, 129 pp, Zbl.552.00007

Brushlinskaya, N.N. [1961] Qualitative integration of a system of *n* differential equations in a region, containing a singular point and a limit cycle. *Dokl. Akad. Nauk SSSR 139*, 9–12, English transl.: Sov. Math., Dokl. 2, 845–848 (1961), Zbl.168,338

Brushlinskaya, N.N. [1965a] Generation of limit cycles and tori at transcritical values of the Reynolds number. *Usp. Mat. Nauk 20*, No. 5, 259–260

Brushlinskaya, N.N. [1965b] On the behavior of the equations of hydrodynamics as the Reynolds number passes through a critical value. *Dokl. Akad. Nauk SSSR 162*, 731–734, English transl.: Sov. Math. Dokl. 6, 724–728 (1965), Zbl.151,144

Brushlinskaya, N.N. [1971] A finiteness theorem for families of vector fields in a neighborhood of a singular point of Poincaré type. *Funkts. Anal. Prilozh. 5*, No. 3, 10–15; English transl.: Funct. Anal. Appl. 5, 177–181 (1972), Zbl.243.34008

Bruter, C.P., Aragnol, A., Lichnerowicz, A.D., (Eds.) [1983] *Bifurcation theory, mechanics and physics.* J. Reidel, Dordrecht, 388 pp, Zbl.516.00020

Bunimovich, L.A., Pesin, Ya.V., Sinai, Ya.G., Yakobson, M.V. [1985] *Ergodic theory for smooth dynamical systems.* Itogi Nauki Tekh., Ser. Sovrem. Problem. Mat., Fundam. Napr. *2*, 113–232, English transl.: Dynamical systems 2, Encyclopedia of Math. Sciences 2, Zbl.678.00021

Bykov, V.V. [1977] On the birth of periodic motions from a separatrix-contour of a three-dimensional system. *Usp. Mat. Nauk 32*, No. 6, 213–214, Zbl.376.34026

Bykov, V.V. [1980] On bifurcations of dynamical systems, near to systems with a separatrix-contour containing a saddle-focus. In: *Methods of the qualitative theory of differential equations*, pp. 44–72, E.A. Leontovich-Andronova (ed.), Gor'kov. Gos. Univ., Gor'kij

Bylov, B.F., Vinograd, R.Eh., Grobman, D.M., Nemytskij, V.V., [1966] *Theory of Lyapunov exponents.* Nauka, Moscow, 576 pp, Zbl.144,107

Campanino, M., Epstein, H. [1983] On the existence of Feigenbaum's fixed point, Comm. Math. Phys. 79, 261–302

Chenciner, A., [1981] Courbes invariantes non normalement hyperboliques au voisinage d'une bifurcation de Hopf dégénerée de difféomorphismes $\mathbf{R}^2$. *C.R. Acad. Sci. Paris*, Sér. 1 *292*, 507–510, Zbl.475.58016

Chenciner, A., [1985] Bifurcations de points fixes elliptiques II. Orbites périodiques et ensembles de Cantor invariants. *Invent. Math. 80*, 81–106, Zbl.578.58031

Chenciner, A., Iooss, G. [1979] Bifurcations de tores invariantes. *Arch. Ration. Mech. Anal. 69*, 109–198, Zbl.405.58033.

Chentsova, N.N. [1982] Investigation of a model system of quasi-stochastic relaxation oscillations. *Usp. Mat. Nauk 37*, No. 5, 205–206; English transl.: Russ. Math. Surv. 37, No. 5, 164–165 (1982), Zbl.506.58020

Chentsova, N.N. [1983] Investigation of a quasi-stochastic regime of an auto-generator on a tunnel diode. In: *Methods of the qualitative theory of differential equations* pp. 95–117, E.A. Leontovich-Andronova (ed.), Gor'kov. Gos. Univ., Gor'kij

Chow, S.-N., Hale, J.K. [1982] *Methods of bifurcation theory.* Springer-Verlag: New York, Heidelberg, Berlin, 515 pp, Zbl.487.47039

Chow, S.-N., Hale, J.K., Mallet-Paret, J. [1975] Applications of generic bifurcations. *Arch. Ration. Mech. Anal. 59*, 159–188; *62* (1976), 209–235, Zbl.328.47036 and Zbl.346.47050

Collet, P., Eckmann, J.P. [1980] *Iterated maps on the interval as dynamical systems.* Birkhäuser, Boston, 248 pp, Zbl.458.58002

Dangelmayr, G., Armbruster, D. [1983] Classification of $Z(2)$-equivariant imperfect bifurcations with corank 2. *Proc. Lond. Math. Soc.*, III. Ser. 46, 517–546, Zbl.517.58027

Dorodnitsyn, A.A. [1947] Asymptotic solution of van der Pol's equation. *Prikl. Mat. Mekh. 11*, No. 3, 313–328, Zbl.34,356

Dufour, J.P. [1979] Stabilité simultanée de deux fonctions. *Ann. Inst. Fou-Fourier 29*, No. 1, 263–282, Zbl.364.58007

Dumortier, F., Roussarie, R., Sotomayor, J. [1985] Generic 3-parameter families of vector fields on the plane, unfolding a singularity with nilpotent linear part. The cusp case of codimension 3. Preprint, 56 pp., published in: Ergodic Theory, Dyn. Syst. 7, 375–413 (1987), Zbl.608.58034

Feigenbaum, M.J. [1978] Quantitative universality for a class of nonlinear transformations. *J. Stat. Phys. 19*, No. 1, 25–52, Zbl.509.58037

Feigenbaum, M.J. [1980] Universal behavior in nonlinear systems. *Los Alamos Science 1*, No. 1, 4–27, Zbl.556.34046

Fenichel, N. [1971] Persistence and smoothness of invariant manifolds for flows. *Indiana Univ. Math. J. 21*, 193–226, Zbl.246.58015

Field, M.J. [1984] Unfolding of a class of SO(2)-invariant diffeomorphisms. Univ. of Sidney, Preprint, 69 pp.

Field, M.J. [1985] Unfolding equivariant diffeomorphisms. Univ. of Sidney, Preprint, 71 pp.

Gapanov-Grekhov, A.V., Rabinovich, M.I. [1984] Nonlinear physics, stochacity and structures. *Physics in the XXth century. Developments and perspectives*, Nauka, Moscow, 219–280

Gaspard, P. [1984] Generation of a countable set of homoclinic flows through bifurcation in multidimensional systems. *Bull. U. Sci, V. Ser., Acad. R. Belg. 70*, No. 2, 61–83, Zbl.555.58021

Gavrilov, N.K. [1973] Three-dimensional dynamical systems that have a structurally unstable homoclinic contour. *Mat. Zametki 14*, No. 5, 687–696; English transl.: Math. Notes 14, 953–957 (1974), Zbl.351.58004

Gavrilov, N.K. [1978] On some bifurcations of a state of equilibrium with one zero and a pair of pure imaginary eigenvalues. In: *Methods of the qualitative theory of differential equations*, pp. 33–40, E.A. Leontovich-Andronova (ed.), Gor'kov. Gos. Univ., Gor'kij

Gavrilov, N.K. [1980] On bifurcations from a state of equilibrium with two pairs of pure imaginary eigenvalues. In: *Methods of the qualitative theory of differential equations*, pp. 17–30, E.A. Leontovich-Andronova (ed.), Gor'kov. Gos. Univ., Gor'kij

Gavrilov, N.K., Shil'nikov, L.P. [1972] Three-dimensional dynamical systems that are close to systems with a structurally unstable homoclinic curve. I., *Mat. Sb.*, Nov. Ser. *88*(130), No. 8, 475–492; English transl.: Math. USSR, Sb. 17, 467–485 (1973), Zbl.251.58005

Gavrilov, N.K., Shil'nikov, L.P. [1973] Three-dimensional dynamical systems that are close to systems with a structurally unstable homoclinic curve. II., *ibid, 90* (130), No. 1, 139–156, 167; English transl.: Math. USSR, Sb. 19, 139–156 (1974), Zbl.258.58009

Golubitsky, M., Keyfitz, B.L., Schaeffer, D. [1981] A singularity theory analysis of a thermal chain branching model for the explosion peninsula. *Commun. Pure Appl. Math. 34*, No. 5, 433–463, Zbl.497.58006

Golubitsky, M., Langford, W.F. [1981] Classification and unfoldings of degenerate Hopf bifurcations. *J. Differ. Equations 41*, 375–415, Zbl.442.58020

Golubitsky, M., Schaeffer, D. [1979a] Imperfect bifurcations in the presence of symmetry. *Commun. Math. Phys. 67*, 205–232, Zbl.467.58019

Golubitsky, M., Schaeffer, D. [1979b] A theory for imperfect bifurcation via singularity theory. *Commun. Pure Appl. Math. 32*, 21–98, Zbl.409.58007

Golubitsky, M., Schaeffer, D. [1982] Bifurcations with $O(3)$ symmetry including applications to the Bénard problem. *Commun. Pure Appl. Math. 35*, 81–111, Zbl.492.58012

Golubitsky, M., Schaeffer, D. [1985] *Singularities and groups in bifurcation theory. Vol. 1*, Appl. Math. Sci. 51, Springer-Verlag: New York, Heidelberg, Berlin, 1985, 436 pp. Zbl.607.35004

Golubitsky, M., Stewart, I. [1985] Hopf bifurcation in the presence of symmetry. *Arch. Ration. Mech. Anal. 87*, No. 2, 107–165, Zbl.588.34030

Gomozov, E.B. [1976] Equivalence of families of diffeomorphisms in a class with finite smoothness. *Vestn. Khar'kov. Univ.* 134, Mat. Mekh. *41*, 95–104, Zbl.444.58006

Gonchenko, S.V. [1980] Doubling bifurcations in systems that are close to systems with a structurally unstable homoclinic curve. In: *Methods of the qualitative theory of differential equations*, pp. 31–43, E.A. Leontovich-Andronova (ed.), Gor'kov. Gos. Univ., Gor'kij, 227 pp.

Gonchenko, S.V. [1983] On stable periodic motions in systems close to systems with a structurally unstable homoclinic curve. *Mat. Zametki 33*, No. 5, 745–755; English transl.: Math. Notes 33, 384–389 (1983), Zbl.519.58029

Gonchenko, S.V. [1984] On some basic bifurcations in a two-parameter family of systems with homoclinic curves. Gor'kov. Gos. Univ., Gor'kij, 1984, 26 pp.

Gradshtejn, I.S. [1953] Application of A.M. Lyapunov's theory of stability to the theory of differential equations with a small coefficient in the derivatives. *Mat. Sb.*, Nov. Ser. *32* (74), No. 2, 263–286, Zbl.50,91

Grebogi, C., Ott, E., Yorke, J.A. [1983] Crises, sudden changes in chaotic attractors, and transients to chaos. *Physica 7D*, 181–200, Zbl.561.58029

Guckenheimer, J. [1973] One parameter families of vector fields on two-manifolds. In: *Dynamical Systm.*, Proc. Symp. Univ. Bahia, Salvador, 111–127, Zbl.285.58007

Guckenheimer, J. [1984] Multiple bifurcation problems of condimension 2. *SIAM J. Math. Anal. 15*, 1–49, Zbl.543.34034

Guckenheimer, J., Holmes, P.J. [1983] *Nonlinear oscillations, dynamical systems, and bifurcations of vector fields.* Appl. Math. Sci. *42*, Springer-Verlag: New York, Heidelberg, Berlin, 453 pp. Zbl.515.34001

Hartman, P. [1964] *Ordinary Differential Equations.* J. Wiley, New York, 612 pp, 2nd. ed.: Baltimore 1973, Zbl.125,321

Hassard, B.D. Kazarinoff, N.D., Wan, Y.-H. [1981] *Theory and Applications of the Hopf Bifurcation.* Lond. Math. Soc. Lect. Notes Ser. 41, Cambridge Univ. Press, Zbl.474.34002

Hénon, M.A. [1976] A two-dimensional mapping with a strange attractor. *Commun. Math. Phys. 50*, 69–76, Zbl.576.58018

Hirsch, M.W., Pugh, C.C., Shub, M. [1977] *Invariant Manifolds* Lect. Notes Math. 583, Springer-Verlag: New York, Heidelberg, Berlin, Zbl.355.58009

Hopf, E. [1948] A mathematical example, displaying the features of turbulence. *Commun. Pure Appl. Math. 1*, 303–322, Zbl.31,329

Il'yashenko, Yu.S. [1977] The zeros of special Abelian integrals in the real domain. *Funkts. Anal. Prilozh 11*, No. 4, 78–79; English transl.: Funct. Anal. Appl. 11, 309–311 (1978), Zbl.386.58008

Il'yashenko, Yu.S. [1978] Multiplicity of limit cycles that arise in the perturbation of a Hamiltonian equation $\omega' = P_2/Q_1$ in a real and complex domain. *Tr. Semin. Im. I.G. Petrovskogo 3*, 49–60; English transl.: Transl., II. Ser., Am. Math. Soc. 118, 191–202 (1982), Zbl.454.34020

Iooss, G., Joseph, D.D. [1981] *Elementary Stability and Bifurcation Theory.* Springer-Verlag: New York, Heidelberg, London, 1981, Zbl.443.34001

Jonker, L., Rand, D. [1981] Bifurcations in one dimension I. *Invent. Math. 62*, 347–365, Zbl.475.58014

Keener, J.P. [1980] Chaotic behavior in piecewise continuous difference equations, Trans. Amer. Math. Soc. 261, no. 2, 589–604

Khazin, L.V., Shnol', Eh. Eh. [1981] The stability of equilibria in critical and near critical cases. *Prikl. Mat. Mekh.. 45*, No. 4, *Stability of critical equilibrium states.* Akad. Nauk SSSR, Pushchino, 1985, 215 pp.; English transl.: Manchester Univ. Press, 1991

Khorozov, E.I. [1979] Versal deformations of equivariant vector fields for the cases of symmetries of orders 2 and 3. *Tr. Semin. Im. I.G. Petrovskogo 5*, 163–192; English transl.: Topics in modern mathematics, Petrovskij Semin. 5, 207–243 (1985), Zbl. 665. 58030

Khorozov, E.I. [1981–2] Bifurcation of a vector field in a neighborhood of a singular point in case of two pairs of pure imaginary eigenvalues, I. & II. *C. R. Acad. Bulg. Sci. 34*, No. 9, 1221–1224 and *35*, No. 2, 145–152, Zbl.488.34026 and Zbl.519.34021

Kostov, V.P. [1984] Versal deformations of differential forms of degree α on the line. *Funkts. Anal. Prilozh. 18*, No. 4, 81–82; English transl.: Funct. Anal. Appl. 18, 335–337 (1984), Zbl.573.58002

Landau, L.D. [1944] The Problem of Turbulence. *Dokl. Akad. Nauk SSSR 44*, No. 8, 331-314

Lanford, O.E. [1982] A computer assisted proof of the Feigenbaum conjecture. *Bull. Am. Math. Soc., New. Ser. 6*, 427–434, Zbl.487.58017

Lefschetz, S. [1957] *Differential Equations, Geometric Theory*. New York, 390 pp, Zbl.80,64

Leontovich, E.A. [1951] On the generation of limit cycles from separatrices. *Dokl. Akad. Nauk SSSR 78*, No. 4, 641–644, Zbl.42,328

Levi, M. [1981] Qualitative analysis of the periodically forced relaxation oscillations. *Mem. Am. Math Soc. 244*, 1–147, Zbl.448.34032

Lorenz, E.N. [1963] Deterministic non-periodic flow. *J. Atmos. Sci. 20*, 130–141

Louck, J.D., Metropolis, N. [1986] *Symbolic Dynamics of Trapezoidal Maps*. D. Reidel Publ. Co., Dordrecht-Boston, Mass., 1986, vii, 312 pp, Zbl.668.58040

Luk'yanov, V.I. [1982] Bifurcations of dynamical systems with saddlenode separatrix loops. *Differ. Uravn. 18*, No. 9, 1493–1506; English transl.: Differ. Equations 18, 1049–1059 (1983), Zbl.497.34032

Luk'yanov, V.I., Shil'nikov, L.P. [1978] On some bifurcations of dynamical systems with homoclinic structures. *Dokl. Akad. Nauk SSSR 243*, No. 1, 26–29; English transl.: Sov. Math., Dokl. 19, 1314–1318 (1978), Zbl.415.58013

Majer, A.G. [1939] Rough mappings of the circle into itself. Uchen. Zap., Gor'kov. Gos. Univ., No. 2, 215–229

Malkin, I.G. [1952] *Theory of stability of motions*. Gostekhizdat, Moscow-Leningrad, 432 pp, Zbl.48, 328

Malta, I.P., Palis, J. [1981] Families of vector fields with finite modulus of stability. In: *Dynamical Systems and Turbulence*, 1980, Lect. Notes Math. 898, Springer-Verlag: New York, Heidelberg, Berlin, 212–229, Zbl.482.58023

Manneville, P., Pomeau, Y. [1980a] Intermittent transition to turbulence in dissipative dynamical systems, *Comm. Math. Phys.* 74, no. 2, 189–197

Manneville, P., Pomeau, Y. [1980b] Different ways to turbulence in dissipative dynamical systems. *Physica 1D*, 219–224

Markley, N.G. [1969] The Poincaré-Bendixon theorem for the Klein bottle. *Trans. Am. Math. Soc. 135*, 159–165, Zbl.175,501

Marsden, J.E., McCracken. M. [1976] *The Hopf bifurcation and Its Applications*. Appl. Math. Sci. 19, Springer-Verlag: New York, Heidelberg, Berlin, xiii, 408 pp, Zbl.346.58007

May, R.M. [1975] Biological populations obeying difference equations: stable points, stable cycles and chaos. *J. Theoretical Biology 51*, 511–524

Medvedev, V.S. [1980] On a new type of bifurcations on manifolds. *Mat. Sb., Nov. Ser. 113* (155), No. 3, 487–492; English transl.: Math. USSR, Sb. 41, 403–407 (1982), Zbl.468.58014

Mel'nikov, V.K. [1962] A qualitative description of resonance phenomena in nonlinear systems. Dubna, OINF., Preprint R-1012, 17 pp.

de Melo, W., Palis, J. jun. [1980] Moduli of stability for diffeomorphisms. In: *Global Theory of Dynamical Systems* Proc. Int. Conf., Evanston/Ill. 1979, Lect. Notes Math. 819, Springer-Verlag: New York, Heidelberg, Berlin, 318–339, Zbl.445.58020

de Melo, W., Palis, J. jun. [1982] *Geometric theory of dynamical systems: An introduction*, Transl. from the Portuguese by A.K. Manning. Springer-Verlag: New York, Heidelberg, Berlin, xii, 198 pp., Zbl.491.58001

de Melo, W., Palis, J. jun., van Strein, S.J. [1981] Characterizing diffeomorphisms with modulus of stability one. In: *Dynamical Systems and Turbulence*, Proc. Symp., Coventry 1980, Lect. Notes Math. 898, Springer-Verlag: New York, Heidelberg, Berlin, 266–285, Zbl.482.58022

Metropolis, N., Stein, M.L., Stein P.R. [1973] On finite limit sets for transformations on the unit interval. *J. Comb. Theory*, Ser. *A15*, 25–44, Zbl.259.26003

Milnor, J. [1985] On the concept of attractor. *Commun. Math. Phys. 99*, 177–195, Zbl.595.58028

Mishchenko, E.F., Rozov, N. Kh. [1975] *Differential equations with a small parameter multiplying the highest derivative and relaxation oscillations*. Nauka, Moscow, 248 pp; English transl.: Plenum Press, New York, London (1980), Zbl.482.34004

Mittelmann, H.D., Weber, H.(eds.) [1980] *Bifurcation Problems and Their Numerical Solutions*. Birkhäuser, Basel, 243 pp., Zbl.429.00010

Myrberg, P.J. [1958, 1959, 1963] Iteration der reellen Polynome zweiten Grades. *Ann. Acad. Sci. Fenn. Ser. AI*, 256, 268, 336 (3), 10, 13, 18 pp, Zbl.105,58, Zbl.106,283

Nejmark, Yu.I. [1959] On some cases of dependence of periodic motions on parameters. *Dokl. Akad. Nauk SSSR 129*, No. 4, 736–739, Zbl.89,296

Nejshtadt, A.I. [1978] Bifurcations of the phase portrait of a system of equations, arising from the problem of loss of stability of an auto-oscillation near to a 1 : 4 resonance. *Prikl. Mat. Mekh. 42*, No. 5, 830–840; English transl.: J. Appl. Math. Mech. 42, 896–907 (1979), Zbl.419.58013

Nejshtadt, A.I. [1984] On the division of motions in systems with rapidly varying phase. *Prikl. Mat. Mekh. 48*, No. 2, 197–204; English transl.: J. Appl. Math. Mech. 48, 133–139 (1984) Zbl.571.70022

Nejshtadt, A.I. [1985] Asymptotic investigation of the loss of stability by an equlibrium as a pair of eigenvalues slowly cross the imaginary axis. *Usp. Mat. Nauk 40*, No. 5, 190–191

Nemytskij, V.V., Stepanov, V.V. [1949] *Qualitative theory of differential equations.* OGIZ, Moscow-Leningrad, 448 pp., Zbl.41,418; English transl.: Princeton Univ. Press; reprint, Dover, New York

Newhouse, S. [1974] Diffeomorphisms with infinitely many sinks. *Topology 13*, 9–18, Zbl.275.58016

Newhouse, S. [1980] Asymptotic behavior and homoclinic points in nonlinear systems. Nonlinear dynamics. *Ann. Acad. Sci. USA 357*, 292–299, Zbl.467.58020

Newhouse, S., Palis, J. jun. [1976] Cycles and bifurcation theory. *Astérisque 31*, 44–140, Zbl.322.58009

Newhouse, S., Palis, J. jun, Takens, F. [1976] Stable arcs of diffeomorphisms. *Bull. Am. Math. Soc. 82*, No. 3, 499–502, Zbl.339.58008

Newhouse, S., Palis, J. jun, Takens, F. [1983] Bifurcation and stability of families of diffeomorphisms. Publ. Math., *Inst. Hautes Etud. Sci. 57*, 5–71, Zbl.518.58031

Nitecki, Z. [1971] *Differentiable Dynamics. An introduction to the orbit structure of diffeomorphisms.* M.I.T. Press, Cambridge, Mass.-London xv, 282 pp., Zbl.246.58012

Nozdracheva, V.P. [1982] Bifurcation of a noncoarse separatrix loop. *Differ. Uravn. 18*, No. 9, 1551–1558; English transl.: Differ. Equations *18*, 1098–1104 (1983), Zbl.514.58030

Palis, J. jun. [1971] Ω-explosions. *Proc. Am. Math. Soc. 27*, 85–90, Zbl.207,544

Palis, J. jun. [1978] A differentiable invariant of topological conjugacies and moduli of stability. *Astérisque 51*, 335–346, Zbl.396.58015

Palis, J. jun., Pugh, C.C [1975] Fifty Problems in Dynamical Systems. In: Lect Notes Math. 468, Springer-Verlag: New York, Heidelberg, Berlin, 345–353, Zbl.304.58011

Peixoto, M.M. [1962] Structural stability on two-dimensional manifolds. Topology *1*, 101–120, Zbl.107,71

Petrov, G.S. [1984] The number of zeros of complete elliptic integrals. *Funkts. Anal. Prilozh. 18*, No. 2, 73–74; English transl.: Funct. Anal. Appl. 18, 148–149 (1984), Zbl.547.14003

Petrowskii, I.G. [1934] Über das Verhalten der Integralkurven eines Systems gewöhnlicher Differentialgleichungen in der Nähe eines singulären Punktes. *Mat. Sb.* Nov. Ser. *41*, 107–155, Zbl.9,352

Poincaré, H. [1987] *Les méthodes nouvelles de la mécanique céleste.* Vols. I, I, III, Gauthier-Villars, Paris, vol. I (1892), 385 pp.; vol. II (1893), viii + 479 pp.; vol. III (1899), 414 pp, Zbl.651.70002

Pontryagin, L.S. [1957] The asymptotic behavior of systems of differential equations with a small parameter multiplying the highest derivatives. *Izv. Akad. Nauk SSSR*, Ser. Mat. *21*, No. 5, 605–626, Zbl.78,80

Pontryagin, L.S., Rodygin, L.V. [1960] Approximate solution of a system of differential equations with a small parameter multiplying the highest derivatives. *Dokl. Akad. Nauk SSSR 131*, No. 2, 255–258, Sov. Math., Dokl. 1, 237–240 (1960), Zbl.117,48

Rabinowitz, P. (ed.) [1977] *Applications of bifurcation theory.* Proc. Symp. Wisc., October, 1976, Academic Press, 389 pp, Zbl.456.00014

Reyn, J.W. [1980] Generation of limit cycles from separatrix polygons in the phase plane. In: Geometric approaches to differential equations, Proc. 4th Scheveningen Conf., 1979, Lecture Notes Math. 810, Springer-Verlag, Berlin-Heidelberg, New York, 264–289; Zbl437.34025

Robbin, J.W. [1984] Unfoldings of discrete dynamical systems. *Ergodic Theory Dyn. Syst. 4*, No. 3, 421–486, Zbl.529.58017

Robinson, C. [1983] Bifurcation to infinitely many sinks. *Commun. Math. Phys. 90*, 433–459, Zbl.531.58035

Roussarie, R. [1986] On the number of limit cycles which appear by perturbation of a separatrix loop of planar vector fields. Bol. Soc. Bras. Mat. *17*, No. 2., 67–101, Zbl.628.34032

Ruelle, D., Takens F. [1971] On the nature of turbulence. *Commun. Math. Phys. 20*, 167–192 & *23*, 343–344, Zbl.233.76041 and Zbl.227.76084

Sacker, R. [1964] On invariant surfaces and bifurcations of periodic ordinary differential equations. Doctoral Thesis, New York Univ., New York, 76 pp.

Sacker, R. [1965] A new approach to the perturbation theory of invariant surfaces. *Commun. Pure Appl. Math. 18*, 717–732, Zbl.133,355

Samborskij, S.N. [1985] Limit trajectories of singularly perturbed differential equations. *Dokl. Akad. Nauk Ukr. SSR*, Ser. A. 1985., No. 9, 22–25, Zbl.572.34050

Samovol, V.S. [1982] Equivalence of systems of differential equations in a neighborhood of a singular point. *Tr. Moskov. Mat. O.-Va 44*, 213–234; English transl.: Trans. Mosc. Math. Soc. 1983, No. 2, 217–237 (1983), Zbl.526.34025

Sanders, J.A., Cushman, R. [1985] Abelian integrals and global bifurcations. In: *Dynamical systems and bifurcations*, 1125, Springer-Verlag: New York, Heidelberg, Berlin, 129 pp. 87–98, Zbl.559.34028

Sattinger, D.H. [1973] *Topics in stability and bifurcation theory*. Lect. Notes Math. 309, Springer-Verlag: New York, Heidelberg, Berlin 1973, 190 pp, Zbl.248.35003

Sattinger, D.H. [1978] Group representation theory, bifurcation theory and pattern formation. *J. Funct. Anal. 28*, 58–101, Zbl.416.47027

Sattinger, D.H. [1979] *Group Theoretic Methods in Bifurcation Theory*. Lect. Notes Math. 762, Springer-Verlag: New York, Heidelberg, Berlin, 241 pp, Zbl.414.58013

Schaeffer, D.G., Golubitsky, M. [1981] Bifurcation analysis near a double eigenvalue of a model chemical reaction. *Arch. Ration. Mech. Anal. 75*, 315–347, Zbl.522.35010

Semenov, N.N. [1934] *Chain reactions*. Goskhimizdat, Moscow-Leningrad, 555 pp.

Serebryakova, N.N. [1963] A qualitative investigation of a system of differential equations in the theory of oscillations. *Priklad. Mat. Mekh. 27*, No. 1, 160–166; English transl.: J. Appl. Math. Mech. 27, 227–237 (1963), Zbl.143,312

Shapiro, A.P. [1974] Mathematical models in concurrences. In: *Upravlenie i Informatsiya/Control and Information*, Far East Scientific Center (DVNC), Acad. Nauk SSSR, Vladivostok, No. 10, 5–75

Sharkovskij, A.N. [1964] Coexistence of cycles of a continuous map of the line into itself, Ukr. Mat. Zh. 16, 61–71

Shil'nikov, L.P. [1963] Some cases of generation of periodic motions from singular trajectories. *Mat. Sb.*, Nov. Ser. *61*(103), No. 4, 443–466, Zbl.121,75

Shil'nikov, L.P. [1966] Generation of a periodic motion from a trajectory exiting from an equilibrium of saddle-node and re-entering it. *Dokl. Akad. Nauk SSSR 170*, No. 1, 49–52; English transl.: Sov. Math. Dokl. 7, 1155–1158 (1966), Zbl.161,288

Shil'nikov, L.P. [1967] Existence of a countable set of periodic motions in four-dimensional space in an extended neighborhood of a saddle-focus. *Dokl. Akad. Nauk SSSR 172*, No. 1, 54–57; English transl.: Sov. Math., Dokl. 8, 54–58 (1969), Zbl.155,418

Shil'nikov, L.P. [1968] The generation of a periodic motion from a trajectory, which is doubly asymptotic to an equilibrium of saddle type. *Mat. Sb.*, Nov. Ser. 77, (119), No. 3, 461–472; English transl.: Math. USSR, Sb. 6, 427–438 (1970), Zbl.165,419

Shil'nikov, L.P. [1969] A new type of bifurcation of many-dimensional dynamical systems. *Dokl. Akad. Nauk SSSR 189*, No. 1, 59–62; English transl.: Sov. Math., Dokl. 10, 1368–1371 (1970) Zbl.219,200

Shil'nikov, L.P. [1970] On the question of the structure of an extended neighborhood of a rough saddle-node equilibrium. *Mat. Sb. Nov.* Ser. *81* (123), No. 1, 92–103; English transl.: Math. USSR, Sb. 10, 91–102 (1970), Zbl.193,53

Shil'nikov, L.P. [1977] Poincaré's theory of dynamical systems with homoclinic curves. *VII Internationale Konferenz über nichtlineare Schwingungen, Bd. 1, 2*, Akademie Verlag, Berlin, 279–293, Zbl.421.34051

Shil'nikov, L.P. [1980] Bifurcation theory and the Lorenz model, pp. 317–355 of the Russian Trans. *Bifurcations giving birth to a cycle and its applications*, Mir, Moscow, 368 pp. of *The Hopf bifurcation and its applications* by J.E. Marsden & M. McCracken, Springer-Verlag: New York, Heidelberg, Berlin, 1976, Zbl.346.58007

Shishkova, M.A. [1973] Investigation of a system of differential equations with a small parameter in the highest derivatives. *Dokl. Akad. Nauk SSSR 209*, No. 3, 576–579; English transl.: Sov. Math., Dokl. 14, 483–487 (1973), Zbl.289.34083

Shoshitajshvili, A.N. [1975] Bifurcations of topological type of singular points of vector fields that depend on parameters. *Tr. Semin. im. I.G. Petrovskogo*, 1975, No. 1, 279–309, Zbl.333.34037

Sinai, Ya.G. [1979] Stochasticity of dynamical systems. *Nonlinear waves*. Nauka, Moscow, 192–212

Sinai, Ya.G., Shil'nikov, L.P. (Eds.) [1981] *Strange Attractors*. Collection of Articles. Mir, Moscow, 253 pp.

Smale, S. [1967] Differentiable dynamical systems. *Bull. Am. Math. Soc. 73*, 747–808, Zbl.202,552

Sotomayor, J. [1973a] Generic bifurcation of dynamical systems. In: *Dynamical Syst*. Proc. Symp. Univ. Bahia, Salvador 1971, 561–582, Zbl.296.58007

Sotomayor, J. [1973b] Structural stability and bifurcation theory. In: *Dynamical Syst.*, Proc. Symp. Univ. Bahia, Salvador 1971, 549–560, Zbl.293.34081

Sotomayor, J. [1974] Generic one parameter families of vector fields on two dimensional manifolds. *Publ. Math., Inst. Hautes Etudes Sci. 43*, 5–46, Zbl.279.58008

Takens, F. [1971] Partially hyperbolic fixed points. *Topology 10*, 133–147, Zbl.214,229

Takens, F. [1974] Forced oscillations and bifurcations. *Comm. Math. Inst. Rijksuniversiteit Utrecht 3*, 1–59

Takens, F. [1976] Constrained equations: a study of implicit differential equations and their discontinuous solutions. In: *Struct. Stab., Theor. Catastr., Appl. Sci.; Proc. Conf. Seattle*, Lect. Notes Math. 525, 143–234, Springer-Verlag: New York, Heidelberg, Berlin, Zbl.386.34003

Takens, F. [1979] Global phenomena in bifurcations of dynamical systems with simple recurrence. *Jahresber. Dtsch. Math.-Ver. 81*, 87–96, Zbl.419.58012

Tikhonov, A.N. [1952] Systems of differential equations containing small parameters multiplying the derivatives. *Mat. Sb.*, Nov. Ser. *31* (73), No. 3, 575–586, Zbl.48,71

Turaev, D.V., Shil'nikov, L.P. [1986] Bifurcations of quasiattractors torus-chaos. In: *Math. Mech. turbulence (modern nonlinear dynamics in appl. to turbulence simulation), Coll. Sci. Works*, Kiev 1986, 113–121, Zbl.625.58008

Vajnberg, M.M., Trenogin, V.A. [1969] The theory of branching of differential equations. Nauka, Moscow, 527 pp; English transl.: Noordhoff Int. Publ, Netherlands (1974), Zbl.186,208

van der Pol, B. [1926] On relaxation oscillations. *Phil. Mag. 2*, Ser. 7, 978–992

Vasil'eva, A.B. [1952] On differential equations containing small parameters with the derivatives. *Mat. Sb.*, Nov. Ser. *31* (73), No. 3, 587–644, Zbl.48,71

Vol'pert, A.I. [1983] Wave solutions of parabolic equations. Preprint, O.I.Kh.F., Akad. Nauk SSSR, Chornogolovka, 48 pp.

Vol'pert, V.A. [1982] Bifurcations of nonstationary regimes of propagating waves. Preprint, O.I.Kh.F., Akad. Nauk SSSR, Chornogolovka, 2, 62 pp.

Voronin, S.M. [1982] Analytic classification of pairs of involutions and its application. *Funkts. Anal. Prilozh. 16*, No. 2, 21–29; English transl.: Funct. Anal. Appl. 16, 94–100 (1982), Zbl.521.30010

Vul, E.B., Sinai, Ya.G., Khanin, K.M. [1984] Feigenbaum's universality and the thermodynamical formalism. *Usp. Mat. Nauk 39*, No. 3, 3–37; English transl.: Russ. Math. Suv. 39, No. 3, 1–40 (1984), Zbl.561.58033

Yakovenko, S.Yu. [1984] On real zeros of a class of Abelian integrals arising in the theory of bifurcations. In: *Methods of the qualitative theory of differential equations* Interuniv Collect., Gork'ij 175–185

Yudovich, V.I. [1965] An example of the birth of a second stationary or periodic flow upon loss of stability of a laminar flow of a viscous incompressible fluid. *Prikl. Mat. Mekh. 29*, No. 3, 453–467; English transl.: J. Appl. Math. Mech. 29, 527–544 (1965), Zbl.148,223

Zel'dovich, Ya.B. [1941] On the theory of heat liberation. *Zh. Tekh. Fiz. 9*, No. 6, 493–508

Zharov, M.I., Mishchenko, E.F., Rozov, N.Kh. [1981] Some special functions and constants that arise in the theory of relaxation oscillation. *Dokl. Akad. Nauk SSSR 261*, No. 6, 1292–1296; English transl.: Sov. Math., Dokl. 24, 672–675 (1981), Zbl.504.34031

Zheleztsov, N.A. [1958] On the theory of discontinuous oscillations in systems of the second order. *Izv. Vyssh. Ucheb. Zaved., Radiofiz. 1* No. 1, 67–78

Żołądek, H. (Zholondek, Kh.) [1983] On versality of a family of symmetric vector vector fields on the plane. *Mat. Sb.*, Nov. Ser. *120* (162), No. 4, 473–499; English transl.: Math. USSR, Sb. 48, 463–492 (1984), Zbl.516.58032

Żołądek, H. [1987] Bifurcation of a certain family of planar vector fields tangent to axes. J. Differ. Equations. *67*, No. 1, 1–55, Zbl.648.34068

Zvonkin, A.K., Shubin, M.A. [1984] Nonstandard analysis and singular perturbations of ordinary differential equations. *Usp. Mat. Nauk 39*, No. 2, 77–127, English transl.: Russ. Math. Surv. 39, No. 2, 69–131 (1984), Zbl.549.34055

Additional References

(Added in the Translation)

[1*] Davidov, A.A. (1985), Normal form of an implicit differential equation in a neighborhood of a singular point, Funkts. Anal. Prilozhen. *19*, no. 2, 1–10

[2*] Il'yashenko, Yu.S. (1991), The concept of a minimal attractor and the maximal attractors of partial differential equations of Kuramoto-Shivashinsky type, Chaos 1, no. 2, 168–173

[3*] Il'yashenko, Yu.S., Yakovenko, S.Yu. (1991), Finitely smooth normal forms of local families of diffeomorphisms and vector fields, Usp. Mat. Nauk *46*, no. 1, 3–39

[4*] Il'yashenko, Yu.S., Yakovenko, S.Yu., Finitely smooth classification of local families of vector fields and diffeomorphisms and their functional moduli, to appear in: *Nonlinear Stokes Phenomena*, Yu.S. Il'yashenko (editor)

[5*] Kolesov, A.Yu., Mishchenko, E.F. (1988), Asymptotics of relaxation oscillations, Mat. Sbornik *137*, no. 1, 3–18

[6*] Nejshtadt, A.I. (1987), On the delay of the loss of stability by dynamic bifurcations, Diff. Uravneniya *23*, no. 12, 2060–2067

[7*] Nejshtadt, A.I. (1988), On the delay of the loss of stability by dynamic bifurcations, II. Diff. Uravneniya *24*, no. 2, 226–233

[8*] Teperin, I.V. (1990), The asymptotics of relaxation oscillations in the case of two slow variables, in: Methods of qualitative and bifurcation theory, Nizhnyi Novogorod, 19–33

[9*] Turaev, D.V., Shil'nikov, L.P. (1986), On the bifurcations of homoclinic figure-eights of a saddle with negative saddle value, Dokl. Akad. Nauk SSSR *290*, no. 6

II. Catastrophe Theory

V.I. Arnol'd

Translated from the Russian
by N.D. Kazarinoff

Contents

Сначала мысль, воплощена
В поэму сжатую поэта,
Как дева юная, темна
Для невнимательного света;
Потом, осмелившись, она
Уже увертлива, речиста,
Со всех сторон своих видна,
Как искушенная жена
В свободной прозе романиста;
Болтунья старая, затем
Она, подъемля крик нахальный,
Плодит в полемике журнальой
Давно уж ведомое всем.
*Е.А. Баратынский**

§ 1. Basic Concepts

1.1. Catastrophes and Bifurcations. The term *catastrophe* was introduced by R. Thom (1972) in order to denote a qualitative change in an object as the parameters upon which it depends change smoothly. This term, replacing the previously used terms *bifurcation*, *perestroika*, and *metamorphosis*, gained wide popularity after Zeeman (1976) suggested the use of the name *catastrophe theory* to unite singularity theory, bifurcation theory and their applications.

It is difficult for a mathematician to agree that the introduction of a new term, unaccompanied by the discovery of any new facts, is a significant achievement. However, the success of "cybernetics", "attractors" and "catastrophe theory" illustrates the fruitfulness of word creation as scientific research. This method, by the way, was known long ago; even Poincaré consciously used it.

"It is difficult to believe," he said, "what a large economy of thought can be achieved by a well-chosen word. Often it is sufficient to invent one new word, and this word becomes an achievement." Moreover, according to Poincaré, "Mathematics is the art of giving one and the same name to different things." "A fact ... acquires its meaning only from that moment when a more penetrating thinker notices its similarity [to something else], which he brings to light and denotes symbolically by some term or other." (all four sentences are from *Science et Méthode*, in the chapter "The future of mathematics", (1908; pp 296–301).

Introducing the term "catastrophe theory", in the 1970s Thom and Zeeman gave wide publicity to the accumulated achievements of H. Whitney's mathe-

* In the beginning a thought embodied / In the laconic lines of a poet / Is like a young maiden, too enigmatic / For inattentive high society to notice; / Then, daring, all "her" sides can be seen, / Now shifty, now eloquent / Like an experienced wife / In the free prose of a novelist / The thought is visible; / After that, an old chatterbox, / Raising an impudent cry / Long since known to all / "She" procreates polemics in magazines.
E.A. Baratynskij (1780–1844)

matical theory of singularities and A.A. Andronov's theory of bifurcations of dynamical systems. Thom was the first to realize the great potentialities for applications of the theory established by Whitney (1955), but his treatment of the subject was so mixed up and unclear that the underlying simple, general ideas remained unknown to a wide public until Zeeman in a series of popular articles advertised a revolution in mathematics "comparable to Newton's invention of mathematical analysis". Beginning with this moment, catastrophe theory started down the path of cybernetics: this name was reserved mainly for speculations of a near-scientific and pseudo-scientific character, while serious works on singularity theory, bifurcation theory and their applications were usually related by their authors (including Thom and Zeeman) to the corresponding special fields.

1.2. Catastrophes and Singularities. The basic idea of catastrophe theory (in Zeeman's treatment) is the following. Let us consider any system, smoothly depending on its parameters, and let us assume that the parameters defining the states of the system are divided into two groups: *internal* and *external.* It is assumed that a dependence exists among the parameters. However, the values of the internal parameters are not uniquely determined by the values of the external ones. Geometrically the states of the system are described by points in the product of the manifolds of values of the internal and external parameters. The meaning of the dependence is that this point always lies in some subset of the product space. It is assumed that this subset is a smooth submanifold in general position in the product space, and that its dimension is equal to the dimension of the space of external parameters. Let us consider the mapping of this submanifold onto the manifold of the external parameters. The theory of singularities yields information on the critical points and values of this mapping for submanifolds in general position. Thom and Zeeman's program consists of using this information for the study of "catastrophes", that is, jumps of the system from one state to another under changes of parameters.

1.3. Zeeman's Machine. Consider some elastic structure, for example, "*Zeeman's catastrophe machine*", illustrated in Fig. 1. This machine consists of a wheel, rotating about a fixed axis, and two springs attached at a point on the rim of the wheel: one spring has its farther end fixed in the plane of the wheel; the other has its farther end attached to the point of a pencil so that, as the pencil moves, it traces a curve on a sheet of paper in this same plane.

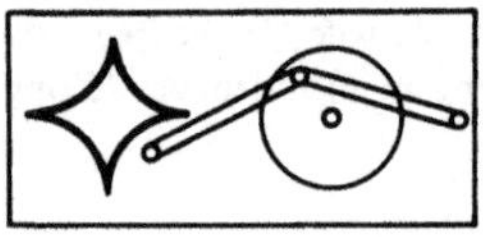

Fig. 1. Zeeman's catastrophe machine

As one changes the position of the pencil (two external parameters), one can observe, generally speaking, that the wheel rotates, smoothly responding to the changes of the parameters; but that in some cases the wheel changes its position with a jump. Such jumps occur for exceptional ("bifurcation") positions of the pencil: on the sheet of paper they form a "*catastrophe curve*" with four cusps. If the pencil intersects this curve while moving, a catastrophe (jump of the wheel) may or may not occur, depending upon the prehistory of the motion.

The theory of singularities allows one to explain the behavior of Zeeman's machine, and to predict in which cases a "catastrophe" will occur and in which cases it will not.

The state space of this machine is 3-dimensional (the two external parameters define the position of the pencil, one internal parameter defines the rotation angle of the wheel). The potential energy of the system is a function of these three parameters (periodic in the rotation angle). For fixed values of the external parameters the system minimizes its potential energy (locally). The dependence that arises between the values of the internal and external parameters is described in 3-dimensional space by a 2-dimensional *surface of equilibria* (see Fig. 2). The surface of equilibria is formed by the critical points of the potential energy, considered as a function of the internal parameter for fixed values of the external ones. These critical points, taken at all possible values of the external parameters, form a surface in the product space.

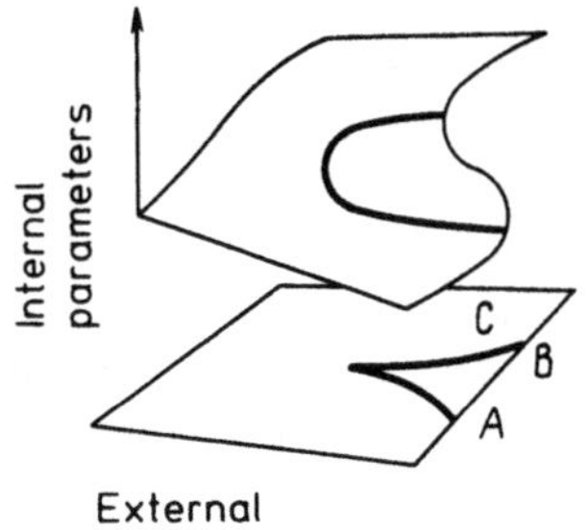

Fig. 2. A model of a "pleat"

The surface of equilibria of Zeeman's machine is smooth, and is situated in a generic way relative to the projection onto the plane of the external variables. Therefore, the only singularities of the projection map are Whitney's folds and pleats (or cusps); namely, 4 pleat points, which project to the cusps of the catastrophe curve.

As the values of the external parameters approach the catastrophe curve, the critical points of the potential energy, considered as a function on the circle, undergo a metamorphosis; see Fig. 3. Upon intersecting the catastrophe curve at a generic point, two critical points of the potential energy merge – a local maximum and a local minimum. The system, in a stable equilibrium state at a point where the potential energy is a local minimum, remains in this state up to the moment of bifurcation. At that moment the critical point becomes unstable, and the system jumps to another equilibrium state (corresponding to another minimum of the potential energy). Thus, whether or not there will be a jump upon intersecting the catastrophe curve depends upon which local minimum the potential energy of the system lies at before the intersection takes place.

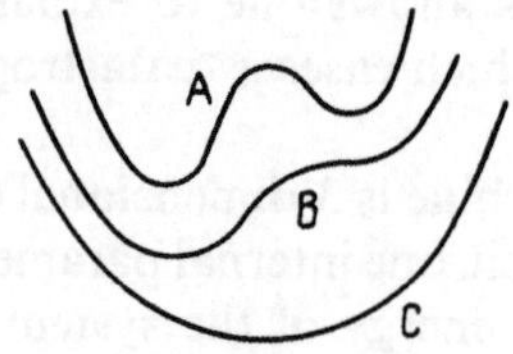

Fig. 3. Metamorphoses of minima

At a cusp point of the catastrophe curve, three 'sheets' of the surface of equilibria merge: the two extreme ones correspond to local minima and the middle one to a local maximum of the potential energy. Knowing from Whitney's theory how these sheets pass into one another near the pleat point, one can easily predict the jumps that occur for different paths around the cusp points of the catastrophe curve.

1.4. Models of Catastrophes. Numerous other models can be investigated following the procedure used above to analyze the jumps of Zeeman's machine. Zeeman considers with special enthusiasm those cases in which the correctness of the choice of parameters, the existence of a dependence among them, and, even more, the smoothness and genericity of the corresponding manifold (the generalized surface of equilibria of his machine) are all problematical. In these cases both the premises and the deductions have the character of metaphors rather than being models in the sense of mathematical physics. As an example, let us consider (basically following Zeeman (1974)) an investigation of the activity of a creative person, for example, a scholar.

In this model the scholar is characterized by three parameters: technique (or technical proficiency) (T), enthusiasm (E), and results (R). From observation it is known that they are not independent. Their dependence is represented by a surface in the 3-dimensional space with coordinates T, E, R. Since, at present, we know nothing about this surface, we can suppose that it is smooth and in general position. In this case, the singularities of the projection of the surface along the R-axis onto the (T, E)-plane will be folds and pleats. It is asserted that the pleat, located as shown in Fig. 4, satisfactorily describes the observed phenomena.

Indeed, if enthusiasm is small, results grow gradually as technique increases. If enthusiasm is sufficiently large, they may increase with a jump as technique increases. The region of great results, whither the jumping point lands, is denoted

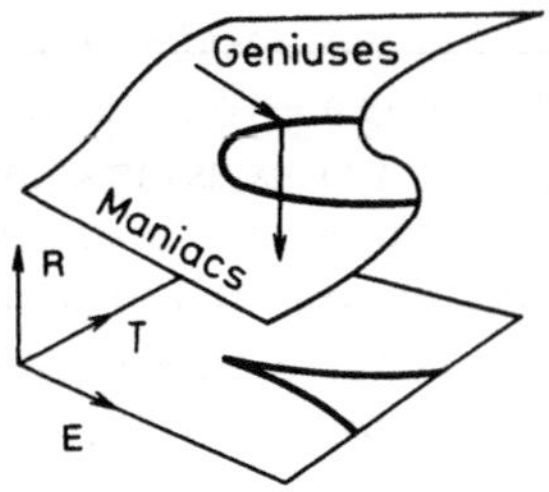

Fig. 4. The "scholar" model

by the word "geniuses" in the figure. On the other hand, the growth of enthusiasm, when not reinforced by a corresponding growth in technique, leads to a catastrophic drop in results. The region of low results, representing the point to which the scholar falls, is denoted in the figure by the word "maniacs".

It is instructive that a genius and a maniac may both have the same enthusiasm and technique, and differ only in their results (and their prehistory).

Hundreds of models of this type have been constructed. Usually, the simplest nontrivial Whitney singularity (a pleat) and three parameters are involved: one internal and two external ones. Moreover, one of the latter (in our example, enthusiasm) is "splitting" in the sense that its increase splits the different possible values of the internal parameter coexisting for a fixed value of the external parameter (results of a genius or a maniac, the equilibria of Zeeman's machine, etc). The second external parameter is called "normal" (in our example it is technique).

1.5. The Verification of Models. Although in "nonphysical" situations (for example, in biology, economics, sociology) models of the type described cannot claim to be especially trustwothy, efforts have been made to verify them through experiments. For example, in analysis of rebellions by prisoners, the rôles of external parameters are played by the strictness of the regime (the normal parameter) and alienation (the splitting parameter), and the rôle of the internal parameter is played by unrest. If alienation is sufficiently great, then a small increase in the strictness of the regime arouses a catastrophic growth of unrest (leading to a rebellion), and a small lessening of the strictness of the regime leads to an equally sudden calmness. Inasmuch as both jumps are observable, one may verify whether or not the corresponding points lie on a catastrophy curve in the strictness-alienation plane which has the characteristic sharp point. However, the observed values of the variables in this model (as in many others) admit, along with the above interpretation, many others, showing no worse agreement with the observations.

To criticize models of this kind is too easy. I will now give here only one example of reasoning, externally like the one introduced above, but leading to an obviously incorrect outcome (this example was communicated to me by A.T. Fomenko).

1.6. An Inadequate Model. Let us consider some functional, for example the area of the minimal surface spanning a given contour, as a function of parameters upon which the contour depends. Let us assume that there are two parameters and that the contour depends upon them in a sufficiently general way. Since the minimal surface is not unique, its area is not a single-valued function of the parameters. However, there is, of course, a dependence between the area and the parameters. It is represented by some surface in 3-dimensional space – the product of the plane of parameters with the axis of area.

This surface depends on the two-parameter family of contours being considered. It is possible to imagine that for typical families this surface is smooth and that its projection onto the plane of parameters does not have any singularities other than folds and pleats. Under this assumption the bifurcation set must be a curve with isolated cusps.

The mistake in this reasoning becomes evident if we look at this problem from another point of view. The surface under study is the graph of a many-valued "minimum function" (a *minimum function* of a family of functions on a manifold is defined to be a function that assigns to a value of the parameters of the family the minimum of the function on the manifold corresponding to this parameter value; for the construction of a many-valued minimum function the minima are replaced by the critical values).

It is known from theory of singularities that the graphs of many-valued minimum functions of typical families of smooth functions are typical *Legendre singularities*, that is, singularities of typical wave fronts or graphs of Legendre transformations of smooth functions.

Fig. 5. The swallowtail

In 3-dimensional space the typical Legendre singularities of fronts are the *cuspidal edge* (of semi-cubical type) and the *swallowtail* (see Fig. 5). Therefore, the graph of the area of the minimal surface (as a many-valued function of the parameters of a typical two-parameter family of contours) is unlikely to have singularities different from swallowtails. (Of course, for justification, the transversality theorem that shows that the areas of surfaces spanning contours of typical families form a typical family of functions on a manifold of surfaces would be needed. This theorem, as far as I know, has not yet been proved; cf. Poston (1972), Fomenko and Tuzhilin (1986) and M. Beeson and A. Tromba, "The cusp catastrophe of Thom and bifurcation of minimial surfaces," *Manuscr. math.*, *46* (1984), 200–220.

In any case, a model in which the surface has a swallowtail is more convincing than a model with a pleat (moreover, with respect to the bifurcation curves and the number of merging preimages of the projection these models do not differ, if one does not count a coincidence of minima attained at widely separate points as a bifurcation).

1.7. Adequate Models. In more serious models one deals with more or less direct applications of the theory of singularities of well defined smooth maps (for example, Zeeman's machine and other models with a "physical" origin). In these cases the predictions of the theory of singularities are verified excellently by experiments.

In most of the models of this type the number of parameters is not large (ordinarily, there are one internal and two external parameters so that the singularities which occur are Whitney's pleats). It is instructive that, in many cases, the results obtained using the theory of singularities were already known earlier in the corresponding area of physics, and curves with semi-cubical cusps in the space of parameters can be discovered in a large number of research papers, whose authors have explicitly discussed and used considerations of genericity formalized in the theory of singularities, and explicitly pointed out the universality of the dependencies they uncovered.

Since smooth mappings and their singularities are all around us, what is surprising is not that some of them were discussed before the creation of singularity theory, but rather how many of them were not discovered, or were not so widely known as they deserved to be. Without claiming completeness, I discuss below some outstanding papers whose authors have investigated singularities, bifurcations and catastrophes in generic systems arising in various areas of knowledge.

§ 2. The Theory of Catastrophes Before Poincaré

2.1. Evolutes and Caustics, Evolvents and Fronts. In 1654 Huygens (1673) constructed the theory of evolutes and involutes of plane curves. He noted that an involute or an evolute of a smooth curve has cusps of semi-cubical type. The *evolute* of a smooth curve is a caustic (the envelope of the family of rays orthogonal to the curve); see Fig. 6. (Leonardo da Vinci (see Bennequin (1986)) had already investigated caustics; the term was apparently introduced by Tschirnhaus (1682)). The family of involutes of the caustic in Fig. 6 is shown in Fig. 7. It is the family of wave fronts corresponding to the same family of rays. (All of this was also discovered by Huygens.) Thus, Huygens essentially discovered the stability of cusps on caustics and wave fronts. Nowadays these singularities are connected with the pleats of the corresponding smooth mappings, and they belong to the

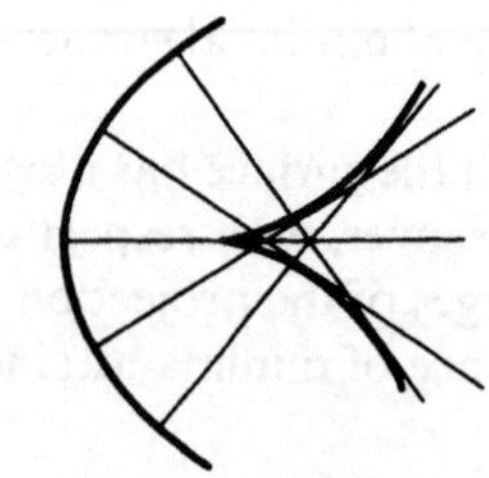

Fig. 6. The caustic of a system of rays

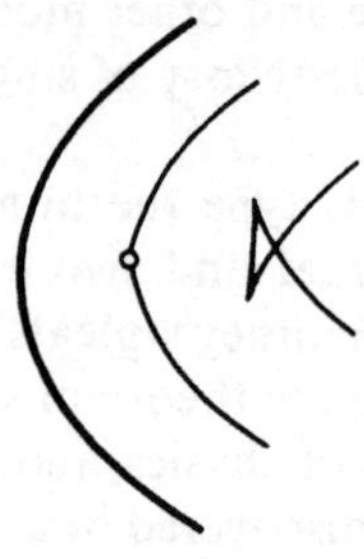

Fig. 7. Singularities of wave fronts

most important applications of catastrophe theory. Fig. 7 was discussed by Cayley (1868b).

2.2. Families of Functions in the Work of Hamilton and His Successors. Hamilton (1828–1837) noted that geometrical optics is conveniently described with the help of a family of functions (the "characteristic function" of Hamilton amounts to the optical length of a path, considered as a family of functions of the terminal point, parametrized by the initial point).

Jacobi in his *Lectures on dynamics* (1866) drew the caustic of the central field of extremals on an ellipsoid that is nearly a sphere; see Fig. 8. He proved that the caustic must have cusps (his proof originated in essentially topological considerations about conjugate points, now included in "Morse theory"), and asserted that a more refined analysis (which, unfortunately, he omitted) shows that there are four such points for an ellipsoid. Jacobi's reasoning has a general character, and it does not use the specifics of the ellipsoid (except for the calculation of the number of cusps). In contemporary terms, one may say that Jacobi investigated the singularities of a special family of functions depending on two parameters, and convinced himself that a typical bifurcation curve ("catastrophe curve") has cusps of semi-cubical type.

2.3. Points of Inflection and Swallowtails. Beginning with the work of Monge (1807) on the "application of analysis to geometry" the geometers of the 19th century systematically studied singularities of curves and surfaces dual to generic

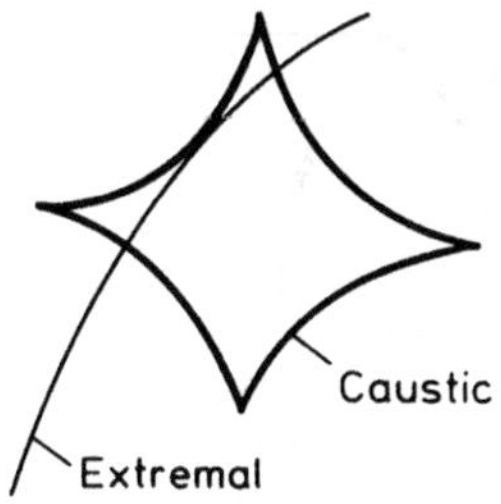

Fig. 8. The caustic of the geodesics of an ellipsoid

smooth curves and surfaces. Plücker (1839) knew well that the only singularities of curves dual to germs of generic smooth curves are cusps of semi-cubical type (the duals of the inflection points of the original curve). In Salmon's textbook (1862) the analogous classification of inflection points of surfaces is analysed in detail.

The swallowtail, which appears instead of the semi-cubical cusp in 3-dimensional space, was thoroughly studied in those times, and one may find it illustrated in standard algebra texts, for example Weber's (1898), who cites Kronecker (1878), or in Brill's catalogue (1892) of mathematical models.

2.4. The Umbrella and Umbilic Singularities of Caustics. Cayley (1852) studied typical singularities of mappings of a surface in 3-dimensional space (in algebraic geometry these singularities are called *pinch points* or *vertices*). The image of the surface in a neighborhood of a critical value is locally diffeomorphic to the *Whitney-Cayley umbrella* $y^2 = zx^2$; see Fig. 9. Moreover, Cayley (1870, 1873) also considered the geometry of the family of equidistants and the *focal set* of a triaxial ellipsoid (that is, the caustics of the system of rays orthogonal to the ellipsoid; see Fig. 10). Besides cuspidal edges of semi-cubical type, one meets here the singularity D_4, the purse (hyperbolic umbilic); see Fig. 11. These correspond to *umbilic points* on an ellipsoid (points where both principal curvatures coincide). Therefore, Thom later called the D_4 and D_5 singularities umbilics. Cayley (1859) also investigated the singularities of the envelope of the family of planes normal to the radius vector at points of an ellipsoid; in contemporary terms these are Legendre singularities.

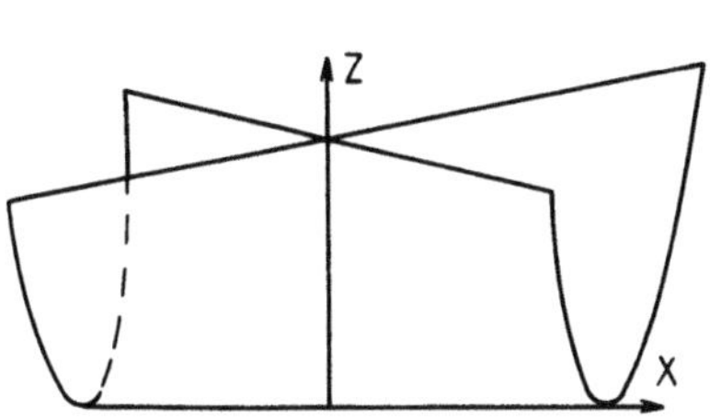

Fig. 9. The Whitney-Cayley umbrella

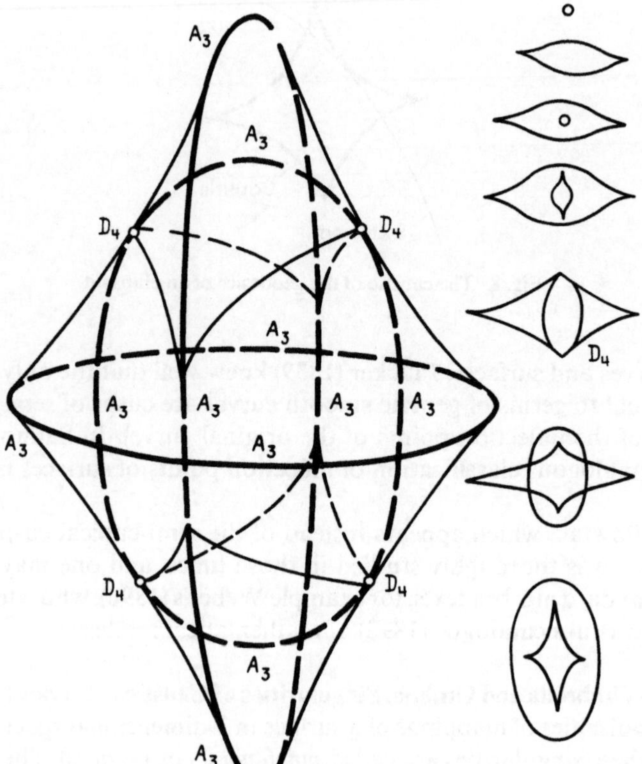

Fig. 10. The focal surface of an ellipsoid

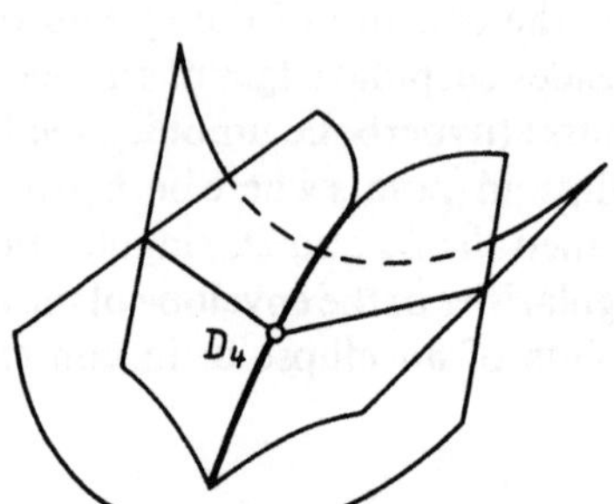

Fig. 11. The purse

From the point of view of catastrophe theory, Cayley investigated perestroikas[1] of critical points in 3-parameter families of functions of two variables (the focal set consists of those points of space, the distance to which has a degenerate critical point on the ellipsoid). He undoubtedly understood that the family he was studying was generic. (Cayley explicity formulated the problem on the topology of families of level curves of a generic smooth function (1868a), and thus arrived at the beginning of Morse theory.) But it seems that he did not know that the singularities he was investigating exhaust (up to complex equivalence) all the cases found in typical families.

2.5. Transversality. Algebraic geometers of the last century systematically and consciously used considerations of genericity, throwing aside exceptional objects, correctly calculating the codimensions of the degeneracies of these objects. In the complex case with which they were basically involved, that method is especially fruitful, since the bifurcation sets, having real codimension exceeding 1, do not divide the parameter space into parts. Thanks to this, for the calculation of discrete invariants, such as topological type, it is sufficient to examine one nondegenerate object. (Often enough, as the nondegenerate object one takes a small perturbation of a strongly degenerate one – this also stimulated the study of singularities.)

From the point of view of contemporary terminology, algebraic geometers used transversality theorems as if they were obvious. The algebraic analog of Sard's theorem (1942), which is the key to this circle of questions, was proved by Bertini (1882).

I note, incidentally, that the "enumerative" methods of the singularity theory of the algebraic geometers of the 19th century anticipated contemporary investigations in many respects. Moreover, from the point of view of catastrophe theory, the algebraic geometry of projective varieties is the far advanced investigation of singularities of a special type (homogeneous), and the *raison d'être* of algebraic geometry is that the initial sections of the Taylor series of smooth functions are polynomials ("... to a great extent the 'local' results already contain the global ones" (A. Grothendieck (1960)), from his address at the International Congress of Mathematicians, Edinburgh, 1958).

The opposite point of view is also possible, of course; according to this view the theory of singularities is included in algebraic geometry as the investigation of degeneracies of algebraic sets.

Returning to the considerations of genericity used by algebraic geometers, I note that they (both in the algebraic and in the smooth setting) went beyond the somewhat restrictive confines of the formulations later proposed by Thom (1956). For example, it is often necessary by a small perturbation of a map to achieve the transversality of this map, not to a submanifold of the image space, but to a

[1] The word metamorphosis was used as a translation of the Russian "perestroika" until recently. In Russian "Morse surgery" was always called "perestroika of Morse". Nowadays, when we have an international word "perestroika", we don't need to substitute "metamorphosis" for it.

map into this space. Let

$$X \underset{f}{\rightarrow} Y \underset{g}{\leftarrow} X,$$

then one can achieve transversality in

$$J^k X \xrightarrow[j^k f]{} J^k Y \xleftarrow[j^k h]{} J^k X,$$

with h arbitrarily close to g without changing f (which is not an embedding; see Arnol'd (1967)). Transversality of f to g means $Tf(T_x X) + Tg(T_z Z) = T_y Y$ for $f(x) = y = g(z)$ and similarly for jets. (For the formulations of theorems establishing the possibility of such a perturbation, see the details in Landis (1981).) As far as I know these simple generalizations of the transversality theorems (without jets and/or with them) are not present in the textbooks. They do *not* follow from Thom's theorems since f (or $j^k f$) *should not be an embedding.*).

§ 3. The Theory of Bifurcations in the Work of Poincaré

3.1. The Classification of Singularities and Normal Forms. Poincaré already used "putting into general position" almost as is done today. For example, his classification of the singular points of generic vector fields on the plane (saddle, node, focus) is the prototype for many classifications in catastrophe theory. The theory of bifurcations of periodic solutions, also created by Poincaré, anticipated the general theory of bifurcations, not only with respect to results, but also (and particularly) with respect to methods. Already in his dissertation Poincaré (1879) (solving a problem of classification of differential equations posed by Darboux, who was the discover of the normal form of a symplectic structure) created the general method of normal forms, which leads to the classification of catastrophes if one applies it to functions instead of to differential equations.

3.2. The Preparation Theorem, Finite Determinacy and Versal Deformations. For his construction of a theory of bifurcations Poincaré systematically used the Preparation Theorem of Weierstrass, which Poincaré proved in his dissertation, apparently independently. Subsequently, he derived from it a fundamental finiteness theorem (Lemma IV on p. 14 of his dissertation (1879) and Section 33 in Chap. 1 of *Les nouvelles méthodes de la Méchanique céleste* (1892)):

If $\varphi: (\mathbb{C}^p \times \mathbb{C}^p, 0) \to (\mathbb{C}^p, 0)$ *is a holomorphic germ and* $z = 0$ *is a root of finite multiplicity of the equation* $\varphi(0, z) = 0$, *then the system* $\varphi = 0$ *is equivalent to* $\psi = 0$, *where* ψ *are polynomials in* z.

In contemporary terms this means, first, the finite V-determinacy[2] of a holomorphic germ of finite multiplicity and, second, the existence of a V-versal deformation of it with a finite number of parameters. Morever, such a deformation is obtained by adding to the left-hand side of each equation the sum of

[2] In Mather's terminology (1968) "contact finite determinacy" (however, Poincaré proved more, since he did not use diffeomorphisms of z-space).

all monomials in z, of degrees less than the multiplicity of the root z, with undetermined coefficients (since Poincaré's proof estimates from above the degrees of the polynomials ψ mentioned in his lemma by the multiplicity of the root).

A theory of bifurcations based upon this lemma is extensively developed in *Les nouvelles méthodes de la Méchanique céleste* (Poincaré, 1892). For example, in Section 51, Vol. I, Poincaré considers the birth of a cycle from a singular point (in a more complex situation than the Hopf bifurcation of the catastrophe theorists); in Section 37, Vol. I, Poincaré derives an equation for the periodic solutions generated upon the birth of a cycle from one of the closed orbits upon perturbing a one-parameter family of closed orbits; in Chap. 28, Vol. III, he sets forth the theory of period-doubling (see Fig. 12), etc.

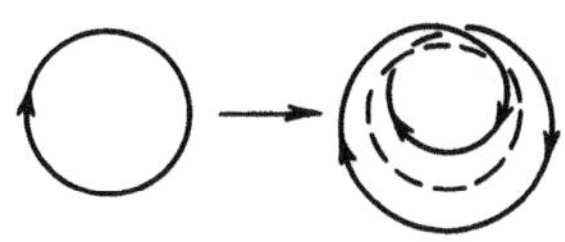

Fig. 12. The period-doubling bifurcation of a cycle

3.3. Poincaré and Contemporary Mathematics. Unfortunately, the simple texts of Poincaré are difficult for mathematicians raised upon set theory. Poincaré would have said, "Pete washed his hands", where a contemporary mathematician would simply write instead, "There exists a $t_1 < 0$ such that the image of the point t_1 under the natural mapping $t \to (\text{Pete}(t))$ belongs to the set of people with dirty hands and a $t_2 \in (t_1, 0]$ such that $\text{Pete}(t_2)$ belongs to the complement of the set mentioned above." Apparently, this is why many of his ideas have remained unnoticed by later generations. For example, cohomology theory and de Rham's theorem remained hidden in Poincaré's work until É. Cartan made sense of an obscure page in it, according to Leray (Poincaré, 1953).

The exceptions are, probably, only Birkhoff and his students Morse and Whitney. R. Thom in his address on Smale's work at the International Congress of Mathematicians in Moscow in 1966 (Thom, 1968) said that Smale was almost the only mathematician who had read Poincaré and Birkhoff.

In *The mathematical heritage of Henri Poincaré*, published recently by the American Mathematical Society (Browder, 1982), one may even read that Poincaré did not know what a manifold is! Actually, the definition of a (real) smooth manifold is set forth in detail in his *Analysis Situs* (Poincaré, 1895). This "naive" definition is the following (in contemporary terms): A *manifold* is a submanifold of Euclidean space (considered up to diffeomorphism); (a complex manifold is a complex structure on a real manifold).

3.4. Naive and Abstract Definitions. Of course, Poincaré's "naive" definition is different from the contemporary one although, by one of Whitney's theorems, the two are equivalent. However, if one thinks about it, the meaning of the contemporary "abstract" definition, together with Whitney's theorem, amounts to the conclusion that natural attempts to generalize Poincaré's naive conception of a manifold do not, in fact, produce any new objects!

The situation here is the same as for abstract groups, which are obtained from the "naive" ones (groups of transformations) by forgetting the set being transformed. The abstract definition is necessary only if one wants to verify that the algebraists' natural attempts to generalize naive groups do not lead to any new concepts. I offer one more example of this kind: the history of Church's thesis: natural-seeming attempts to generalize the naive conception of an algorithm did not lead to any new objects.

This axiomatizing in attempts to generalize inevitably arises in the process of working out concepts, but then it is better to forget about it – and, in particular, when teaching, it is better to return to the "naive" definition of a manifold, given by Poincaré. (Poincaré himself discussed in detail the advantage in teaching methodology of using naive definitions of the circle and of fractions: *Science et méthode*, Ch. II (Mathematical definitions and teaching), pp. 355–356 (Poincaré, 1908).

Along with submanifolds of Euclidean space, Poincaré studied other constructions as well, for example, gluing a manifold together out of charts from an atlas (at this point it would have been necessary to prove that all objects obtained by such gluings can be realized as submanifolds, i.e., Whitney's theorem, but Poincaré apparently considered this to be obvious). As for such manifolds in *Analysis Situs*, among other things, Poincaré studied in detail fibrations with base a circle and fibers tori, which are carriers of the first example of Anosov flows, proposed by Smale, to whom Thom pointed out this example.

3.5. Catastrophe Theory in the Work of Poincaré. Returning to catastrophe theory, I shall point out two more lesser-known examples of the "application" of its ideas in Poincaré's work: the study of singularities of caustics and "cut loci", and singularities of solutions of the Hamilton-Jacobi-Bellman equations (see Fig. 13) in the article *Sur les lignes géodesiques des surfaces convexes* (Poincaré, 1905a), and the study of singularities in the dependence of multiple integrals on parameters (the chapter on the expansion of a perturbation function in *Les nouvelles méthodes de la Méchanique céleste* (Poincaré, 1892), in Section 98, Vol. I).

Ending this discussion of the theory of bifurcations constructed by Poincaré on the basis of the preparation theorem and the theorem on versal deformations which he proved, I note that this theory is literally applicable only to bifurcations from equilibria or periodic orbits, but not to bifurcations of the differential equations themselves.

Versal deformations of differential equations defined by vector fields on the line with a degenerate singular point were found only in 1984:

$$\dot{x} = x^{\mu+1} + \lambda_1 x^{\mu-1} + \cdots + \lambda_\mu + \lambda_0 x^{2\mu+1}$$

(a theorem of V.P. Kostov (1984), who also investigated the deformations of

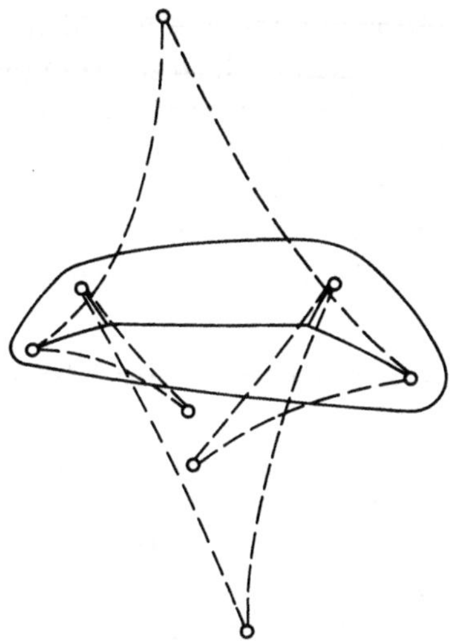

Fig. 13. Singularities of the distance to a curve

forms $f(dx)^\alpha$, $\alpha \in \mathbb{C}$ (vector fields correspond to $\alpha = -1$)). The higher-dimensional case requires a transition to homeomorphisms or diffeomorphisms of finite smoothness.

3.6. Analyticity and Smoothness. An argument often put forward by "catastrophists" in discussions of the relationship between their theory and classical results, is that Poincaré's theory of bifurcations and the similar later theories of algebraic geometers relate to analytic or holomorphic objects, whereas now theorems on finite determinacy, versality, and stability can also be proved for infinitely (or even finitely) differentiable objects, which "substantively widens" the possibilities for applications.

The equivalence of a smooth object to its normal form, i.e. the possiblity of safely throwing away tails of Taylor series, actually creates several conveniences. For example, one may use algebraic geometry (say, the theory of quadratic forms, replacing a function in the neighborhood of a critical point by a quadratic form). However, one should not overvalue the applied worth of the results depending on throwing away tails. For example, the Morse Lemma (transforming a function to a standard quadratic form by a diffeomorphism) does not give a greater amount of topological information on the behavior of the function near a critical point than the transformation to normal form of its 2-jet by a linear transformation, and transformation at the level of the k-jet, for arbitrary k, is apparently sufficient for all practical purposes.

The contemporary development of singularity theory has shown that even in theoretical mathematical respects, the smooth (C^∞) theory in the majority of the simplest problems repeats the analytic theory. For example, the smooth analog of the preparation theorem was pointed out by Thom and proved by Malgrange

(1962–3). However, problems are not rare in which the smooth and analytic classifications are completely different. Such, for example, is the simplest problem on breaking of symmetry, which leads to the classification of pairs of involutions of the line with a common fixed point. The smooth classification repeats the formal classification here, whereas there exist formally equivalent, but analytically different classes depending upon arbitrary holomorphic functions (S.M. Voronin (1981, 1982) Écalle (1975), Dufour (1977)).

§ 4. The Theory of Bifurcations in the Work of A.A. Andronov

4.1. The Point of View of Function Space. "The main thing we must do with the differential equations of physical models is to investigate what is possible and what it is necessary to change in them", said Poincaré (*La valeur de la science*, Part 2, Chapter V (Analysis and physics), p. 222 (1905b)). In 1931, following this prescription, A.A. Andronov (1933) came out with a vast program (also set forth in the preface to Andronov and Khajkin, 1937) which differs from the contemporary program of catastrophe theorists only in that the qualitative theory of differential equations and Poincaré's theory of bifurcations take the place of Whitney's theory of singularities of differentiable mappings, as yet uncreated at his time.

From the mathematical point of view, the basis of Andronov's approach was to consider a whole class of admissible models of the phenomenon investigated at the same time, instead of a single model. That class, in fact, is usually a function space, since the functions that enter into the right-hand sides of the equations (for example, the characteristics of the nonlinear elements of an electric circuit) are usually known only approximately. In some cases the function space comes down to a finite-dimensional space of parameters, but here also the exact values of the parameters are usually unknown. It is above all a question of studying systems which are typical (in the given class of models) and parameter values corresponding to typical systems.

4.2. Structural Stability. A.A. Andronov reasons in the following way. Since the exact values of the parameters in a model are unknown, a conclusion drawn from the mathematical investigation of a model is worth the confidence of the practical worker only to the extent that this conclusion is stable with respect to small changes of the parameters. For example, a conclusion about the periodicity of a steady-state regime of motion of a system is well founded only if the corresponding periodic solution of the equations of the model is preserved under all small changes in the model (within the corresponding function space); here, of course, small changes in the periodic solution itself and in its period are allowed.

In formalizing these ideas, A.A. Andronov arrived at the *concept* of *structural stability* (in his terminology *roughness*) of properties of the system being investigated; the class of objects equivalent to the given one in relation to the properties of interest to us must be open in the corresponding function space of objects.

Andronov applied these ideas, above all, to dynamical systems and to orbital topological equivalence. In the 2-dimensional case, on the plane or on the sphere, he, together with L.S. Pontryagin, succeeded in proving the structural stability of generic systems (Andronov and Pontryagin (1937)). However, Andronov's program itself was more general and concerned any classification of arbitrary objects of analysis, for example, functions, fields, bifurcations, etc., for which he at once gave examples (see Andronov (1933), Andronov and Khajkin (1937)).

Depending on the classification being studied, the partition of the function space into classes can be very complex. For example, it can be discrete in some regions of the function space or parameter space and continuous in others; see Fig. 14. Indeed, this is how things stand for the partition of the space of dynamical systems with a more than 2-dimensional phase space into topological orbital equivalence classes; this was cleared up only in the sixties, principally thanks to Smale (1966).

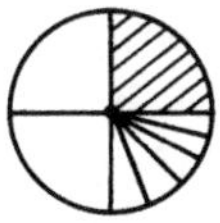

Fig. 14. Partition into classes

One should emphasize that the concept of roughness (structural stability) appears in Andronov's work as both a general physical and a general mathematical idea. The objects being studied need not necessarily be dynamical systems, and the classification need not necessarily be topological. In Andronov's terminology, the classification we use is determined by the questions we ask about a system, and the requirement of structural stability of models forbids our asking questions that are too precise on the qualitative behaviour of systems in those cases when a small change of the model changes the answers to these questions. This means a useful "qualitative" classification partitions the function space into discrete (and not continuous) parts.

I note such a classification has now been found in the theory of singularities of differentiable mappings (thanks to the efforts of Thom (1964), A.N. Varchenko (1975, 1974), Mather (1973), Looijenga (1974), Wirthmüller (see Gibson *et al.* (1977), and others), but is unknown in the theory of dynamical systems.

4.3. Bifurcation Sets. According to the arguments set forth above, nongeneric objects need not necessarily be met, and one may neglect them in an initial analysis (for example, the singular points of a generic vector field are nodes, foci and saddles, but not centers at all; and therefore the investigation of centers may be put off as less important).

This reasoning of Poincaré lessens the value of so many results of mathematical analysis, where, traditionally, special attention is given to the more difficult but rarely encountered degenerate cases, that it is dangerous to mention it. (There are also several bases of support for the opposite point of view, that degenerate cases are the most interesting since they stand out from the general mass of typical cases, which is Thom's remark.)

Matters stand differently, however, if one does not consider an individual object, but rather a family of objects that depend on parameters. For example, in a 1-parameter family a degeneration of codimension one, corresponding to a hypersurface (a submanifold given by one equation) of the function space under consideration, becomes unremovable; in a 2-parameter family a degeneration of codimension two is unremovable, etc.

The surface just mentioned is called a *bifurcation set*. Bifurcation sets are the walls that divide the function space of objects under consideration into regions consisting of equivalent generic objects (it is assumed, of course, that the classification considered is discrete).

A 1-parameter family of objects is represented by a curve (a mapping of the axis of parameter values into the function space). Such a curve may intersect the bifurcation set transversally (at a nonzero angle). The point of intersection corresponds to a bifurcation value of the parameter. As the parameter changes, at the moment it passes through a bifurcation value, the class of the object (the number and type of singular points, etc.) changes – the object undergoes a "qualitative change" or bifurcation.

Fig. 15. Transversal intersection

A nearby curve intersects the bifurcation set as before in a nearby point (Fig. 15); therefore, the bifurcation considered is unremovable under a small perturbation of the family (although the bifurcation takes place in the perturbed family at a slightly different value of the parameter). A nontransversal intersection of the curve with the bifurcation set is not found in generic 1-parameter families (in particular, the curve representing the family does not intersect the singular set of the bifurcation variety).

In generic 2-parameter families, one already unavoidably finds degenerations of codimension two (but not of codimension 3), in generic k-parameter families degenerations of codimension k.

Thus, *degenerations of finite codimension k are found ever more rarely as the codimension becomes higher, and their investigation must include the study of bifurcations in generic k-parameter families, in which these degeneracies are unremovable.*

In his theory of bifurcations, Poincaré only dealt with equilibrium states and periodic solutions. Andronov (1933) gave bifurcation theory its contemporary form; he began to deal with bifurcations of objects of a more general nature. In particular (and mainly), he dealt with bifurcations of phase portraits (that is, bifurcations of the topological structure of the partition of the phase space into curves).

Formally, Andronov was the first to investigate bifurcations, (in families with a finite number of parameters) up to changes of variables, fibered over the parameter space, just as is now usually done in contemporary singularity and catastrophe theory.

4.4. Degrees of Nonroughness. The general considerations set forth above on the classification of higher-order degenerations were formalized by Andronov and Leontovich (1938) in the form of a peculiar calculus of "degrees of nonroughness", or "degrees of structural instability".

Let there be given a discrete partition of the function space under study into various classes of objects: open regions of structurally stable (rough) objects and a bifurcation set. A point of the bifurcation set is said to be of *degree of nonroughness* 1 (or *first degree of structural instability*) if and only if all the nearby points of the bifurcation set lie within the same class of the partition. Further degrees of nonroughness are defined by induction: for example, points of degeneration of *degree of nonroughness* 2 are those points of the boundary of the set of points of degree of nonroughness 1, a whole neighborhood of which in the boundary mentioned belongs to a single class of the original partition, etc.

This definition, since it does not use a priori specific properties of the bifurcation set, except for its topology and the original partition into classes, anticipates the modern theory of stratifications of Whitney, Thom, Varchenko and Mather. From the contemporary theory, in particular, it follows that in the problems of singularity theory (but not of the theory of dynamical systems) all objects, except for a set of infinite codimension in the function space, are distributed according to degrees of nonroughness. (A priori this is not at all obvious, and, indeed, for the topological orbital classification of multi-dimensional dynamical systems it is not even true.)

Generally speaking, the degree of nonroughness need not coincide with the codimension of the corresponding stratum of the bifurcation set (since the dimension can change by more than 1 upon passing to the following stratum). This is so, for example, in the case of the classification of linear operators by rank. Thus, the degree of nonroughness of an operator of rank $n - 2$, acting on an n-dimensional space, is equal to 2, while the codimension of the set of operators of rank $n - 2$ in the space of all operators is equal to 4.

It is interesting, however, that this phenomenon apparently does not occur in the problem of classifying the critical points of functions up to smooth changes in the independent variables.

All known strata of the natural stratification of the space of germs of holomorphic functions at a critical point may be joined with the stratum of nondegenerate quadratic forms by a chain of contiguous strata, the dimensions of whose consecutive links descend by 1. (The conjecture that such a chain exists for any stratum has been discussed since 1972 and is still unproved.)

4.5. Structural Stability and Deformational Stability. From A.A. Andronov's point of view, bifurcation theory, in essence, deals with the splitting of the function space of all objects (models) of the type being studied into classes up to some kind or other of equivalence (usually, topological). Assertions about bifurcations in generic families that depend on finitely many parameters are obtained as consequences of this investigation. Namely, in sufficiently good cases the bifurcation set in function space has the structure of a local direct product of its section by a finite-dimensional subspace (transversal to the stratum of the bifurcation set to which the point being studied belongs) and an infinite-dimensional manifold of finite codimension (equal to the codimension of the stratum and the dimension of the cross-section), along which "nothing essential changes."

In this case the generic family with finitely many parameters is also a transversal section of the indicated stratum. The bifurcation set in function space leaves a trace (the preimage) in parameter space called the *bifurcation diagram* of the family. This bifurcation diagram in the situation being studied has the same local structure as the trace of the bifurcation set on the special cross-section described above (up to multiplication by a smooth manifold, if the number of parameters is larger than the codimension of the stratum). In this way, there arise universal hypersurfaces (not depending upon the family if it is generic) with singularities (the bifurcation diagrams) in finite-dimensional spaces.

The point of view of the contemporary theory of singularities and catastrophes is opposed in some sense to Andronov's appoach. Instead of the function space, these theories consider generic families with finitely many parameters immediately, and, instead of describing a neighborhood in function space of the object being studied, one studies families with finitely many parameters that contain this object. In particular, the structural stability of an object (a whole neighborhood belongs to the same equivalence class) is replaced by *deformational stability* (each deformation of the object with finitely many parameters is trivial).

In exactly the same way, the structural stability of a deformation of an object (each nearby family is locally equivalent to the family which gives the deformation) is replaced by *versality* (each deformation of the same object with finitely many parameters is equivalent to a deformation induced from the given one).

For most classifications of the contemporary theory of singularities, the difference pointed out above turns out to be illusory, that is, the deformational stability of an object implies its structural stability, and versality implies the structural stability of a deformation.

However, this fact is not evident a priori, and its proof in concrete classifications is generally not simple. One can imagine a situation where an equivalence class contains a neighborhood of an object in each finite-dimensional cross-section through it, but does not contain any neighborhood in the whole infinite-dimensional function space. I think, therefore, that in the future it will be necessary to return from the finitely-many-parameters formulation to a direct description of the partition of function space as A.A. Andronov and as his decendants proposed (see, for example, the investigation of the partition of the function space of diffeomorphisms of the circle into itself in my diploma thesis in 1959 (Arnol'd (1961)).

4.6. The Bifurcation Which Gives Birth to a Cycle. Catastrophe theorists have chronically underestimated the contribution of A.A. Andronov and his school to the theory of bifurcations. An especially outstanding example of this appears in the history of one of the most important bifurcations in dynamical systems, a bifurcation giving birth to a cycle. Already in the invited address "Mathematical problems of the theory of self-oscillations", given in November of 1931, Andronov (1933) pointed out that in typical one-parameter families of vector fields on the plane at isolated values of the parameter from an equilibrium state (smoothly changing with the parameter) a limit cycle is born or dies, the radius of which is proportional to the square root of the difference between the value of the parameter and its critical value. The birth or death of a cycle is accompanied by a *loss of stability of* the equilibrium state: a *soft* loss of stability, when a stable cycle is born (see Fig. 16), a *hard* loss of stability when a dying unstable cycle gives its instability to the equilibrium state (see Fig. 17).

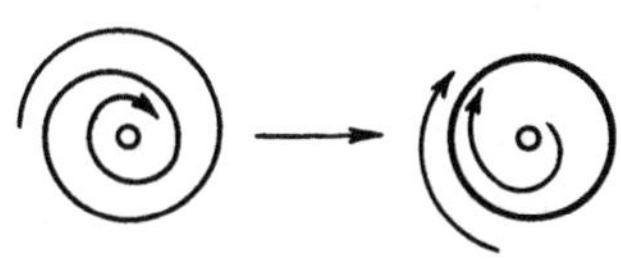

Fig. 16. A bifurcation generating a cycle

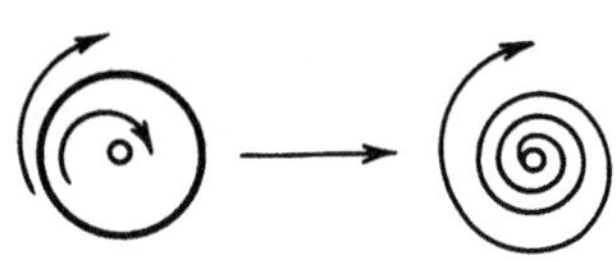

Fig. 17. A bifurcation killing a cycle

I emphasize that he dealt with a fully defined bifurcation of the phase portrait, occurring in typical one-parameter families in one of two standard ways (up to a homeomorphism of the phase portrait that depends continuously upon the parameter).

The theory of this bifurcation was set forth in the book by Andronov and Khajkin (1937), and detailed proofs are given in the article by Andronov and Leontovich (1939), where he also dealt with another bifurcation occurring in typical one-parameter families, the birth or death of a cycle at the moment of intersection of two separatrices exiting and entering a saddle.

In catastrophe theory a bifurcation giving birth to a cycle upon loss of stability by an equilibrium state is usually called a *Hopf bifurcation*, since E. Hopf (1942) investigated cycles being born also in the multi-dimensional case. In fact, Hopf's work is related to the theory of bifurcations in Poincaré's sense and not in Andronov's, since Hopf considered only a cycle and not the whole phase portrait. Moreover, Poincaré's theory contains all that is necessary for investigating the birth of a cycle (although, formally, in his *Les nouvelles Méthodes de la Méchanique céleste* he investigated only the more complex problems on the birth of cycles from equilibrium states of Hamiltonian systems).

The investigations (Brushlinskaya (1961), Nejmark (1959), and Shoshitajshvili (1975)) carried out later on bifurcations of a phase portrait upon birth of a cycle from an equilibrium state showed that, in a typical one-parameter family of multi-dimensional systems, the same phenomena take place as in Andronov's 2-dimensional theory. By a homeomorphism of phase space that continuously depends upon the parameter, the system can be reduced to a local normal form, which is the direct product of Andronov's standard system

$$\dot{z} = (i + \varepsilon)z \pm z^2\bar{z}$$

and the standard saddle

$$\dot{x} = x, \quad \dot{y} = -y.$$

(The dimensions of the x, y-spaces are arbitrary. For example, in the situation of loss of stability the variable x is absent.)

4.7. Delayed Loss of Stability. Up to now, we have been dealing with changes of the phase portraits depending on a parameter that does not change with time. In the theory of relaxation oscillations, those cases in which a parameter changes slowly with time (with speed proportional to a small parameter ε) are also important. Over a time of order $1/\varepsilon$ the parameter changes by a finite amount, and a stable (attracting) focus, "frozen" at a fixed value of the parameter of the system, may change into an unstable one, for example, softly. Upon "unfreezing" (i.e., upon taking into account the slow change of the parameter) the following startling effect emerges in an analytic system: the system remains near the already unstable equilibrium state for a time of order $1/\varepsilon$ during which a cycle that is born manages to grow to finite size (not small as $\varepsilon \to 0$), and only after this does the system jump onto the cycle, so that the whole process of loss of stability looks as if it were hard.

This phenomenon was first discovered by M.A. Shishkova (1973) in a model example. Recently, A.I. Nejshtadt (Uspekhi Mat. Nauk, 40 (1985), No. 5, 300–301; see also Diff. Uravn. 23 (1987), no. 12, 2060–2067; ibid. 24 (1988), no. 2, 226–233) proved that just such a delayed loss of stability is also typical for general systems

$$\dot{x} = f(x, y, \varepsilon)$$

$$\dot{y} = \varepsilon g(x, y, \varepsilon)$$

with f and g analytic, if the "fast motion" $\dot{x} = f(x, Y, 0)$, $Y = \text{const.}$ has an equilibrium $x = X(Y)$ that loses its stability as a pair of eigenvalues transversally crosses the imaginary axis into the right half-plane, owing to a slow change in Y with speed $\varepsilon g(X(Y), Y, 0)$.

It is interesting to note that the delay effect is strongly connected with analyticity. For finitely, or even infinitely, smooth systems the delay time is small compared to the time $1/\varepsilon$ of essential change of the slow variables y (the delay time in finitely smooth systems is of order $C(1/\varepsilon)\sqrt{(\varepsilon \ln \varepsilon)}$, and increasing the degree of smoothness increases only the constant C).

The mechanism of delay in crossing over is explained in that the fast damping of oscillations near the stable focus attracts the phase curve so close to it that, even after the loss of stability, quite a long time is required to move a finite distance away from the focus that has lost stability. Analogous delay phenomena are also met in other situations, for example, when a cycle loses its stability to a period-doubled cycle.

Returning to the ordinary case of birth of a cycle, in which the parameter (fixed slow variable) does not change with time, I note that this bifurcation should correctly be called an *Andronov bifurcation.*

4.8. The Pleat in the Work of A.A. Andronov. From what has been said above, it is obvious that A.A. Andronov's work contained almost completely the ideology of the contemporary bifurcation theory of the catastrophe theorists, and the unique thing that he did not possess was theory of singularities later developed by Whitney. It is interesting to note, however, that the singularities of mappings of a surface onto the plane (not up to diffeomorphisms as with Whitney (1955) but up to homeomorphisms) were known to Andronov.

Projections of the pleat are drawn in *The theory of oscillations* by A.A. Andronov, A.A. Vitt and S. Eh. Khajkin (1959) on p. 832, Fig. 569 of paragraph 2, § 11, Ch. X to describe the relaxation oscillations of an electronic rectifier. Namely, the surface projected is the surface of slow motions in 3-dimensional phase space; the horizontal direction of the projection is along the phase curves of fast motion. In the drawing the jump curve, on which the jump of the slow motion from one sheet onto the other takes place, is delineated; see Fig. 18. A closely related situation is investigated in § 5, Ch. IV (Andronov, Vitt and Khajkin (1959)), Figs. 175 and 176, which are reproduced in our Figs. 19 and 20. Here the disposition of straight lines is studied ("two external parameters" in the terminol-

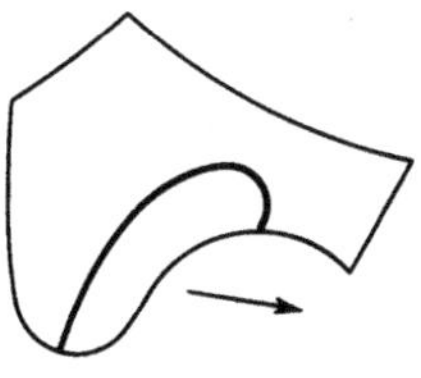

Fig. 18. The pleat according to Andronov-Khajkin-Vitt (A.A. Andronov *et al.* (1959), Fig. 569, Chapter X, § 11)

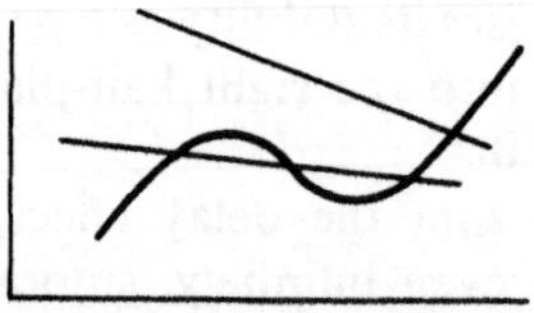

Fig. 19. Bifurcations of stationary regimes (A.A. Andronov *et al.* (1959), Fig. 175, Chapter IV, § 5)

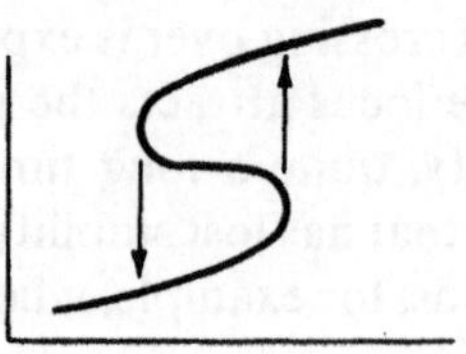

Fig. 20. Relaxation oscillations (A.A. Andronov *et al.* (1959), Fig. 176, Chapter IV, § 5)

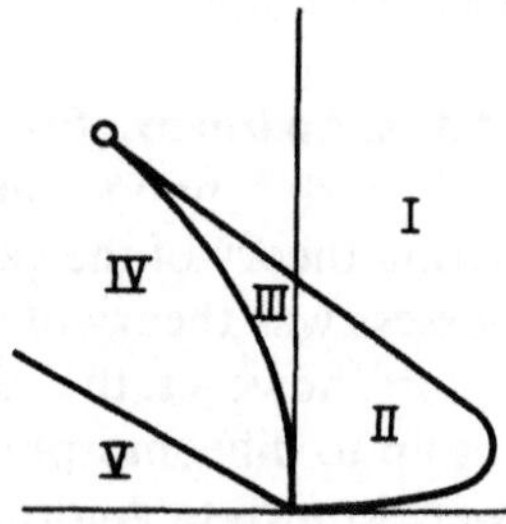

Fig. 21. A bifurcation diagram (from *The Theory of Indirect Regulation* by Andronov, Bautin and Gorelik (1946))

ogy of the catastrophe theorists) in relation to a generic curve with a point of inflection.

Numerous bifurcation diagrams of two-parameter and three-parameter families can be found, for example, in an article by A.A. Andronov, N.N. Bautin and G.A. Gorelik "A theory of indirect regulation" (1946). The cusp here (Fig. 14 of the article is reproduced as our Fig. 21) is related to a special, and not fully smooth, system, but the authors note that the structure of the parameter space is "in a qualitative way not specific" for the case they consider.

§ 5. Physicists' Treatment of Catastrophes Before Catastrophe Theory

5.1. Thermodynamics. Physicists have always made use of constructions more or less equivalent to catastrophe theory in investigations of concrete problems. In this sense, catastrophe theory may be compared with mathematical analysis.

Without the help of analysis, Huygens was able to solve the majority of problems solved by Newton. But such solutions required the genius of Huygens while, nowadays, the same problems may be solved with the help of analysis by any student. In exactly the same way, mastery of the technique of singularity theory allows one to obtain results automatically that otherwise require inventiveness and substantial efforts of imagination, simultaneously extending them to more complicated situations where "elementary" methods would lead to vast calculations. Porteous (1974) in an article "Nobel prizes for catastrophes" points out many examples of acknowledged physical achievements, whose authors, independently of one another, used ideas formalized later in singularity theory.

These ideas were systematically used in thermodynamics from the time of J.C. Maxwell and, especially, J.W. Gibbs. The perestroika (Fig. 22) of the isotherms of van der Waals' equations of state is a typical example of an application of the geometry of the pleat. An analysis of the asymptotics in a neighborhood of the critical point quickly leads to the understanding that this geometry is independent of the exact form of the equations of state. This fact was well known in the time of Maxwell, and is mentioned in most thermodynamics textbooks (for example, in the book of L.D. Landau and E.M. Lifshits (1964, § 84)).

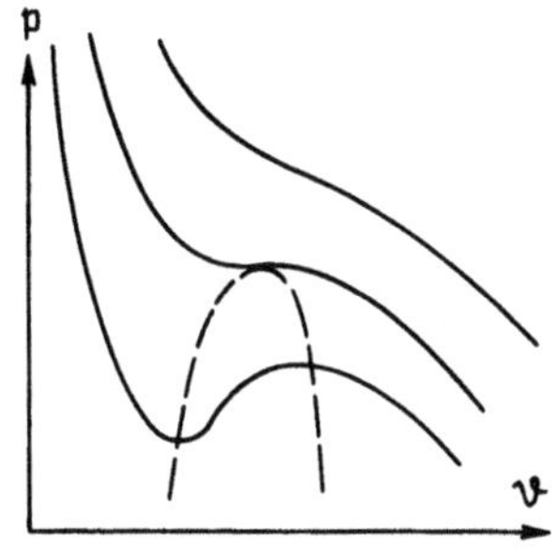

Fig. 22. van der Waals's diagram

Maxwell's idea was to draw the horizontal portion of an isotherm so that the areas of its lunes lying above and below it coincide. This equality implies that the transition from one of two competing minima of the potential to the other occurs at that moment when the second one becomes lower. Therefore, the equality just pointed out is called *Maxwell's convention* in catastrophe theory, too. A specific bifurcation diagram (formed of those values of the parameters at which the potential has two minima of equal height) corresponds to it. Therefore, even in the more general situation of the theory of singularities (and also for complex-valued functions), the set of values of the parameters at which a function of a family has several critical points with a common critical value is called the *Maxwell stratum*. For example, for a two-parameter family the typical singularity of the critical value as a (multivalued) function of the parameters is the swallowtail, and the Maxwell stratum corresponds to the curve of self-intersection on the swallowtail. In the parameter plane the projection of this curve has the form of a ray, tangent at its cusp to a semi-cubical parabola (the projection of the cuspidal edge of the swallowtail) formed by those values of the parameters at which the function has a degenerate critical point (Fig. 23).

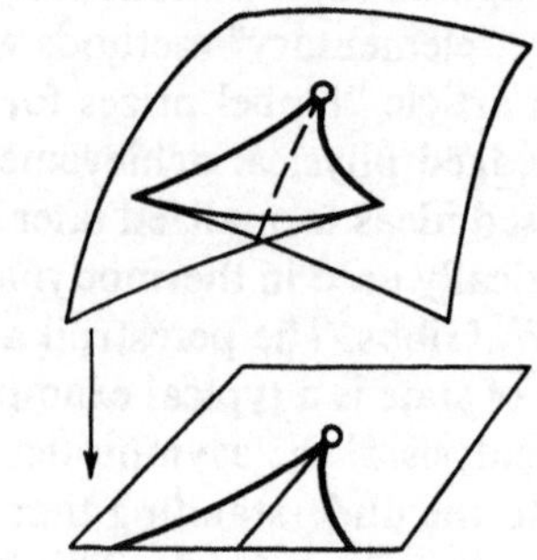

Fig. 23. A caustic and the Maxwell stratum

In typical two-parameter families, three equal critical values may also occur (for isolated values of the parameters), and $k + 1$ in k-parameter families. This consequence of the transversality theorem is called *Gibbs' phase rule* in thermodynamics (the discovery of the necessity of giving rigorous proofs of similar facts is due to mathematicians of a later period).

Physicists, chemists, and mineralogists (the last in connection with investigations of processes of crystallization of magma) were not afraid to go considerably farther in the geometrical investigation of singularities of generic multi-parameter families. For example, in a series of papers of F.A. Schreinemakers (1917) at the time of World War I, one can find a detailed geometrical analysis of singularities of curves and surfaces of equilibria of multi-phase systems. Fig. 24, taken from this work, illustrates the set of critical values of a pleat on the (p, T)-plane, in which the dotted lines single out the part that corresponds to "nonphysical" states. The author specially notes that MSm is a single curve with a cusp S (the branches MS and mS are called *spinodal curves*).

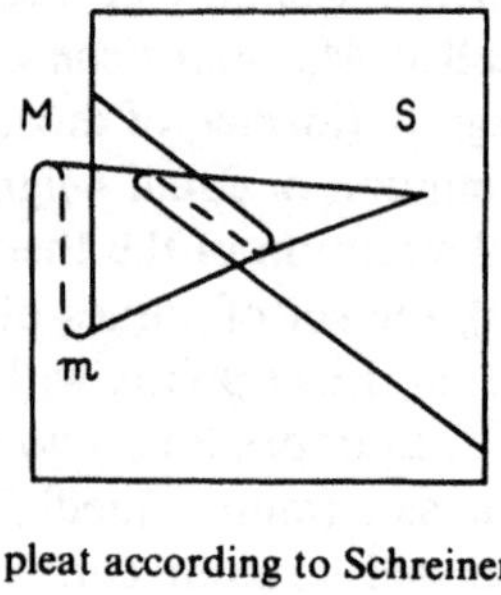

Fig. 24. The pleat according to Schreinemakers (1917)

The graphs of Legendre transformations of nonconvex generic smooth functions have the same singularities as generic wave fronts: both are realizations of the same *Legendre singularities* (Arnol'd (1975)).

Legendre singularities can also be described as singularities of the critical value of a function that depends upon parameters, considered as a many-valued function of the parameters.

Thermodynamic quantities can often be obtained from one another by a Legendre transformation; therefore, in work on thermodynamics, various Legendre singularities may occur. The connection of this circle of questions with contact structure was already described by Gibbs (1873).

5.2. Thermal Explosions. In N.N. Semenov's theory of thermal explosions, (1929) transitions that are jumps from one reaction mode to another ("ignition" and "extinction") arise under a smooth change of a system with, for example, one-dimensional phase space (the phase variable describing the concentration of one of the chemical substances).

On the plane of the values of the parameter and the phase variable, the stationary regimes form a smooth curve; however, its projection onto the axis of the parameter has singularities. As the parameter changes, one of the stable equilibria disappears, merging with an unstable equilibrium, and the system has to jump to a new regime with sharply different characteristics ("self-ignition"); see Fig. 25.

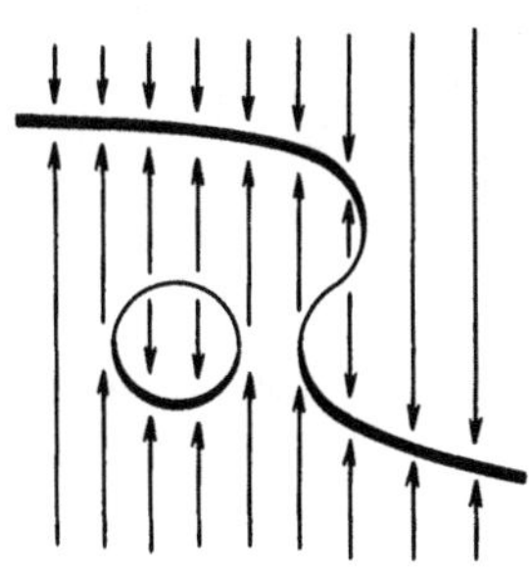

Fig. 25. Morse perestroika of a curve of equilibria

The analysis in this situation carried out by N.N Semenov and his followers differs from the corresponding arguments of contemporary catastrophe theory only in terminology ("explosion" instead of "catastrophe"); universality, i.e., the independence of the character of the phenomena observed from the specifics of the problem was, from the very beginning, evident to everyone concerned with these questions.

Along with the analysis of generic systems in the theory of combustion, their perestroikas as other parameters of the problem are varied were also investigated. For example, in 1940 Ya.B. Zel'dovich (1941) analyzed the phenomena occurring during Morse perestroikas of equilibrium curves (birth of new islands or their merging with the main curve). In contemporary mathematical theory the analogous analysis was carried out only in recent years (Diener, 1984).

5.3. Short-Wave Asymptotics. The analysis of a wave field near a caustic led G.B. Airy (1838) to the functions which bear his name (see Fig. 26)

$$Ai(z) = -\int_{-\infty}^{\infty} e^{i(zx - x^3/3)}\,dx.$$

The phase in the exponent is precisely a versal deformation of the function $-x^3/3$ which has a simple degenerate critical point at the origin.

In exactly the same way, the analysis of a field near a cusp point of the caustic led Pearcey to the analogous integral

$$Pe(z_1, z_2) = \int_{-\infty}^{\infty} e^{i(x^4/8 - z_1 x^2/2 + z_2 x)}\,dx,$$

whose phase is a versal deformation of the singularity $x^4/8$. The level curves of the square of the modulus of Pearcey's function are shown in Fig. 27.

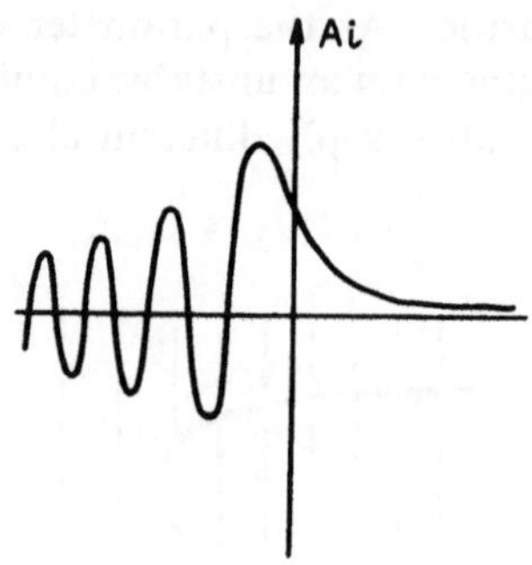

Fig. 26. The Airy function

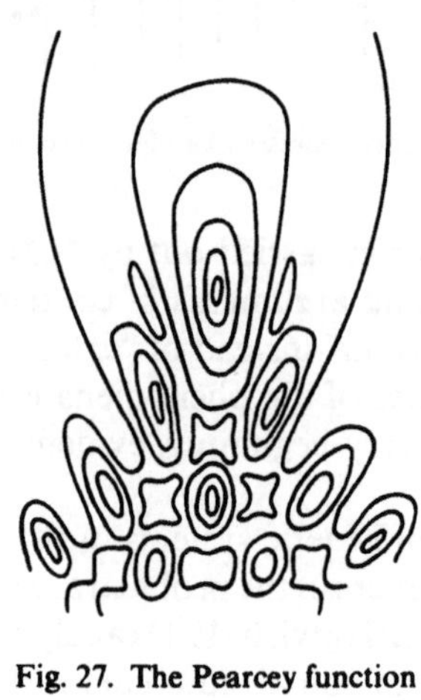

Fig. 27. The Pearcey function

Neither Pearcey nor Airy (nor Fresnel (1818) who, still earlier, looked at the analogous integral for a nondegenerate critical point) rigorously proved that the standard functions they studied correctly describe the high-frequency asymptotics of a generic field. However, they clearly understood the universality of the phenomena they discovered (otherwise it would have been difficult to correctly guess the standard phases). In this connection, it is worthwhile, I believe, to mention that the high-frequency asymptotics found by M.A. Leontovich (1944) and V.A. Fok (1946) for an electromagnetic field near a scattering obstacle still have not been digested by by catastrophe theory (compare Buldyrev and Lanin (1981) and Popov (1984)).

5.4. The Theory of Elasticity. Let us consider the maximal load that an elastic bar can carry, when bent in the form of the arc shown in Fig. 28.

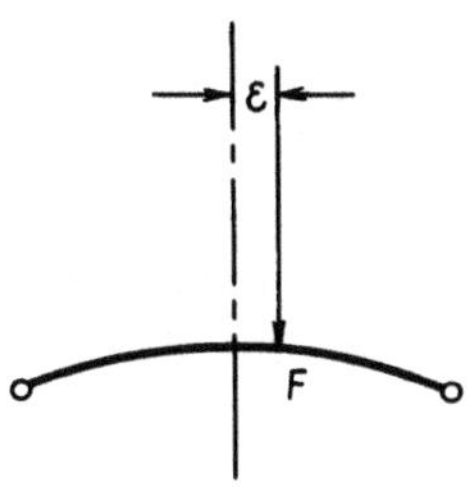

Fig. 28. An off-center load

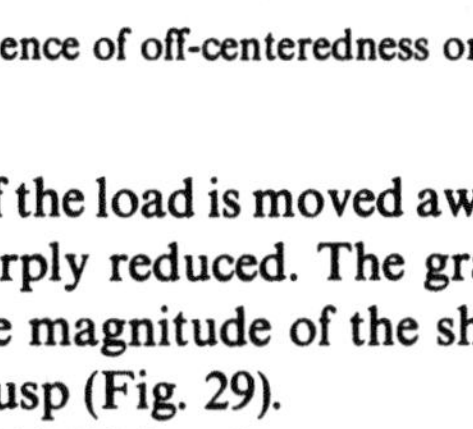

Fig. 29. The influence of off-centeredness on the critical load

If the point of application of the load is moved away from the axis of symmetry, then the maximal load is sharply reduced. The graph showing the dependence of the maximal load upon the magnitude of the shift (of the point of load from the axis of symmetry) has a cusp (Fig. 29).

This result of W.T. Koiter (1945) is a forerunner of many results of the theory of sensitivity of elastic models to imperfections. In contemporary terms the problem reduces to investigating the family of potentials $-x^4 + (F_0 - F)x^2 + \varepsilon x$. The largest value of the parameter F at which this function of x has a minimum is sought. Naturally, one obtains a semi-cubical parabola.

To specialists in the theory of elasticity the universality of this phenomenon, as well as the form of the smooth surface of equilibria, was well known. If the equilibrium surface is projected along the x-axis onto the (F, ε)-plane, it has a pleat over the apex of a semi-cubical parabola. Investigations also took place of systems with a larger number of parameters and having a more complicated geometry of their singularities. For example, during experiments upon elastic constructions, it may be important to carry out loading without dangerous snaps, and for this it is necessary to know the "catastrophe diagram" and the topology of how the sheets of the equilibrium surface are joined over it.

The necessary calculations were carried out without a general theory for finding the correct terms to throw away among those in the Taylor series and which ones it is important to keep, as was done in obtaining the above formula for the potential. That the model system one obtains is actually a normal form (i.e., that all the other terms of the Taylor series can be killed by a suitable choice of the coordinate system) does not have great practical importance.

5.5. The Work of L.D. Landau. Among the physicists who systematically applied catastrophe theory before it was born, one should particularly single out L.D. Landau. In his hands, the art of throwing away inessential terms of Taylor series, retaining the higher-order, but "physically important" terms, yielded many excellent results, now included within catastrophe theory.

Thus, for example, in work done in 1943 on the generation of turbulence, by this method Landau (1943) directly wrote out the equation of the "Hopf bifurcation" for the square A of the amplitude of an oscillation losing its stability:

$$\dot{A} = A(\varepsilon + cA)$$

(it seems that he did not know the work of A.A. Andronov on bifurcations generating cycles).

Landau's *theory of phase transitions of the second kind* reduces to the analysis of bifurcations of critical points of symmetric functions (§§ 138–139 in Landau and Lifshits (1964); Gufan (1982)) (the "kind" indicates the derivative of the function which depends discontinuously upon the parameter). The *Landau curves* (F. Pham, 1967) in the theory of Feynman integrals depending upon parameters, with their stable cusps, are included among the fundamental bifurcation diagrams of contemporary singularity and catastrophe theory.

Of course, the contemporary general theory allows one to investigate with less effort, and more rigor, more complicated singularities (which occur when the number of parameters is larger). However, the simplest and the most frequently encountered singularities have the greatest practical value in most cases: the expenditure of energy in surmounting technical difficulties that stand in the paths of investigation of the more complicated cases, is not always justified by the practical value of the results obtained. To the contrary, the fundamental work of the forerunners of catastrophe theory (those mentioned above as well as many others) retains all its value even now when its mathematical structure has been completely clarified by singularity and bifurcation theory.

§ 6. Thom's Conjecture

6.1. Gradient Dynamics. In the middle of the 1960s R. Thom, the first person to appreciate the value of H. Whitney's theory of singularities for applications, undertook to apply it to problems of biology (namely, to embryology), and broadly to problems of morphogenesis in general. The rough model with which he began was the following: The state of each point of an organism is described by a vector **c** of concentrations of chemicals, and its evolution, to a first approximation, is described by a gradient dynamical system:

$$\dot{\mathbf{c}} = -\mathbf{grad}\, U(\mathbf{c})$$

whose potential function U depends upon four parameters, the three coordinates of the point considered and on the "slow" time (the left-hand side denotes the derivative of **c** with respect to fast time).

The analysis of this model led Thom to the necessity of investigating generic four-parameter families of smooth functions (the dimension of the space to which the vector **c** of concentrations belongs is not restricted). Putting aside (important) questions on the relation of such a model to reality, I shall discuss below the mathematical problems that arise here.

First it was necessary to choose an equivalence relation, up to which the classification is carried out. As is natural in the theory of dynamical systems, Thom chose the topological classificiation, in which two systems are equivalent if their phase portraits are homeomorphic. (For gradient systems it would also have been possible to choose the topological equivalence of phase flows.)

The character of the singular points of a gradient field depends upon the nature of the critical points of its potential function. Therefore it was natural to begin with the investigation of bifurcations of critical point in typical four-parameter families of functions. Analyzing these bifurcations, Thom (1969) carried out their topological classification (with some unimportant, easily corrected errors) and gave names to the singularities encountered. (In my opinion, these were not completely successful choices; for example, one must always expend some effort to remember what an *elliptic umbilic* is and what a *hyperbolic umbilic* is. Thus, I prefer the terms "pyramid" and "purse" which bring to mind the forms of the corresponding caustics; see Fig. 30.)

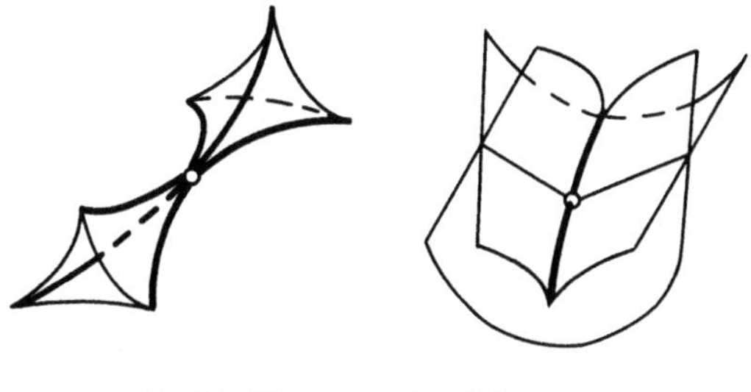

Fig. 30. The pyramid and the purse

Nevertheless, the introduction of unified names (even if for objects met earlier in various theories) was an important step forward.

6.2. The Classification of Critical Points of Functions. Thom did not publish his research on the theory of catastrophes until 1969, and even now it is difficult to say exactly what he proved. In any case, in a paper by Mather which was written in 1968 under Thom's influence (but which also remained unpublished) a theorem is proved which classifies, up to diffeomorphisms, the simplest degenerate critical points of smooth functions (unavoidable in generic four-parameter families[3]). A theorem, close to this one, on singular points of analytic hypersurfaces was published in the same year by G.N. Tyurina (1968)

The classification theorem is not precisely formulated in Thom's article (1969); however, the correct list of potentials was given. Therefore Zeeman, setting forth the formulation and proof of the above mentioned theorem in an unpublished paper (Trotman and Zeeman (1974)), called it *Thom's Theorem*. As far as one can judge from Thom's article, he was concerned with the topological classification of gradient systems and was more interested in how functions and bifurcation diagrams "look" than in their classification up to diffeomorphism[4].

Thom's point of view was that topological perestroikas of a potential must affect the behavior of the gradient system, and consequently bifurcation diagrams in the 4-dimensional space of parameters may be looked at as boundaries separating regions of different behavior. He interpreted those boundaries in the spatial cross-sections of space-time as separators of structures, and he regarded their perestroikas in time as a process of morphogenesis.

The analysis of perestroikas of cross-sections of bifurcation diagrams in 4-dimensional space-time was carried out only in 1974 (Arnol'd (1975)), when adequate methods in singularity theory had been worked out (based upon the connection between bifurcation diagrams and the theory of reflection groups). The perestroikas of wave fronts (bifurcation diagrams of zeros of functions) in 3-dimensional space are illustrated in Fig. 31.

6.3. The Classification of Gradient Systems. By analyzing the bifurcations of critical points of potentials in generic families, Thom thought that he would obtain the topological classification of the local bifurcations in the corresponding gradient dynamical systems. The assertions he announced reduce to the result that in gradient systems that depend generically upon four parameters, one finds only 7 topologically different bifurcations of local phase portraits, corresponding to the 7 types of degeneracy of critical points of potentials described by him.

---

[3] Mather's (1971) classification of the simplest degeneracies of complete intersections, published at the same time, contains many errors, later corrected by M. Giusti (1977).

[4] Apparently he had in view some sort of variant of combinatorial equivalence, never formulated to this day, intermediate between homeomorphism and diffeomorphism.

Fig. 31. Perestroikas of wave fronts

This "*magnificent seven*" was repeatedly played up in subsequent publications of catastrophe theorists (see, for example, Stewart (1975)). However, as early as 1973, J. Guckenheimer (1973) pointed out that in typical families of multidimensional gradient systems there are topological bifurcations that do not reduce to perestroikas of critical points of potentials.

Let us consider, for example, a gradient system on the plane, and let us assume that it has two critical points which are saddles. Then at some bifurcation value of the parameter the exiting separatrix of the higher saddle (higher in potential) may coincide with the entering separatrix of the lower saddle (Fig. 32). This value of the parameter is a bifurcation value from the point of view of the classification of gradient systems, although for the potential nothing special happens at this value of the parameter (it may be transformed by a diffeomorphism into any nearby potential). The bifurcation in Fig. 32 is called the appearance of a *saddle connection*.

The possibility that saddle connections appear as saddles approach one another shows that Thom's assertion about the exactly seven topological bifurcations of local phase portraits of gradient dynamical systems is unfounded. Therefore this assertion (whose proof, by the way, Thom never published) is better called *Thom's conjecture* rather than Thom's Theorem.

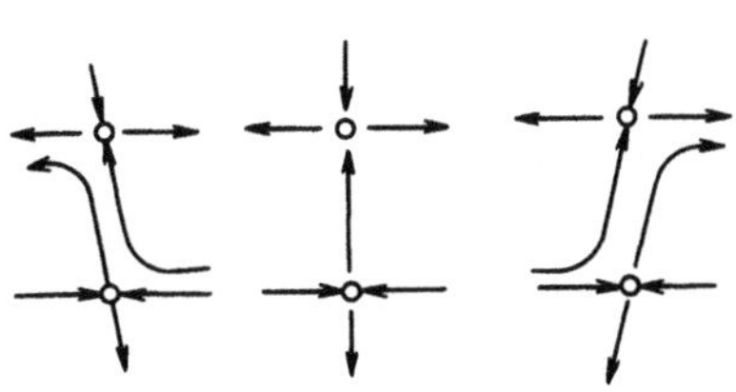

Fig. 32. Saddle connections

6.4. Bifurcations of Gradient Systems. In one-dimensional systems (all of which are gradient systems) saddle connections do not appear. In this case Thom's program for the investigation of bifurcations of gradient systems can be realized.

If the rank of the second differential of a function at a critical point decreases only by one (the so-called A_k series of singularities), then the investigation of the bifurcations of the phase portrait of the corresponding gradient dynamical system in a neighborhood of this point may be reduced to the investigation of bifurcations in a 1-dimensional system (because of the "hyperbolicity" in the transversal direction; see Shoshitajshvili (1975)).

If the number of parameters upon which the potential in a typical family depends is less than three, then the rank of the second differential of the potential at a critical point does not fall by more than one. Therefore, in 1- and 2-parameter generic families, Thom's program can also be realized, and saddle connections cannot appear in the investigation of bifurcations of the local phase portrait in a neighborhood of a critical point of the potential.

On the other hand, in typical 3-parameter families of gradient systems on the plane, saddle connections do appear near a critical point of a potential of type D_4, and they contribute new components to the bifurcation diagrams (to the caustics). Moreover, as G. Vegter (1982) pointed out, different (diffeomorphic) critical points of type D_4 (and different generic metrics on the plane) lead to different (nonhomeomorphic) bifurcation diagrams. From this it follows that, already in the case of three parameters, the number of topologically different bifurcations of local phase portraits of typical gradient systems is a fortiori greater than Thom suggested. Therefore, Thom's conjecture on local bifurcations in gradient systems is not correct in the form in which it was stated above.

6.5. Stating Thom's Conjecture More Precisely. From what was said above, it is clear that the number of topologically different bifurcations of local phase portraits in typical four-parameter families of gradient systems is greater than seven. However, the possibility remains that the number of such bifurcations is nevertheless finite, and, in this improved, more accurate form, Thom's conjecture has not been disproved.

In order to formulate the improved version of the conjecture, let us consider the space of smooth functions on a fixed Riemannian manifold. Suppose a family of such functions is given that smoothly depends upon ℓ parameters. Two such families are said to be *locally topologically gradiently equivalent* at a point if there exists a germ of a homeomorphism of the product of the space of parameters with the Riemannian manifold into itself, fibered over a homeomorphism of the space of parameters, and mapping the phase curves of gradients of functions in the first family (in a neighborhood of the point studied) into the phase curves of gradients of functions of the second family (preserving the direction of motion).

The improved version of Thom's conjecture asserts that for $\ell \leqslant 4$, a typical family of smooth functions on a Riemannian manifold is locally (in a neighborhood of each value of the parameters and each point of the Riemannian mani-

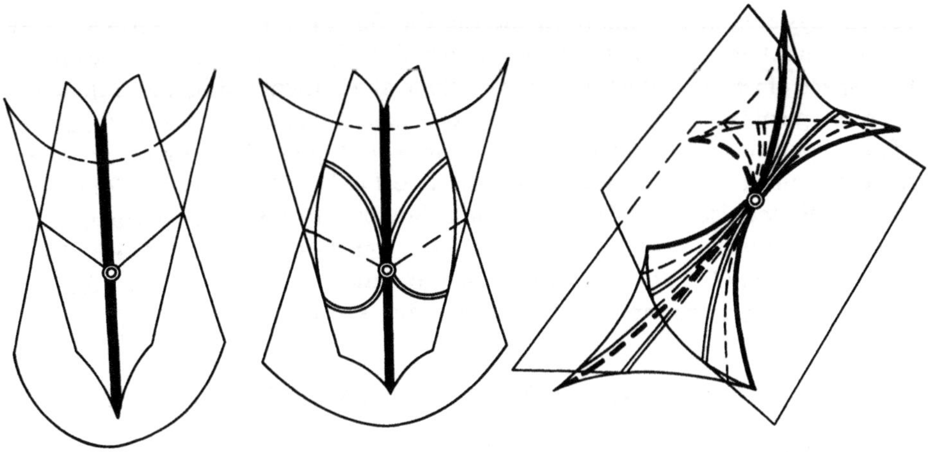

Fig. 33. Bifurcation diagrams of gradient fields

fold) topologically gradiently equivalent to one of a finite number of germs of families (depending only on the number of parameters ℓ, and on the dimension of the manifold).

From this, apparently, it follows that the number of types of local bifurcation diagrams for a fixed number of parameters is finite (up to homeomorphisms). This number of types of diagrams, apparently, is independent of the dimension of the Riemannian manifold.

Thom's conjecture, made more acccurate in this way, has been refuted if the number of parameters is large (of order 10), but it is true for one- and two-parameter families of potentials. The case of three parameters is discussed below.

6.6. Bifurcations of Gradient Systems of Type D_4. The investigation of three-parameter families near a D_4 singularity, where it is first necessary to take saddle connections into account, was begun by G. Vegter (1982). According to B.A. Khesin (1984), in this case, the number of topologically different local bifurcations of gradients is four. Thom distinguished only two types of such bifurcation diagrams (the purse and the pyramid). The expansion of the list occured to take into account saddle connections. Saddle connections sometimes arise, and sometimes not (Fig. 33), depending upon the function and the metric[5]. It is interesting that for the normal form of the D_4-singularity (the potential $x^3 + y^3 + ux + vy + wxy$, depending upon parameters u, v, w on the plane with the standard metric $dx^2 + dy^2$) there are no saddle connections, whereas after a small deformation of either the family or the metric, they may appear (this was already noticed by Guckenhemier (1973)). This means that the standard normal forms of D_4 are not suitable for describing the topologically versal deformations of

[5] Strictly speaking, Vegter considered another equivalence relation (continuous dependence of the transforming homeomorphisms under discussion on the parameters has been proved only for values of the parameters distinct from zero).

gradient systems in the standard metric: the function $x^3 + y^3$ belongs to an exceptional set, from the point of view of gradient dynamics of codimension $\geqslant 4$ in the space of germs of functions on the plane at a critical point, and it does not occur in typical three-parameter families of gradient systems (although functions diffeomorphic to $x^3 + y^3$ do occur). The number of topologically different typical gradient families D_4, according to Khesin (see *Funkts. Anal. Prilozh.* 20 (1986), No. 3, 94–95; see also Itogi Nauki i Tekhniki, Sovremennye Problemy Matematiki, Noveishie Dostizhenya vol. 33, VINITI, 1988 translated in J. Sov. Math., 1990, V. 52, no. 4, 3279–3305), is equal to 4 (potential functions $x^3 + y^3$ for D_4^+, $x^3 - 3xy^2$ for D_4^-, and metric $(11/10)\,dx^2 + dy^2 + p\,dxdy$).[5a]

§ 7. Classifications of Singularities and Catastrophes

7.1. Codimension and Modality. At first glance, the most natural classification principle is classifying by codimension beginning with small codimensions. *To classify objects "up to codimension $\leqslant k$"* means to represent the entire space of objects studied as a finite union of submanifolds of codimensions not greater than k (called *classes*) and a remainder of codimension $\geqslant k + 1$ so that within each class an object's properties that are of interest to us do not change. Then all objects in typical, no more than k-parameter families, belong to our classes: the remaining ones may be avoided by a small perturbation of the family.

For example, suppose the objects are germs of smooth functions of two variables with critical point 0 and critical value 0. Then the classification up to codimension $\leqslant 4$ is formed by the classes of functions locally transformable to functions in the following list (Thom's seven)

A_1	A_2	A_3	A_4	D_4	A_5	D_5
$\pm x^2 \pm y^2$	$x^3 \pm y^2$	$\pm x^4 \pm y^2$	$x^5 \pm y^2$	$x^2y \pm y^3$	$\pm x^6 \pm y^2$	$x^2y \pm y^4$

by a diffeomorphic change of independent variables. Classification up to codimension k differs, generally speaking, from *classification by codimensions of orbits* (or of classes under some equivalence relation). This is so because the orbits may form continuous families. In this case, the appearance of objects whose orbits have codimension k may turn out to be unavoidable under classification up to some codimension less than k.

For the problem considered here, of the classification of functions of two variables, a difference first appears for $k = 7$ (if the number of variables is more than two, then it appears for $k = 6$). Namely, the functions with a singularity of type $x^4 + y^4 + ax^2y^2$ form a class of codimension 7 (it is impossible to avoid the appearance of such singularities by a small perturbation of a generic seven-parameter family) although the codimension of the orbit of each fixed function in the class is not less than eight.

[5a] The coefficient 11/10, forgotten in Khesin's article, is necessary. Normal forms of the gradient families correspond for example to the values $p = \pm 1$ and $-1/2$ for D_4^+, and to the value $p = 1$ for D_4^-.

The codimension of the orbit of a germ of a function at a critical point of multiplicity μ is equal to $\mu - 1$ (the multiplicity μ is indicated in the notation for the function; for example, A_1 is a generic singularity, the codimension of an orbit is equal to zero, etc.).

The germs which can be joined to a given germ by a continuous path consisting of germs of constant multiplicity form the *μ-equivalence* class of germs. For the μ-equivalence class of a holomorphic function:

the codimension of the class

= the codimension of the orbit – the number of moduli.

The number of moduli or the *modality* of an object is the least number m for which some neighborhood of the object can be covered by at most a finite number of at most m-parameter families of orbits. (Recently, V. Serganova and V. Vasil'ev discovered that the theory of matroids implies that the $\mathbb{C}$-modality and the $\mathbb{R}$-modality of a function germ may be different.) In the given case, we are talking about a neighborhood of a jet of the function (in the space of k-jets with critical value zero at the critical point zero), and the finite number referred to in the definition must remain bounded as $k \to \infty$ (which is so for functions with a critical point of finite multiplicity).

It turns out that, algebraically, the most natural classification is not by multiplicity or up to a given codimension, but classification by modality. Namely, for functions of the same modality, other properties also turn out to be similar (their intersection forms, monodromy groups, etc.), and, indeed, they appear together in comparisons with other classification problems.

7.2. Simple Objects. Objects of modality zero are called *simple*. A neighborhood of a simple object is covered by a finite number of orbits. A germ of a function at a critical point is simple if it can be deformed in only a finite number of ways (up to smooth changes of the independent variables).

The list of simple critical points of holomorphic functions of two variables is formed by two series of singularities and three exceptional ones (Arnol'd (1972)):

A_k, $k \geqslant 1$	D_k, $k \geqslant 4$	E_6	E_7	E_8
$x^{k+1} + y^2$	$x^2y + y^{k-1}$	$x^3 + y^4$	$x^3 + xy^3$	$x^3 + y^5$

(if the number of variables is larger, it is necessary to add a nondegenerate quadratic form in the missing variables). This list repeats the list of Weyl groups (the crystallographic groups of Coxeter, generated by reflections) without the multiple links present in their Dynkin diagrams (that is, with 90° and 120° angles between the generating mirrors). The connection between functions and reflection groups is the following. Include a function with a critical point of multiplicity μ in a generic $(\mu - 1)$-parameter family as the function that corresponds to the zero value of the parameter. At a typical value of the parameter close to zero, the critical point splits into μ nondegenerate (Morse) critical points with different critical values. These critical values may be considered as a μ-valued function of

$\mu - 1$ variables (the argument being the parameter of the family). The graph of this μ-valued function lies in a μ-dimensional complex space and is called the *bifurcation diagram of zeros* of the original singularity. For example, the bifurcation diagram of zeros of the singularity A_2 is a semi-cubical parabola, and of A_3 is a swallowtail. The investigation of the geometry of bifurcation diagrams of zeros, their cross-sections, and projections, forms in a technical respect the most important part of catastrophe theory.

It turns out that *the bifurcation diagrams of zeros of the simple singularities are diffeomorphic to the varieties of nonregular orbits of the corresponding reflection groups.* This fact permits the use of techniques developed in the theory of reflection groups in investigations of the bifurcation diagrams of simple singularities. For example, from this, the theory of perestroikas of wave fronts easily follows: the cross-sections of the bifurcation diagrams of a simple singularity by the level hypersurfaces of a generic function are standard (Arnol'd (1976)).

In particular, consider the 3-dimensional space of polynomials

$$x^4 + ax^2 + bx + c.$$

The polynomials having a multiple zero form the bifurcation diagram of zeros of A_3, a swallowtail (Fig. 34). A generic homomorphic function, defined in a neighborhood of the point 0, can be reduced to the normal form a + const by a diffeomorphism that preserves the swallowtail. In particular, a generic surface, passing through the singular point itself, can be reduced by a diffeomorphism that preserves the swallowtail to the vertical coordinate plane $a = 0$, and the family of level lines of a generic function on the surface of the swallowtail is reduced to the family of plane sections of the swallowtail by the vertical planes parallel to $a = 0$.

Where classification by an "incorrect" principle (by codimensions) leads to a chaotic, difficult to understand, hierarchy of more and more complicated bifurcation diagrams, the "correct" classification (by modality) is governed by simple general rules.

7.3. Functional Moduli. In more complex situations of the theory of singularities, along with simple objects and objects of finite modality, one also encounters

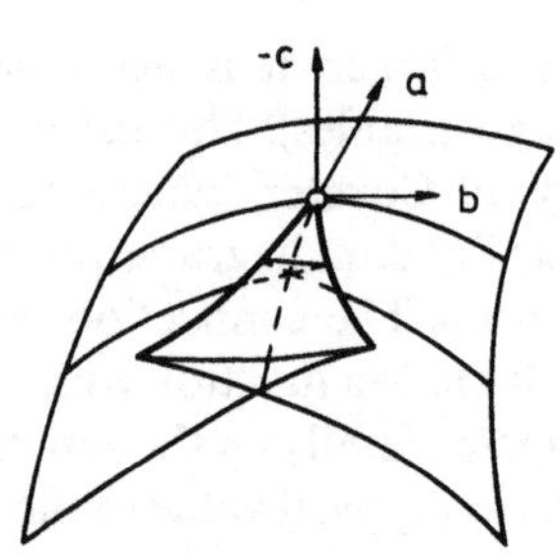

Fig. 34. Sections of the swallowtail

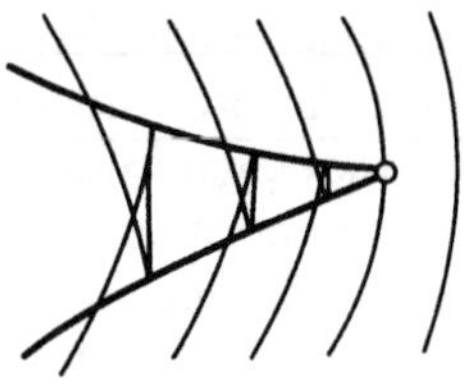

Fig. 35. Integral curves of an implicit differential equation

objects whose moduli are arbitrary functions. This is the situation, for example, in the problem of classification of singularities of smooth mappings of spaces having the same sufficiently large dimension (Wall, 1981). In these cases (as also, moreover, generally in the presence of moduli) the classification quickly becomes difficult to handle.

Let us consider, as an example, the singularities of solutions of an implicit differential equation $F(x, y, p) = 0$, unsolved for the derivative $p = dy/dx$. For generic F, the surface $F = 0$ in the 3-dimensional (x, y, p)-space of 1-jets of functions is smooth, and the only singularities of its projection onto the (x, y)-plane of 0-jets are folds[6] and pleats. The critical values of this projection form the *discriminant curve* of the equation. A pleat corresponds to a cusp of the discriminant curve.

An integral curve of such a generic equation has a cusp at an ordinary point of the discriminant curve. The smaller region bounded by the semi-cubical discriminant curve close to the projection of the pleat is covered three times by the integral curves (Fig. 35): three branches pass through each point. The picture obtained has functional moduli not only with respect to diffeomorphisms, but even with respect to homeomorphisms of the plane.

In fact, if one thinks of the entire family as being parameterized in some nice way, the numbering of the three integral curves passing through a given point of the plane gives three functions of two variables. Two of them can generally be taken as coordinates. The third is then a function of these two coordinates. This function of two variables is defined invariantly up to renumberings of the integral curves, that is, up to choices of three functions of one variable. Whatever one might say about superpositions, there are more functions of two variables than of one, so that there is present here a *functional modulus* in the form of a function of two variables, and in this sense the problem of reduction to normal form is hopeless.

Nevertheless, J. Bruce (1984) proposed a simple, and completely satisfying, description of the systems of integral curves which arise, based upon an original

[6] The investigations of this situation in catastrophe theory (R. Thom (1972), L. Dara (1975) and F. Takens (1976)) basically repeat the work of M. Cibrario (1932), and A.A. Shestakov, and A.V. Pkhakadze (1959) carried out decades earlier.

stabilization: the passage to a space of higher dimension, where the moduli disappear (as a consequence of enlarging the group of admissible transformations). Stabilization here means that each integral curve is lifted from the plane into 3-dimensional space at a height corresponding to its number. The resulting curves form a surface in 3-space diffeomorphic to the swallowtail partitioned by plane sections. This configuration is diffeomorphic to the standard decomposition of the standard swallowtail of Fig. 34 into the curves $a = \text{const}$, and has no moduli at all (with respect to diffeomorphisms of 3-dimensional space).

The original picture in the $\{(x, y)\}$ plane can be obtained from this standard system of curves on the standard swallowtail with the aid of a projection (a mapping of rank two) of 3-dimensional space onto the plane. It is this projection which introduces the functional moduli in the picture of Fig. 35, which may be considered as a 2-dimensional representation of the already rigid (having no moduli) situation of Fig. 34.

Stabilization was successfully applied even earlier in other problems of singularity theory. For example, the perestroikas in a one-parameter family of plane curves near its envelope are described by the standard transversal sections $x + z = c$ of the standard Whitney umbrella $y^2 = zx^2$ (Dufour (1983), Arnol'd (1976); see Fig. 36). The perestroikas of the involutes of a generic planar curve close to an inflection point (Fig. 37) are described by standard sections of the variety of nonregular orbits of the group H_3 of symmetries of the icosahedron (O.P. Shcherbak (1983), Arnol'd (1983), de l'Hospital (1696)).

The perestroikas of trajectories of the slow motion of a generic relaxation system with two slow variables near a point of degeneration of the contact structure on the jump curve, are described by standard sections $x + z = c$ of the *folded Whitney umbrella* $y^2 = z^3x^2$ (Fig. 38).

Apparently, an analogous stabilization also enables one to avoid moduli in the investigation of other situations, especially those in which functional moduli occur.

7.4. The Selection of the Classifying Group. Experience shows that a small change in posing a problem in the investigation of singularities often completely changes the resulting classification. If one makes an unfortunate choice of the

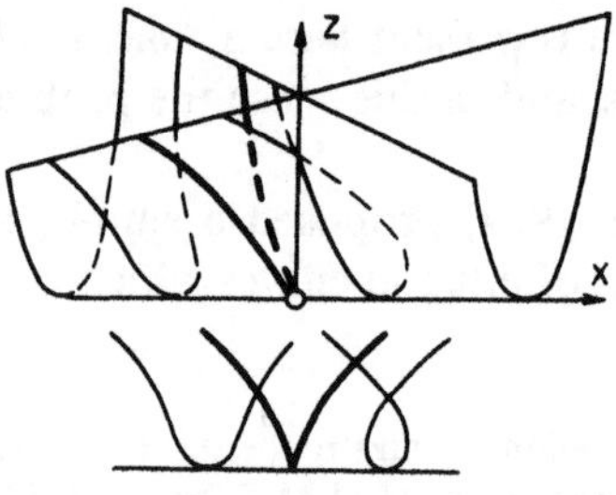

Fig. 36. Sections of the Whitney-Cayley umbrella

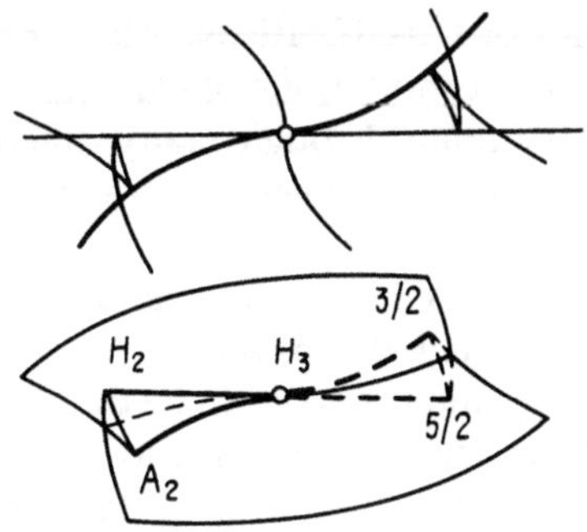

Fig. 37. Involutes of a plane curve and the discriminant of H_3

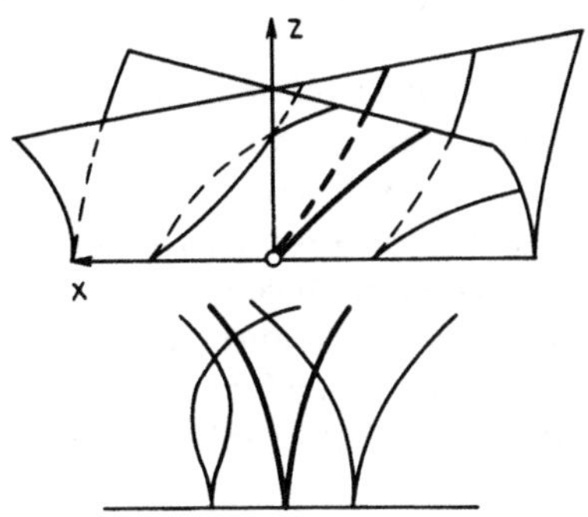

Fig. 38. Sections of the folded umbrella

equivalence relation, a difficult-to-handle chaos results, while a nearby, but fortunate, choice of the equivalence relation may lead to a simple and beautiful, clearly final, classification.

Besides the example given above (the comparison of singularities of small codimension with simple singularities in Sects. 7.1–2), let me mention the classification of the perestroikas of wave fronts. In catastrophe theory this question is usually considered from the point of view of so called (*r*, *s*)*-equivalence*, which leads to a complicated and difficult-to-handle classification (Wassermann (1975 and 1976)).

In the "correct" setting of the problem (Arnol'd (1976)), only a small change of the equivalence group is made, but the results immediately become natural and easy to handle.

In exactly the same way, the equivalence relation of the theory of *boundary singularities* (Arnol'd (1978)) (diffeomorphisms preserving a hyperplane) differs little from the equivalence relation of one of the variants of the theory of "imperfect bifurcations" of M. Golubitsky and D. Schaeffer (1979) (diffeomorphisms preserving a foliation into hyperplanes). However, while the classification of boundary singularities turns out well, for all its usefulness, it is hard to say that the list of imperfect bifurcations is finished.

7.5. Principles for Choice of Classifications. Alas, there are no general rules for the selection of successful formulations of classification problems, and they must be found gropingly, running through many different, to all appearances equally worthy, variants. The following general principles have turned out to be the most fruitful.

A. *A classification is considered more successful, the larger the number of different kinds of problems in which it appears.*

In this sense the most successful turns out to be the A, D, E – classification (see Sect. 7.2), which appears (in a rather mysterious way) in problems of such different natures as the theory of simple singularities, the theory of crystallographic reflection groups, the theory of simple Lie algebras, the theory of simple representations of quivers, and in the theory of regular polyhedra in 3-dimensional Euclidean space (where E_6 corresponds to the tetrahedron, E_7 to the octahedron, and E_8 to the icosahedron).

B. *For the comparison of different classifications one should begin with the investigation of bifurcation diagrams of objects of low codimension.*

Bifurcation diagrams serve as distinctive fingerprints of singularities. Their coincidence in various classification problems allows one to establish natural isomorphisms between hierarchies of objects of different theories.

Thus, it is the coincidence of bifurcation diagrams which has led to the discovery of the connection between singularities of functions on a manifold with boundary, and the Weyl groups (crystallographic Coxeter groups) B, C, F (Arnol'd (1978)).

Example 1. In Fig. 39 the F_4 caustic is illustrated. It is realized in the theory of boundary singularities in the following way (I.G. Shcherbak (1984)).

In 3-dimensional Euclidean space consider a generic surface with boundary. A point of space is called a *caustic point* if the distance to it, as a function on the surface with boundary, is degenerate.

The set of caustic points consists of three components. One is the *focal surface* (the set of centers of curvature) of the original surface. Another is the *focal surface of the boundary*; the third is formed by the normals to the surface at the points of its boundary.

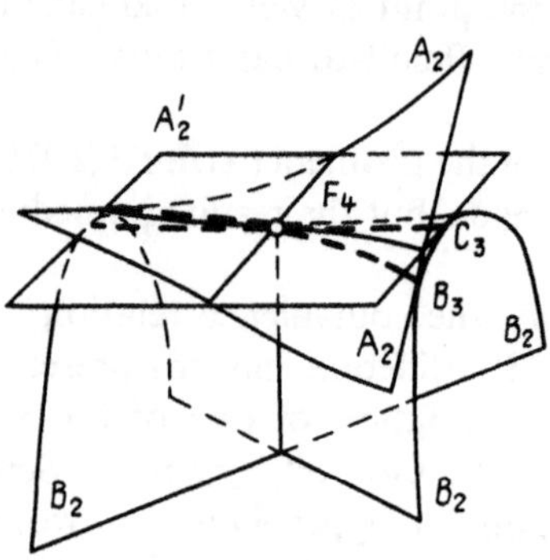

Fig. 39. The caustic F_4

At isolated points, the boundary of a generic surface has the direction of a line of curvature on the surface. It turns out that the caustic set in a neighborhood of the corresponding center of curvature is diffeomorphic to the caustic F_4, illustrated in Fig. 39 (more exactly, a part of it, since in Fig. 39 the centers of curvature of the extension of the surface beyond its boundary are also taken into account). The perestroika F_4 of fronts (equidistants) according to I.G. Shcherbak is portrayed in Fig. 40.

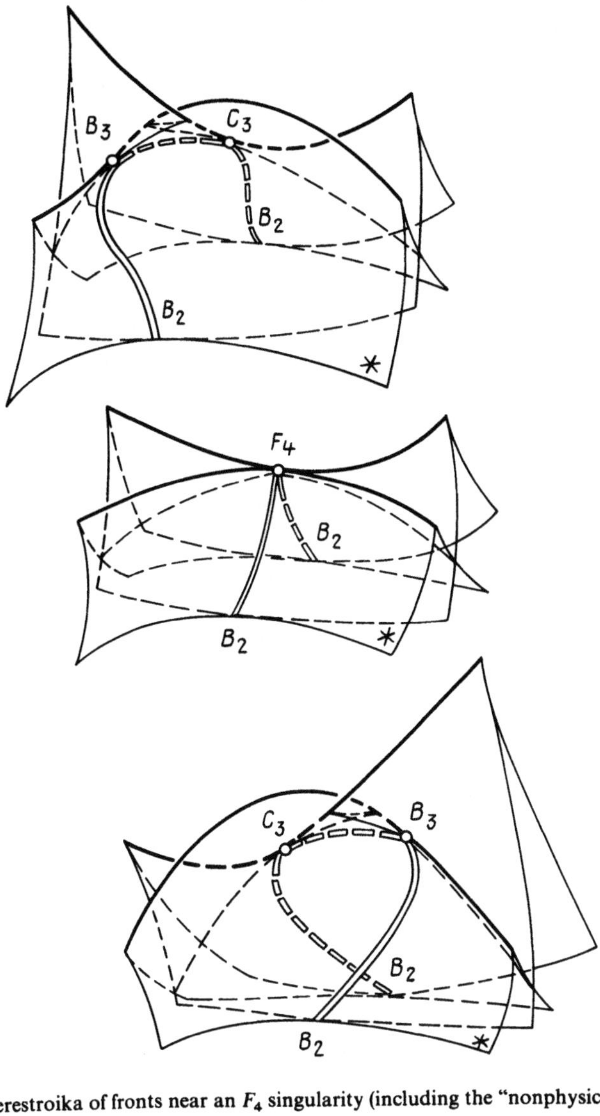

Fig. 40. The perestroika of fronts near an F_4 singularity (including the "nonphysical" part (*) to the right of the curve B_2)

Example 2. Cayley (1859) investigated the singularities of the envelope of the family of planes perpendicular to the radius vectors at the points of an ellipsoid. For the analogous problem on an ellipse, he discovered the same perestroikas of this envelope as the parameter changes as for the family of equidistants (Fig. 41). This "fingerprint" suggests that the singularities of the envelope of the family of planes normal to its radius vectors at the points of a given hypersurface, are Legendre singularities. Legendre singularities can be described in the following way.

The variety of tangent hyperplanes to a hypersurface of (projective) space forms a hypersurface in the dual space, called the *front* of the original hypersurface. The fronts of generic smooth hypersurfaces have singularities, and these are called *Legendre singularities.* It is easy to see that the equidistants or Legendre transformations of generic smooth objects have exactly these singularities.

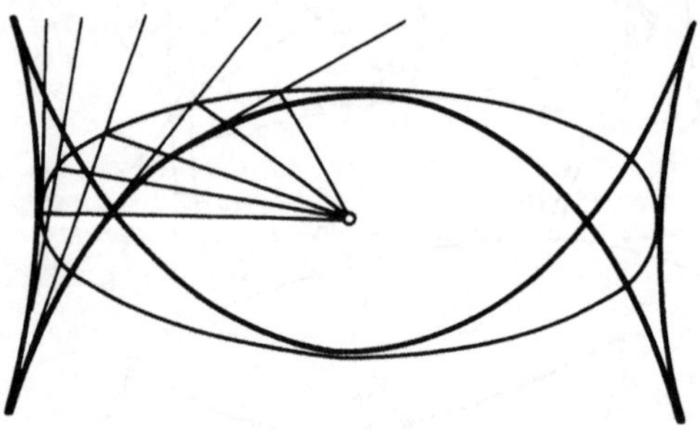

Fig. 41. The envelope of the normals to the radius vectors of the points of an ellipse

The envelope of the family of hyperplanes normal to the radius vectors at the points of a given hypersurface of Euclidean space is, as is easy to see, the front of the inversion of the given hypersurface. From this, it is evident that the "fingerprint" led to a correct answer, and at the same time we obtain a classification of singularities of envelopes of normal hyperplanes.

By an analogous method, the singularities of *pedal loci* – the hypersurfaces formed by the bases of the perpendiculars dropped from the origin to the tangent planes of the original hypersurface (Fig. 42) – also lead to Legendre singularities, for a pedal locus is the inversion of the front of the original hypersurface.

7.6. Recurrence of Singularities. It may be that the most astounding conclusion from the large number of different 'classification' problems in singularity and bifurcation theory is that a comparatively small list of standard forms (the cusp (semi-cubical), the swallowtail, ...) turns out to be universal, and serves in a large number of different theories, between which, at first glance, no connection at all is apparent. Here I shall cite one more recent example of this kind (Arnol'd and Sevryuk (1986); Sevryuk (1986); Devaney (1976)).

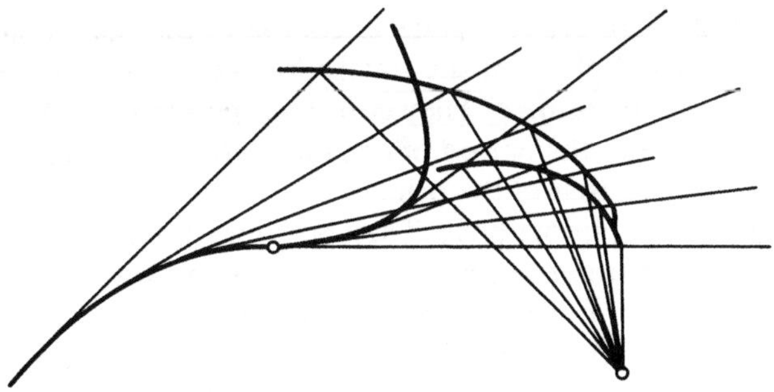

Fig. 42. A singularity of the pedal locus of a curve with an inflection point

In investigating equations of the form

$$u^{(iv)} + Au'' + (u')^2 + u = 0,$$

a number of physicists discovered strange phenomena: for a fixed value of the parameter A, the 4-dimensional phase space of this equation does contain cycles which are not isolated (as should happen for a generic equation), in fact whole one-parameter families of them.

This circumstance is explained by the reversibility of the equation. Namely, the equation is invariant under change of sign of the independent variable together with the involution of phase space (u, u', u'', u''') that changes the signs of the derivatives of odd orders. The phase curves are reflected in the plane of fixed points of the involution (the (u, u'')-plane), and the reflections reverse their direction (Arnol'd (1984b)); that is, phase curves are preserved under the involution and change of sign of the independent variable.

Let us consider the traces of the cycles referred to above on the plane of fixed points of the involution. Each cycle intersects this plane twice, and the set of all such traces forms a curve.

The equilibrium $u \equiv 0$ lies on the fixed plane. All 4 eigenvalues of the system linearized about the equilibrium point are purely imaginary for $A > 2$ (for reversible systems such a configuration of eigenvalues is stable). To the pair of eigenvalues $\pm i\omega$ there corresponds an invariant plane of the linearized equation, filled by the cycles of period $2\pi/\omega$ (of the linear system). This plane intersects the fixed plane of the involution in a straight line, which passes through the origin.

Generally speaking, the original nonlinear equation also has a 2-dimensional invariant manifold, filled out by cycles of a nearby period (this is a generalization of Lyapunov's theorem on Hamiltonian systems which is true also for general reversible systems; here it is required that the ratio of the frequencies not be an integer).

The invariant surface (manifold) described above intersects the plane of fixed points of the involution in a curve, passing through the origin and tangent to the straight line, constructed in the same way as for the linearized equation.

Since, for $A > 2$, there are two pairs of complex conjugate eigenvalues $\pm i\omega_1$ and $\pm i\omega_2$, on the fixed plane of the involution we obtain two curves, passing through 0 in different directions. They consist of points of cycles whose periods on one curve are close to the period of one characteristic oscillation, and on the other curve are close to that of the other.

Now let A approach 2 from above. Then both eigenfrequencies approach each other, and at $A = 2$ there arise two double purely imaginary eigenvalues (with Jordan blocks of order two). Moreover, the directions of the two curves described above approach each other. In our system (and also in any typical family of reversible systems as two purely imaginary eigenvalues merge) it turns out that a standard perestroika of cycles occurs, which can be described in the following way.

When $A > 2$ and is close to 2, both of the curves described above join to form a small figure-eight (see Fig. 43), which at $A = 2$ has contracted to a point. If one places each of these figure-eights in its own plane, then in 3-dimensional space (the product of the fixed plane of the involution and the axis of values of the parameter A) they form a surface. It turns out that this surface is diffeomorphic

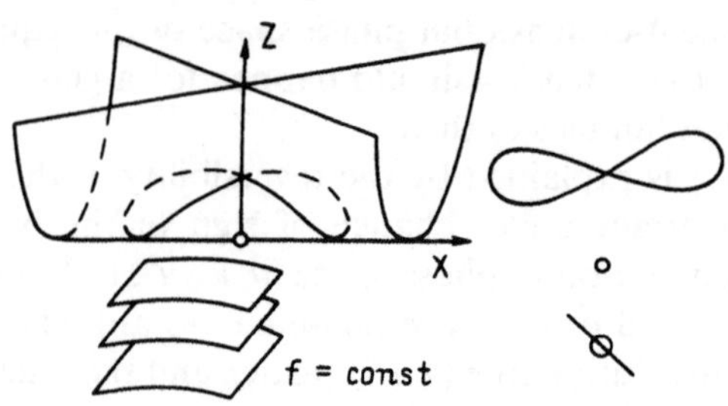

Fig. 43. Nontransversal sections of the Whitney-Cayley umbrella

to the standard Whitney umbrella ($y^2 = zx^2$). Moreover, its foliation into sections $A =$ const. is also standard. Namely, let us consider a generic function $f(x, y, z)$, without critical points, whose restriction to the x-axis has a critical point at 0 (the direction of this axis at 0 is connected invariantly to the umbrella). This function can be reduced near 0 to the normal form $\pm z \pm x^2 +$ const. by a diffeomorphism of three-dimensional space that preserves the umbrella.

Depending on the signs, the foliation of the umbrella into level curves consists of figure eights or of curves of hyperbolic form (Fig. 44). In our example, the case of figure eights is realized, but generally in generic reversible systems both possibility can be realized.

The appearance of the standard foliation of the standard Whitney umbrella in this problem is the unexpected result of rather long calculations. The same two families of curves (figure eights and hyperbolic curves) appear in the theory of Lagrange cobordisms in the following situation. Consider a 2-dimensional Lagrangian manifold, say, an invariant torus of motion in a central force field on the $\{(q_1, q_2)\}$-plane. If one considers the line $q_1 = c$ as the boundary of the plane, then the *Lagrangian boundary* of our manifold is the projection onto the $\{(q_2, p_2)\}$-plane, along the p_1-axis of the curve along which the manifold intersects the 3-dimensional subspace $q_1 = c$ of the 4-dimensional phase space $\{(p_1, p_2, q_1, q_2)\}$ (see Arnol'd (1983)).

At some (critical) values c, the Lagrangian boundary changes structure – out of nothing a figure eight is born, which then is torn apart, as is shown in Figs. 43 and 44. Thus the umbrella and its sections can be obtained by projecting a Lagrangian manifold along the p_1-axis in 3-dimensional (q_1, q_2, p_2)-space and cutting this projection with the planes $q_1 = $ const.

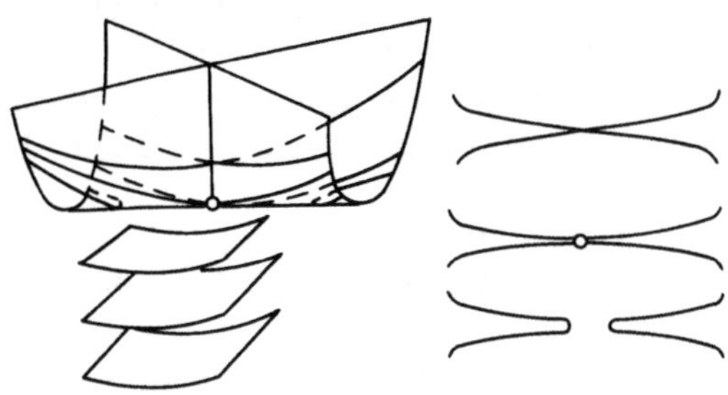

Fig. 44. A second variant of nontransversal intersections

Besides the (1, 1) resonance $\omega_1 = \omega_2$ examined above, upon changing the parameter in a reversible system, other resonances are also possible, $\omega_1 = k\omega_2$ or $(1, k)$ resonance. Upon passing through such resonances, there also occur perestroikas of one of the curves in which cycles intersect the invariant plane of the involution. These perestroikas, recently studied by M.B. Sevryuk, are connected with the theory, constructed by O.V. Lyashko, of critical points of functions on a manifold with a singular boundary (Lyashko (1983)) in the same way as the perestroikas examined above for the resonance $\omega_1 = \omega_2$ are connected with sections of the Whitney umbrella (the role of the umbrella is taken in this theory by the surfaces A_μ of the form

$$x^{\mu+1} \pm y^2 \pm z^2 = 0, \quad k = \mu + 1;$$

see M.B. Sevryuk, *Uspekhi Mat. Nauk*, *40* (1985), No. 5, 235–236 and *Reversible Systems*, Springer Lect. Notes in Math, vol. 1211, Springer Verlag, 1986).

The reasons why the same geometric forms appear in such different problems are not yet completely clear; however, in accord with the general principles of classification pointed out above, one may consider the coincidence discovered as confirmation of the reasonableness of the 'classification' problems considered.

7.7. The Problem of Going Around an Obstacle. Often, upon comparing two classifications by their bifurcation diagrams, it turns out that objects from one of them correspond only to a part of the objects from the other. Thus, the simple critical points of functions do not correspond to all the Weyl groups of simple

Lie groups, but only to groups of the series *A, D, E* (without multiple links in their Dynkin diagrams).

In such cases, one should look for a generalization of the first "classification" problem that furnishes the missing objects. For example, going over to critical points of functions on a manifold with boundary already furnishes all the Weyl groups (except for G_2). *Weyl groups* are the crystallographic groups generated by reflections. They constitute a part of the Coxeter groups – irreducible finite groups generated by Euclidean reflections. The list of Coxeter groups contains, besides the Weyl groups, yet another infinite series $I_2(p)$, (the symmetry groups of *p*-gons) and two special groups, H_3 and H_4. The group H_3 is the symmetry group of the icosahedron in three-dimensional space. The group H_4 consists of the symmetries of a regular polytope in $\mathbb{R}^4$ having 120 vertices. These vertices lie on the sphere $S^3 \approx SU(2)$ and form the binary group of the icosahedron (which covers the group of 60 rotations of the icosahedron under the double covering $SU(2) \to SO(3)$).

The question arises, what singularities correspond to these Coxeter groups? At present such singularities have been found by O.P. Shcherbak (1983) (see also Arnol'd (1984a) [*not* translated in Russian Math. Surveys!]). It has turned out that all of them can be realized in generic variational problems with one-sided constraints. A typical example of such a problem is that of going around an obstacle bounded by a smooth generic surface in 3-dimensional Euclidean space.

The connection between the singularities of the solution of a variational problem and a reflection group is the following. Consider an extremal (not necessarily a minimal one): it consists of segments of geodesics on the surface of the obstacle, and of segments of tangents to it. The length of the extremals from a fixed initial set (for example, a point) to a variable point of space is a (many-valued) function of this terminal point. Let us consider the graph of this many-valued function of "time" or "action". It turns out that for a generic obstacle and initial set, this graph will have singularities only of a standard form. In a neighbor-

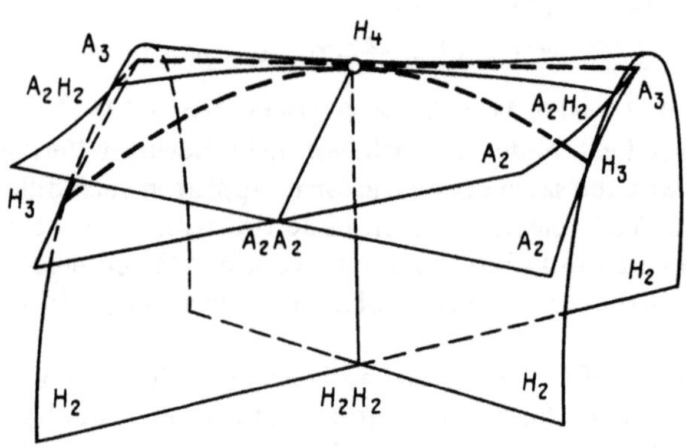

Fig. 45. The caustic H_4

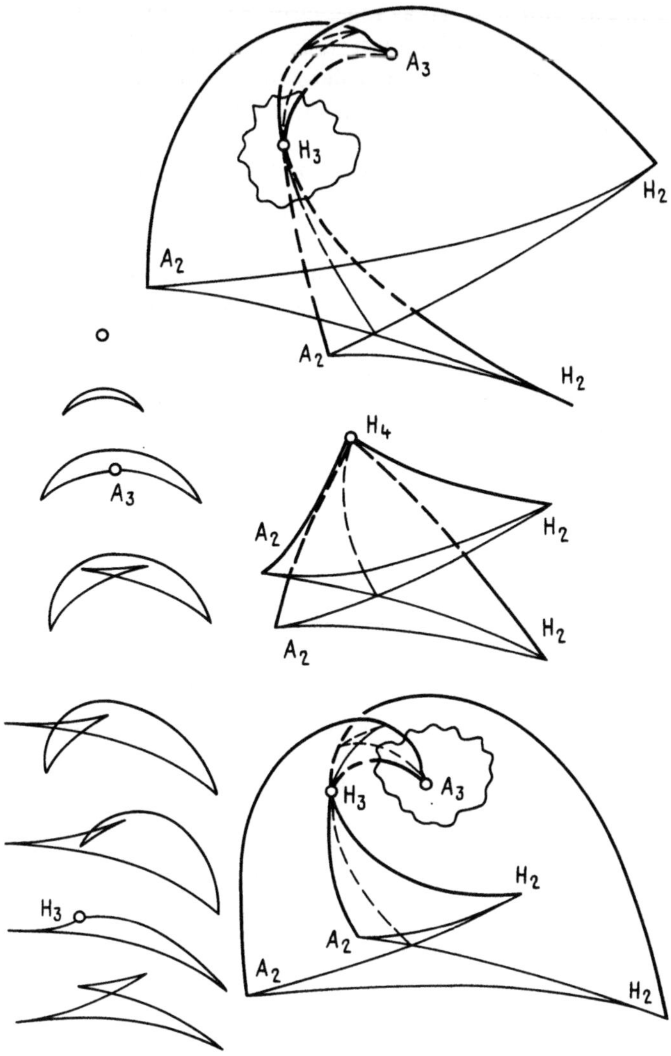

Fig. 46. Perestroika of fronts near an H_4 singularity

hood of some of them, the graphs will be diffeomorphic to the varieties of nonregular orbits of the groups H_3 and H_4.

In particular, the most complicated singularity, H_4, appears at one of the points of a ray tangent to the surface of the obstacle at a parabolic point in an asymptotic direction. The corresponding caustic is drawn in Fig. 45. It consists of a Whitney umbrella H_2 and the folded umbrella A_2, intersecting along the curve A_2H_2 and having a cubic tangency along the curve H_3.

The perestroika of wave fronts corresponding to H_4 is drawn in Fig. 46. The front has two cuspidal edges: A_2 of order 3/2, and H_2 of order 5/2 with singular points A_3 (a swallowtail) and H_3 (a tangency of the two edges). As the front moves, the edge A_2 sweeps out the folded umbrella of the caustic and H_2, the ordinary umbrella. At the moment of perestroika, the points A_3 and H_3 merge, and then appear anew, but a whole new front is obtained from the old one by a symmetry.

Thanks to this theory of O.P. Shcherbak (the details may be found in a posthumous article by Shcherbak, *Wave fronts and Coxeter groups*, Russian Math. Surveys, 1988, No. 3), one is able to apply the apparatus of the theory of reflection groups to the problem of going around an obstacle, and because of this one can investigate in detail the geometry of the corresponding singularities and metamorphoses.

In other cases, however, the theory of singularities turns out to take the lead. For example, the Weyl group F_4 is, in Goryunov's theory (1981) of singularities of projections of complete intersections, the progenitor of a whole series of simple singularities of projections of the curves

$$F_\mu\colon x^2 + y^2 = 0, \quad z = y^p \quad \text{or} \quad xy^q \qquad (\mu = 2p + 1, 2q + 4)$$

onto the z-axis, for which, up to now, analogs have not been found in other "classification" problems.

In the same way, extensive classifications of simple projections of complete intersections of arbitrary dimension onto a space of any dimension (Goryunov (1983)), as well as the near-by investigations of "imperfect bifurcations" (Golubitsky and Schaeffer (1979)), lead to lists (still not put into order) whose analogs have not yet been discovered in other areas of mathematics.

In this way, in the process of its development, singularity theory in turn experiences the ordering influence of unexpectedly discovered connections with areas far from it, or catastrophically plunges into a chaotic piling-up of ever more complicated special cases, unconnected by evident rules.

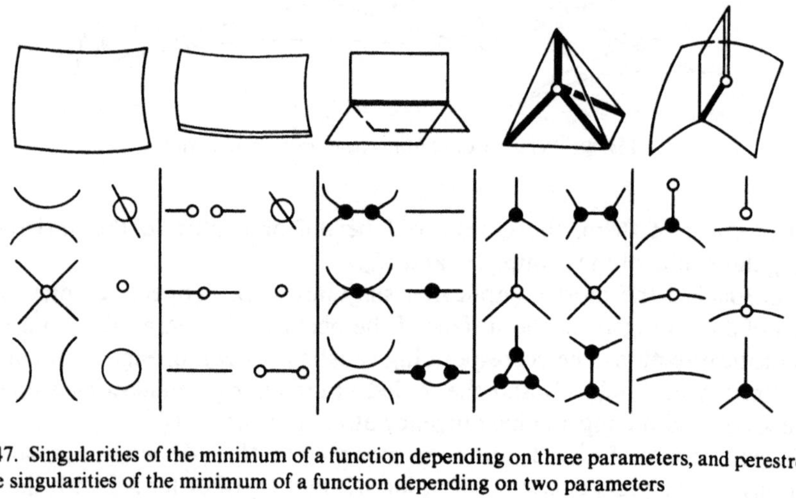

Fig. 47. Singularities of the minimum of a function depending on three parameters, and perestroikas of the singularities of the minimum of a function depending on two parameters

As an example of this genre, in Fig. 47 surfaces are drawn on which the minimum function of a typical three-parameter family of smooth functions on a compact manifold looses its smoothness (following I.A. Bogaevskij), as well as typical perestroikas of curves, playing an analogous role for two-parameter families. The relation of these patterns to those of shock waves, and to their perestroikas is discussed by V.I. Arnol'd (Bifurcation Theory in Mathematics and Mechanics, XVIIth International Congress of Theoretical and Applied Mechanics, Gerenoble, August, 1988).

Recommended Literature

Elements of catastrophe theory are found in the works of Huygens (1673), de l'Hospital (1696), Hamilton (1828–1837), Cayley (1852–1873), Gibbs (1873), and others.

The classical works of Poincaré (1879, 1892) and Andronov (1933–1959) on bifurcation theory, which together with Whitney's work (1955), laid the foundation of the theory of singularities of differentiable maps, contain virtually all the basic ideas of catastrophe theory.

The initial exposition of these ideas is in Thom's book (1972), which is intended primarily for biologists, and is renowned for its incomprehensibility. The later published collection of articles by Thom (1981) already contains a brief summary of the necessary mathematical apparatus of singularity theory, and several (not always convincing) indications of how it should be used.

For example, the verbs "to tear" and "to sew" in Thom's opinion correspond to the swallowtail, while "to break" and "to give birth" correspond to the pleat.

A cycle of articles by Zeeman (1977) is devoted to a broad advertisement of catastrophe theory as a universal method for the construction of models in all branches of knowledge, including medicine and sociology. Just these models, often lightweight in character, were subsequently exposed to harsh criticism. The books of Poston and Stewart (1978) and Gilmore (1981) were devoted to more serious applications to the natural sciences. The mathematical apparatus of singularity theory is briefly described in the textbook of Bröcker and Lander (1975); see also the textbook by Bruce and Giblin (1984). A more detailed exposition of singularity theory can be found in the books of Arnol'd, Varchenko and Gusejn-Zade (1982, 1984); the first volume is devoted to classification (of singularities, critical points of functions, caustics and wave fronts), and the second volume is devoted to monodromy and the asymptotics of integrals (a theory which Gilmore calls the quantum theory of catastrophes). The surveys by Arnol'd (1983), and Arnol'd, Varchenko, Givental and Khovanskij (1984), and vol. 22 in the series Sovremennye Problemy Matematiki, including Goryunov (1983) contain much supplementary material, among which is the theory of Newton polyhedra – the most effective method for shortening the laborious time-consuming computations on which the mathematical apparatus of the theory is based. The exposition of catastrophe theory by Arnol'd (1981–1986) intended for nonspecialists, contains among other things, as does the present article, several results not appearing in other publications.

Detailed bibiographies may be found in the books by Poston and Stewart (1978) and Arnol'd, Varchenko and Gusejn-Zade (1982 and 1984), and a survey of newer works can be found in the lecture of D. Bennequin (1986) at the Bourbaki seminar in November, 1984. See also four recent books by V.I. Arnol'd: (1) *Contact geometry and wave propagation*, Monographie de l'Enseignement Mathématique vol. 34, Geneva, 1992, 52 pp. (2) *The theory of singularities and its applications*, Lezioni Fermiane, Pisa, 1991, 72 pp. (3) *Singularities of caustics and wavefronts*, Mathematics and its applications, Soviet Series vol. 62, Kluwer Acad. Publishers, Amsterdam, 1991, 280 pp. (4) Singularity theory and applications (Arnol'd, ed.), Advances in Soviet Mathematics, vol. 1, Amer. Math. Society, Providence, RI, 1990.

References*

Airy, G.B. [1838] On the intensity of light in the neighborhood of a caustic. *Trans. Cambridge Philos. Soc. 6*, 379–403

Andronov, A.A. [1933] Mathematical problems of the theory of self-oscillations. Paper at The First All-Union Conference on Auto-oscillations. Nov. 1931, *The First All-Union Conference on Auto-oscillations.* GTTI, *Moscow-Leningrad*, 32–71. (Also pp. 85–124 in: *Collected works*, Moscow-Leningrad, Izd. Akad. Nauk SSSR, Moscow, 1956, 538 pp.)

Andronov, A.A., Bautin, N.N., Gorelik, G.A. [1946] Theory of indirect regulation. *Avtomekhanika i Telemekhanika 7*, No 1, 15–41. (Also pp. 317–356 in: *Collected works*, Moscow-Leningrad, Izd. Akad. Nauk SSSR, Moscow, 1956, 538 pp.)

Andronov, A.A., Khajkin, S.Eh. [1937] *Theory of Oscillations.* ONTI, Moscow-Leningrad, 520 pp.; English transl.: Princeton Univ. Press, 1949, ix, 358 pp., Zbl.85,178

Andronov, A.A., Leontovich, E.A. [1938] Sur la théorie de la variation de la structure qualitative de la division du plan en trajectoires. *Dok. Akad. Nauk SSSR* (2) *21*, 423–426 (Also pp. 217–221 in: *Collected works*, Moscow-Leningrad, Izd. Akad. Nauk SSSR, Moscow, 1956, 538 pp.), Zbl.22,22

Andronov, A.A., Leontovich, E.A. [1939] Some cases of dependence of limit cycles on parameters. *Uchen. Zap. Gor'k. Univ.*, No. 6, 3–24 (Also pp. 188–216 in: *Collected works*, Moscow-Leningrad, Izd. Akad. Nauk SSSR, Moscow, 1956, 538 pp.)

Andronov, A.A., Pontryagin, L.S. [1937] Systèmes grossières. *Dokl. Akad. Nauk SSSR 14*, 247–250. (Also pp. 183–187 in: *Collected works*, Moscow-Leningrad, Izd. Akad. Nauk SSSR, Moscow, 1956, 538 pp.), Zbl.16,113

Andronov, A.A., Vitt, A.A. Khajkin, S.Eh. [1959] *Theory of Oscillations.* 2nd ed., Fizmatgiz, Moscow, 916 pp., English transl.: Pergamon Press, New York, 1966, Zbl.85,178

Arnol'd, V.I. [1961] Small divisors. I. On mappings of the circle onto itself. *Izv. Akad. Nauk SSSR*, Ser. Mat. *25*, 21–86; English transl.: Am. Math. Soc., Transl., II. Sect. 46, 213–284 (1965), Zbl.135,426

Arnol'd, V.I. [1967] On a characteristic class, entering into conditions of quantization. *Funkts. Anal. Prilozh 1*, No. 1, 1–14; English transl.: Funct. Anal. Appl. 1, No. 1, 1–13 (1967), Zbl.175,203

Arnol'd, V.I. [1972] Normal forms of functions near to degenerate critical points, Weyl's groups A_k, D_k, E_k and Lagrangian singularities. *Funkts. Anal. Prilozh 6*, No. 4, 3–25; English transl.: Funct. Anal. Appl. 6, 254–272 (1973), Zbl.278.57011

Arnol'd, V.I. [1975] Critical points of smooth functions. *Proc. Int. Congr. Math., Vancouver, 1974*, vol. 1, 19–39, Zbl.343.58002

Arnol'd, V.I. [1976] Wave front evolution and equivariant Morse lemma. *Commun. Pure Appl. Math. 29*, No. 6, 557–582, Zbl.343.58003

Arnol'd, V.I. [1978] Critical points of functions on manifolds with boundary, the simple Lie groups B_k, C_k, F_4 and singularities of involutes. *Usp. Mat. Nauk 33*, No. 5, 91–105; English transl.: Russ. Math. Surv. 33, 99–116 (1978), Zbl.408.58009

Arnol'd, V.I. [1981] *Catastrophe Theory.* Znanie, Moscow, 64 pp., 2nd revised and expanded ed. 1983, 80 pp., English transl.: 2nd revised and expanded edition Springer-Verlag, Berlin, Heidelberg, New York, Tokyo, 1986, XIII, 108 pp., Zbl.517.58002

Arnol'd, V.I. [1983] Singularities of ray systems. *Usp. Mat. Nauk 38*, No. 2, 77–147; English transl.: Russ. Math. Surv. 38, No. 2, 87–176 (1983), Zbl.522.58007

Arnol'd, V.I. [1984a] Singularities in the variational calculus. *Usp. Mat. Nauk 39*, No. 5, 256

Arnol'd, V.I. [1984b] Reversible systems. In: *Nonlinear and turbulent processes*, pp. 1164–1174, Gordon and Breach, New York

---

*For the convenience of the reader, references to reviews in Zentralblatt für Mathematik (Zbl.), compiled using the MATH database, and Jahrbuch über die Fortschritte der Mathematik (Jbuch.) have, as far as possible, been included in this bibliography.

Arnol'd, V.I., Sevryuk, M.B. [1986] Oscillation and bifurcations in reversible systems. In: *Nonlinear Phenomena*. Moscow, Mir. R. Sagdeev Editor p. 31–64

Arnol'd, V.I., Varchenko, A.N., Gusejn-Zade, S.M. [1982, 1984] *Singularities of differential maps I. The classification of critical points, caustics and wave fronts.* Nauka, Moscow, 304 pp.; English transl.: Monographs in Math. 82, Birkhäuser, Boston, Basel, Stuttgart, 1985, x, 382 pp., Zbl.513.58001 *II. Monodromy and asymptotics of integrals*, Nauka, Moscow, 1984, 336 pp.; English transl.: Birkhäuser, Boston, Basel, Stuttgart, 1988, 492 pp., Zbl.545.58001

Arnol'd, V.I., Varchenko, A.N., Givental', A.B., Khovanskij, A.G. [1984] Singularities of functions, wave fronts, caustics, and multidimensional integrals. *Sov. Sci. Rev., Sect. C., Math. Phys. Rev. 4*, 1–92, Zbl.546.58007

Bennequin, D. [1986] Caustique mystique (d'après Arnol'd *et al.*). *Sémin. N. Bourbaki*, 37e ann., vol. 1984/85, Exp. No. 634, *Astérisque* 133/34, 19–56, Zbl.632.58011

Bertini, E. [1882] Sui sistemi lineari, *Rend. Inst. Lombardo 2* (15), 24–29, Jbuch 14, 433

Brill, L. (ed.) [1892] *Catalog mathematischer Modelle für den höheren mathematischen Unterricht.* Darmstadt, 220 pp.

Browder, F.E. (ed.) [1982] *The mathematical heritage of Henri Poincaré.* Proc. Symp. Pure Math, vol 39, Am. Math. Soc., Providence, I. (435 pp.), II (470 pp.), Zbl.512.00008

Bröcker, Th., Lander, L. [1975] *Differential germs and catastrophes.* London Math. Soc. Lect. Notes Ser. 17, Cambridge Univ. Press, 179 pp, Zbl.302.58006

Bruce, J.W. [1984] A note on first order differential equations of degree greater than one and wavefronts' evolution. *Bull. Lond. Math. Soc. 16*, No. 2, 139–144, Zbl.503.34003

Bruce, J.W., Giblin, P.J. [1984] *Curves and singularities: A geometrical introduction to singularity theory.* Cambridge Univ. Press, Cambridge, xii, 222 pp, Zbl.534.58008

Brushlinskaya, N.N. [1961] Qualitative integration of a system of *n* differential equations in a region, containing a singular point and a limit, cycle. *Dokl. Akad. Nauk SSSR 139*, No. 1, 9–12; English transl.: Sov. Math., Dokl. 2, 845–848 (1961), Zbl.168,338

Buldyrev, V.S., Lanin, A.I. [1981] The radiation field of whispering gallery waves on a concave-convex boundary. Zap. Nauchn. Semin. *Leningr. Otd. Mat. Inst. Steklova 104*, 49–65; English transl.: J. Sov. Math. 20, 1776–1787 (1982), Zbl.472.35023

Cayley, A. [1852] On the singularities of surfaces. *Cambridge and Dublin Math. J.* 7 (1852), 166–171 (*The Collected Math. Papers of A. Cayley*, Cambridge Univ. Press, *2* (1889), 28–32)

Cayley, A. [1859] Sur la surface qui est l'enveloppe des plans conduits par les points d'un ellipsoïde perpendicularement aux rayons menés par le centre. *Ann. Mat. pura ed appl.* (*Tortolini*), *2*, 3–14 (pp. 123–133 in: *The Collected Math. Papers of A. Cayley*, Cambridge Univ. Press *8* (1895), 580 pp.)

Cayley, A. [1868a] A "Smith's Prize" paper; solutions, Question 6. *Oxford, Cambridge and Dublin Messenger of Math. 4*, 201–226 (pp. 416–418 in: *The Collected Math. Papers of A. Cayley*, Cambridge Univ. Press *8* (1895), 580 pp.)

Cayley, A. [1868b] A "Smith's Prize" paper; solutions, Question 6. *Oxford, Cambridge and Dublin Messenger of Math. 4*, 201–226 (pp. 420–422 in: *The Collected Math. Papers of A. Cayley*, Cambridge Univ. Press *8* (1895), 580 pp.)

Cayley, A. [1870] A "Smith's Prize" paper; solutions, Question 6. *Oxford, Cambridge and Dublin Messenger of Math. 5*, 182–203 (pp. 474–490 in: *The Collected Math. Papers of A. Cayley*, Cambridge Univ. Press *8* (1895), 580 pp.)

Cayley, A. [1873] On the centro-surface of an ellipsoid. *Cambridge Phil. Trans. 12*, Part I, 319–365 (pp. 4316–365 in: *The Collected Math. Papers of A. Cayley*, Cambridge Univ. Press *8* (1895), 580 pp.), Jbuch 5, 409

Devaney, R.L. [1976] Reversible diffeomorphisms and flows. *Trans. Am. Math. Soc. 218*, 89–113, Zbl.363.58003

Diener, M. [1984] The canard unchained or how fast/slow dynamical systems bifurcate. *Math. Intell. 6*, No. 3, 38–42, Zbl.552.34055

Dufour, J.-P. [1977] Sur la stabilité des d'applications différentiables *Ann. Sci. Éc. Norm.* Super., IV. Ser. *10*, No. 2, 153–174, Zbl.354.58011

Dufour, J.-P. [1983] Familles des courbes planes différentiables. *Topology 22*, No. 4, 449–474, Zbl.521.58012

Écalle, J. [1975] Théorie itérative. Introduction à la théorie des invariants holomorphes. *J. Math. Pure Appl., IX, Ser 54*, 183–258, Zbl.285.26010
Fok, V.A. [1940] The field of a plane wave near to the surface of a conducting body. *Izv. Akad. Nauk SSSR, Ser. Fiz 10*, No 2, 171–186, 399–409
Fomenko, A.T., Tuzhilin, A.A. [1986] Many-valued mappings, minimal surfaces, and soap films. *Vestn. Mosk. Univ.* Ser. I 1986, No. 3, 3–12; English transl.: Mosc. Univ. Math. Bull. 41, No. 3, 1–12 (1986), Zbl.604.58019
Fresnel, A. [1818] Mémoire sur la diffraction de la lumière. *Mém. de l'Acad. des Sci. 5*, 339–353
Gibbs, J.W. [1873] A method of geometrical representation of the thermodynamical properties of substances by means of surfaces. *Trans. Connecticut Acad. 2*, 382–404
Gibson, C.G., Wirthmüller, K., du Plessis, A.A., Looijenga, E. [1976] *Topological stability of smooth mappings*. Lect. Notes Math. 552, Springer-Verlag: New York, Heidelberg, Berlin, Tokyo, 155 pp, Zbl.377.58006
Gilmore, R. [1981] *Catastrophe theory for scientists and engineers*. Wiley-Interscience, New York, 666 pp, Zbl.497.58001
Giusti, M. [1977] Sur les singularités isolées d'intersections complètes quasihomogènes. *Ann. Inst. Fourier 27*, No. 3, 163–192, Zbl.353.14003
Golubitsky, M., Schaeffer, D. [1979] Imperfect bifurcations in the presence of symmetry. *Commun. Math. Phys. 67*, 205–232, Zbl.467.58019
Goryunov, V.V. [1981] The geometry of bifurcation diagrams of simple projections onto a straight line. *Funkts. Anal. Prilozh. 15*, No. 2, 1–8; English transl.: Funct Anal. Appl. 15, 77–82 (1981), Zbl.507.58010
Goryunov, V.V. [1983] Singularities of projections of complete intersections. In: Itogi Nauki Tekh., Ser. Sovrem. Probl. Mat. *22*, 167–206, Zbl.555.58009
Grothendieck, A. [1960] The cohomology theory of abstract algebraic varieties. In: Proc. Int. Congr. Math. 1958, 103–118, Zbl.119,369
Guckenheimer, J. [1973] Bifurcation and catastrophe. In: *Dynamical Syst.*, Proc. Symp. Univ. Bahia, Salvador 1971, 95–109, Zbl.287.58005
Gufan, Yu.M. [1984] *Structured phase transitions*. Nauka, Moscow, 304 pp.
Hamilton, W.R. [1828, 1830, 1831, 1837] Theory of systems of rays. Irish Acad. Trans. *15*, 69–174; *16*, 4–62; *16*, 93–125; *17*, 1–144
Hopf, E. [1943] Abzweigung einer periodischen Lösung von einer stationaren Lösung eines Differentalsystems. *Ber. Math.-Phys. Sachs. Acad Wiss.*, Leipzig, *95*, 3–22, Zbl.63,32. Also pp. 134–154 in: Marsden, J.E. and McCracken, M., *The Hopf bifurcation and Its Applications*. Appl. Math. Sci. 19, Springer-Verlag: New York, Heidelberg, Berlin, Tokyo, Hong Kong 1976, xiii, 408 pp, Zbl.346.58007
Huygens, C. [1673] *Horologium Oscillatorium*. Paris
Jacobi, G. [1866] *Vorlesungen über Dynamik*. G. Reimer, Berlin, 578 pp. (Vorles. 6)
Koiter, W.T. [1945] *On the stability of elastic equilibrium*. Dissertation, Delft, 833 pp.
Kostov, V.P. [1984] Versal deformations of differential forms of degree α on the line. *Funkts. Anal. Prilozh. 18*, No. 4, 81–82; English transl.: Funct. Anal. Appl. 18, 335–337 (1984), Zbl.573.58002, see also Math USSR Izvestiya, Vol. 37 (1991) No. 3 p. 525–537
Kronecker, L. [1878] Über Sturm'sche Funktionen. *Monatsber. Dt. Acad. Wiss.*, Berlin, 95–121, Jbuch 10, 51
Landau, L.D. [1944] The Problem of Turbulence. *Dokl Akad. Nauk SSSR 44*, No. 8, 339–342
Landau, L.D. [1959] On analytic properties of vertex parts in quantum field theory. *Nucl. Phys. 13*, 181–192, Zbl.88,220
Landau, L.D., Lifshits, E.M. [1984] *Statistical physics*. Nauka, Moscow, 568 pp; English transl.: Pergamon Press, New York etc. 1958, 484 pp. Zbl.80,197 (Zbl.659.76001)
Landis, E.E. [1981] Tangential singularities, *Funkts. Anal. Prilozh. 15*, No. 2, 36–49; English transl.: Funct. Anal. Appl. 15, 103–114; Zbl.493.58005
Leontovich, M.A. [1944] On one method of solution of problems on the diffraction of waves. *Izv. Akad. Nauk SSSR Ser. Fiz. 8*, No. 1, 16–22
de l'Hôspital, G.F. [1768] *Analyse des infiniment petits pour l'intelligence des lignes courbes*. Moutard, Paris, 380 pp.

Looijenga, E.J.N. [1974] Structural stability of families of C^∞ functions and the canonical stratification of $C^\infty(N)$. Preprint, Inst. Hautes Etudes Sci.
Lyashko, O.V. [1983] Classification of critical points of functions on manifolds with singular boundary. *Funkts. Anal. Prilozh. 17*, No. 3, 26–38; English transl.: Funct. Anal. Appl. *17*, 187–193 (1983), Zbl.598.58010
Malgrange, B. [1962–3] Le théorème de préparation en géométrie differentiable. *Sémin. H. Cartan, 15* (1962/63), No. 11, No. 12, No. 13, No. 22 (1964); Zbl.119,285
Mather, J.N. [1968] Finitely determined map germs. *Publ. Math. Inst. Hautes Études Sci. 35*, 127–156, Zbl.159,250
Mather, J.N. [1969] Right equivalence. Preprint, Univ. of Warwick
Mather, J.N. [1971] Stability of mappings: VI. The nice dimensions. In: *Proc. Liverpool Singularities-Symp. I*, C.T.C. Wall (ed.), pp. 207–253, Lect. Notes Math. 192, Springer-Verlag: New York, Heidelberg, Berlin, Tokyo, Hong Kong, 319 pp., Zbl.211,561
Mather, J.N. [1973] Stratifications and mappings In:*Dynamical Syst.*, Proc. Symp. Univ. Bahia, Salvador 1971, 195–232 (1973), Zbl.286.58003
Maxwell, J.C. [1875] On the dynamical evidence of the molecular constitution of bodies. *Nature 11*; also in: *The scientific papers*, Maxwell, J.C., Part II, 1927
Monge, G. [1807] *Application de l'Analyse à la Géométrie.* Bernard, Paris, 416 pp.
Nejmark, Yu. I. [1959] On some cases of dependence of periodic motions on parameters. *Dokl. Akad. Nauk SSSR 129*, No. 4, 736–739, Zbl.89,296
Pearcey, T. [1946] The structure of an electromagnetic field in the neighborhood of a cusp of a caustic. Phil. Mag. *37*, 311–317
Pham, F. [1967] Introduction à l'étude topologique des singularités de Landau. *Mém. Sci. Math. 164*, Gauthier-Villars, Paris, 143 pp.
Plücker, J. [1839] *Theorie der algebraischen Kurven*, Bonn
Poincaré, H. [1879] *Sur les propriétés des fonctions définies par les équations aux différences partielles.* Thèse, Gauthier-Villars, Paris, 93 pp.
Poincaré, H. [1887] *Les méthodes nouvelles de la Mécanique céleste.* Vol. I, II, III, Gauthier-Villars, Paris, 385 pp, 479 pp., 414pp., Zbl.651.70002
Poincaré, H. [1895] *Analysis situs. J. École Polytech. 2*(1), 1–23, Jbuch 26, 541
Poincaré, H. [1905a] Sur les lignes géodésiques des surfaces convexes. *Trans. Am. Math. Soc. 6*, 237–274, Jbuch 36, 669
Poincaré, H. [1905b] *La valeur de la science.* Flammarion, Paris, 278 pp; German transl.: Leipzig (1906), Jbuch 37, 56
Poincaré, H. [1908] *Science et méthode.* Flammarion, Paris, 308 pp.
Poincaré, H. [1953] *Oeuvres.* Vol. 6, p. 184, Gauthier-Villars, Paris, 540 pp.
Popov, M.M. [1984] A whispering gallery in the neighborhood of an inflection point on the boundary. Asymptotics of the wave field. *Zap. Nauch. Semin. Leningr. Otd. Mat. Inst. Steklova 140*, 151–166, English transl.: J. Sov. Math. 32, 205–214 (1986, Zbl.557.35097
Porteous, I.R. [1974] Nobel prizes for catastrophes. *Manifold* 15, 34–36
Poston, T. [1972] The Plateau problem. In *Summer College on global Analysis*, ICTP, Trieste
Poston, T., Stewart, I.N. [1978] *Catastrophe Theory and its Applications.* Pitman, London, 491 pp, Zbl.382.58006
Salmon, G. [1862] *A treatise on the geometry of three dimensions.* Hodges and Smith, Dublin, 465 pp.
Sard, A. [1942] The measure of critical values of differential maps. *Bull. Am. Math. Soc. 48*, 883–890
Schreinemakers, F.A. [1917] In – mono – and divariant equilibria – 24. *Koninkijke Acad. van Wettenschappen te Amsterdam 19*, 1203–1205
Semenov, N.N [1934] *Chain reactions.* Goskhimtekhizdat, Leningrad, 555pp.
Sevryuk, M.B. [1986] *Reversible systems.* Lect. Notes Math. 1211, Springer-Verlag: New York, Heidelberg, Berlin, Tokyo, 1986, 319 pp, Zbl.661.58002
Shcherbak, I.G. [1984] The focal set of a surface with boundary and caustics of groups, generated by reflections, B_k, C_k, F_4. *Funkts. Anal. Prilozh. 18*, No. 1, 90–91; English transl.: Funct. Anal. Appl. 18, 84–85 (1984), Zbl.544.58002
Shcherbak, O.P. [1983] Singularities of families of involutes in a neighborhood of a reflection point

of a curve and the group H_3, generated by reflections. *Funkts. Anal. Prilozh. 17*, No. 4, 70–72; English transl.: Funct. Anal. Appl. *17*, 301–303 (1983), Zbl.534.58011

Shishkova, M.A. [1973] Investigation of a system of differential equations with a small parameter in the highest derivatives. *Dokl. Akad. Nauk SSSR 209*, No. 3, 576–579; English transl.: Sov. Math. Dokl. 14, 483–487 (1973) Zbl.289.34083

Shoshitajshvili, A.N. [1975] Bifurcations of topological types of vector fields near singular points. *Tr. Semin. im. I.G. Petrovskogo*, No. 1, 279–309, Zbl.333.34037

Smale, S. [1966] Structurally stable systems are not dense. *Am. J. Math. 88*, 491–496, Zbl.149,200

Stewart, I.N. [1975] The seven elementary catastrophes. *The New Scientist 68*, 447–454

Thom, R. [1956] Une lemme sur les applications différentiables. *Bol. Soc. Math. Mex.*, II, Ser. 1, 59–71, Zbl.75, 322

Thom, R. [1964] Local properties of differentiable mappings. In: *Differ. Analysis, Bombay Colloq.* 1964, pp. 191–206, Zbl.151,320

Thom, R. [1968] Sur les travaux de Stephen Smale. In: *Proc. Int Congr. Math. Moscow, 1966*, pp. 25–28, Mir, Moscow, 728 pp.

Thom, R. [1969] Topological models in biology. *Topology 8*, No. 3, 313–335, Zbl.176,505

Thom, R. [1972] *Stabilité structurelle et morphogénèse.* Benjamin, New York, 362 pp., Zbl.294.92001

Thom, R. [1981] *Modèles mathématiques de la morphogénèse.* C. Bourgois, Paris, 318 pp, 347.58003; English transl.: J. Wiley, New York etc. (1983)

Trotman, D.J.A., Zeeman, E.C. [1974] The classification of elementary catastrophes of codimension 5. *Lect. Notes, Univ. of Warwick*, 56 pp.

Tschirnhaus, W. [1682] *Acad. Sci., Paris*

Tyurina, G.N. [1968] On the topological properties of isolated singularities of complex spaces of codimension one. *Izv. Akad. Nauk SSSR, Ser. Mat. 32*, 605–620; English transl.: Math. USSR., Izv. 2, 557–571 (1969) Zbl.176,509

Varchenko, A.N. [1975] Versal topological deformations. *Izv. Akad. Nauk SSSR, Ser. Mat. 39*, 294–314; English transl.: Math. USSR, Izv. 9 (1975), 277–296 (1976), Zbl.333.32005

Varchenko, A.N. [1976] Local topological properties of smooth mappings. *Izv. Akad. Nauk SSSR, Ser. Mat. 38*, 1037–1090; English transl.: Math. USSR, Izv. 8 (1974), 1033–1082 (1975), Zbl.313.58009.

Vegter, G. [1982] Bifurcations of gradient vector fields. *Astérique*, No. 98–99, 39–73, Zbl.526.58034

Voronin, S.M. [1981] Analytical classification of germs of conformal maps $(C, 0) \to (C, 0)$ with identical linear parts. *Funkts. Anal. Prilozh. 15*, No. 1, 1–17; English transl.: Funct. Anal. Appl. 15, 1–13 (1981), Zbl. 463.30010

Voronin, S.M. [1982] Analytical classifications of pairs of involutions. *Funkts. Anal. Prilozh. 16*, No. 2, 21–29; English transl.: Funct. Anal. Appl. 16, 94–100 (1982), Zbl.521.30010

Wall, C.T.C. [1981] Finite determinacy of smooth map germs. *Bull. Lond. Math. Soc. 13*, 481–539, Zbl.451.58009

Wassermann, G. [1975] Stability of unfoldings in space and time. *Acta Math. 135*, 57–128, Zbl.315.58010

Wassermann, G. [1976] (r, s)-stable unfoldings and catastrophe theory. In: *Structural stability, the theory ofscience: Proceedings of the conference held at the Battelle Seattle Res. Center*, P.J. Hilton (ed.), pp. 253–262 Lect. Notes Math. 525 Springer-Verlag: New York, Heldelberg, Berlin, Tokyo, Hong Kong, 408 pp, Zbl.335.58009

Weber, H. [1898] *Traite d'algèbre supérieure.* Gauthier-Villars, Paris, *84*, 283 pp.

Whitney, H. [1955] On singularities of mappings of Euclidean spaces. I. Mappings of the plane into the plane, *Ann. Math.*, II. Ser. *62*, 374–410, Zbl.68,371

Zeeman, E.C. [1976] Catastrophe theory: a reply to Thom. In: *Dynamical Syst. Warwick 1974*, A. Manning (ed.), Lect. Notes Math, 468, Springer-Verlag: New York, Heidelberg, Berlin, Tokyo, 405 pp., Zbl.307.58009

Zeeman, E.C. [1977] *Catastrophe theory: selected papers 1972–77.* Addison-Wesley, Reading, Mass., 675 pp., Zbl.398.58012

Zel'dovich, Ya.B. [1941] On the theory of thermo-stress. Exothermic reactions in jets. I. II. *Zh. Tekh. Fiz. 11*, No. 6, 493–508

Author Index

Subject Index

GPSR Compliance

The European Union's (EU) General Product Safety Regulation (GPSR) is a set of rules that requires consumer products to be safe and our obligations to ensure this.

If you have any concerns about our products, you can contact us on ProductSafety@springernature.com

In case Publisher is established outside the EU, the EU authorized representative is:

Springer Nature Customer Service Center GmbH
Europaplatz 3
69115 Heidelberg, Germany

Batch number: 10682213

Printed by Printforce, the Netherlands